Polymer–Inorganic Nanostructured Composites Based on Amorphous Silica, Layered Silicates, and Polyionenes

Kostyantyn M. Sukhyy
Ukrainian State University of Chemical Engineering, Ukraine

Elena A. Belyanovskaya
Ukrainian State University of Chemical Engineering, Ukraine

A volume in the Advances in Chemical and Materials Engineering (ACME) Book Series

IGI Global
PUBLISHER of TIMELY KNOWLEDGE

Published in the United States of America by
IGI Global
Engineering Science Reference (an imprint of IGI Global)
701 E. Chocolate Avenue
Hershey PA, USA 17033
Tel: 717-533-8845
Fax: 717-533-8661
E-mail: cust@igi-global.com
Web site: http://www.igi-global.com

Library of Congress Cataloging-in-Publication Data

Names: Sukhyy, Kostyantyn M., 1976- author. | Belyanovskaya, Elena A., 1972- author.
Title: Polymer-inorganic nanostructured composites based on amorphous silica, layered silicates, and polyionenes / by Kostyantyn M. Sukhyy, and Elena A. Belyanovskaya.
Description: Hershey, PA : Engineering Science Reference, an imprint of IGI Global, [2022] | Includes bibliographical references and index. | Summary: "The intent of this monograph is to further the development of physical and chemical principles of technology for polymer-inorganic nanostructured composites based on amorphous silica, layered silicates and polyionenes to use in the creation of composites for technical purposes"-- Provided by publisher.
Identifiers: LCCN 2022024058 (print) | LCCN 2022024059 (ebook) | ISBN 9781799897286 (h/c) | ISBN 9781799897293 (s/c) | ISBN 9781799897309 (ebook)
Subjects: LCSH: Polymeric composites. | Silicates.
Classification: LCC TA455.P58 S85 2022 (print) | LCC TA455.P58 (ebook) | DDC 620.1/92--dc23/eng/20220728
LC record available at https://lccn.loc.gov/2022024058
LC ebook record available at https://lccn.loc.gov/2022024059

This book is published in the IGI Global book series Advances in Chemical and Materials Engineering (ACME) (ISSN: 2327-5448; eISSN: 2327-5456)

British Cataloguing in Publication Data
A Cataloguing in Publication record for this book is available from the British Library.

All work contributed to this book is new, previously-unpublished material.
The views expressed in this book are those of the authors, but not necessarily of the publisher.

For electronic access to this publication, please contact: eresources@igi-global.com.

Advances in Chemical and Materials Engineering (ACME) Book Series

ISSN:2327-5448
EISSN:2327-5456

Editor-in-Chief: J. Paulo Davim University of Aveiro, Portugal

MISSION

The cross disciplinary approach of chemical and materials engineering is rapidly growing as it applies to the study of educational, scientific and industrial research activities by solving complex chemical problems using computational techniques and statistical methods.

The **Advances in Chemical and Materials Engineering (ACME) Book Series** provides research on the recent advances throughout computational and statistical methods of analysis and modeling. This series brings together collaboration between chemists, engineers, statisticians, and computer scientists and offers a wealth of knowledge and useful tools to academics, practitioners, and professionals through high quality publications.

COVERAGE

- Electrochemical and Corrosion
- Polymers Engineering
- Biomaterials
- Fracture Mechanics
- Statistical Methods
- Artificial Intelligence Methods
- Metallic Alloys
- Coatings and surface treatments
- Numerical Techniques
- Multifuncional and Smart Materials

The Advances in Chemical and Materials Engineering (ACME) Book Series (ISSN 2327-5448) is published by IGI Global, 701 E. Chocolate Avenue, Hershey, PA 17033-1240, USA, www.igi-global.com. This series is composed of titles available for purchase individually; each title is edited to be contextually exclusive from any other title within the series. For pricing and ordering information please visit http://www.igi-global.com/book-series/advances-chemical-materials-engineering/73687. Postmaster: Send all address changes to above address.

Titles in this Series

For a list of additional titles in this series, please visit:
www.igi-global.com/book-series/advances-chemical-materials-engineering/73687

Development, Properties, and Industrial Applications of 3D Printed Polymer Composites
R. Keshavamurthy (Dayananda Sagar College of Engineering, India) Vijay Tambrallimath (Manipal Institute of Technology, Manipal Academy of Higher Education, India) and J. Paulo Davim (University of Aveiro, Portugal)
Engineering Science Reference • © 2023 • 300pp • H/C (ISBN: 9781668460092) • US $270.00

The Classes of Higher Dimensional Polytopes in Chemical, Physical, and Biological Systems
Gennadiy Vladimirovich Zhizhin (Russian Academy of Natural Sciences, Russia)
Engineering Science Reference • © 2022 • 366pp • H/C (ISBN: 9781799883746) • US $215.00

Advances in the Modelling of Thermodynamic Systems
Elhoucine Essefi (University of Gabes, Tunisia) and Ibtissem Jendoubi (Faculty of Sciences of Bizerte, Tunisia)
Engineering Science Reference • © 2022 • 309pp • H/C (ISBN: 9781799888017) • US $270.00

Handbook of Research on Tribology in Coatings and Surface Treatment
Amirhossein Pakseresht (FunGlass – Centre for Functional and Surface Functionalized Glass, Alexander Dubček University of Trenčín, Slovakia) and Omid Sharifahmadian (FunGlass – Centre for Functional and Surface Functionalized Glass, Alexander Dubček University of Trenčín, Slovakia)
Engineering Science Reference • © 2022 • 470pp • H/C (ISBN: 9781799896838) • US $315.00

Advanced Manufacturing Techniques for Engineering and Engineered Materials
R. Thanigaivelan (Muthayammal Engineering College, India) N. Rajan (Vinayaga Mission Kripananda Variyar Engineering College, India) and T.G. Argul (St. Mary's Engineering College, India)
Engineering Science Reference • © 2022 • 351pp • H/C (ISBN: 9781799895749) • US $270.00

IGI Global
PUBLISHER of TIMELY KNOWLEDGE

701 East Chocolate Avenue, Hershey, PA 17033, USA
Tel: 717-533-8845 x100 • Fax: 717-533-8661
E-Mail: cust@igi-global.com • www.igi-global.com

Table of Contents

Preface

The present monograph is focused on methods of obtaining hybrid organic-inorganic nano-structured materials and application of hybrid organic-inorganic nanostructured materials such as adsorptive heat energy storage and humidity sensors. Influence of gel formation on properties of organic-inorganic polymers is considered.

The first chapter is focused on state of the art and formation of sol-gel technology of obtaining multifunctional hybrid silicate materials with technically improved properties. The crucial steps and new approaches to establish the mechanism of formation of noncrystalline systems (sols and products of sol-gel synthesis) are substantiated to be study of the fractal structure of sols, gels and synthesized silicate nanocomposites. The use of sol-gel technology is proved to be especially relevant for the formation of new silicate materials and coatings required for power engineering. New products of sol-gel synthesis such as catalytic coatings for gas sensors and next-generation electrochemical devices are shown to be extremely necessary in transport engineering, as well as to ensure the safety of harmful industries, mining, agriculture and everyday life.

The second chapter is devoted to basic principles of process control of sol – gel synthesis of new class of organic-inorganic nanostructured materials on basis of tetraethoxysilane and polyionenes doped by orthophosphoric acid with improved functional properties. The chemical compositions improvement results from complex investigation of morphology, physical and chemical properties. Polyionenes are shown to increase the rate of hydrolytical polycondensation reaction of tetraethoxysilane. Increasing the polyionene molecular weight leads to increasing rate of gel formation process. The polyionenes in silicophosphate sols are stated to prevent crystallization processes occurred in silicophosphate xerogels with silica phosphates and pyrophosphates.

In the third chapter, results of the study of evolution of organic-inorganic nanocomposites during gel formation from sols based on TEOS, orthophosphoric acid and a number of polyionenes are given. Schemes of multilevel fractal aggregation of inorganic nanocomposite and organic-inorganic nanocomposites are suggested according to SAXS data. Formation of surface fractal aggregates is shown to occur at all structural levels. Mass fractal aggregates are generated on the second structural level (100 – 300 nm). A generation of the initial silicophosphate composite starts from the formation of practically spherical aggregates with order of 12 nm which forms compact lattice of the inorganic polymer on the kind of mass fractal with average particle size of 80 nm on the higher scale level. On the third level this structure occurs to be reorganized with forming of surface fractal clusters with very smooth surface and average particle size of 470 nm. Formation of surface fractal is supposed to strongly affect the protonic conductivity nanocomposites.

In the fourth chapter correlation of fractal structure and protonic conductivity of solid inorganic and organo-inorganic phosphorosilicate nanocomposites which were obtained by the sol-gel method is shown. The optimal ratio of H_3PO_4: $Si(OEt)_4$ is stated to be 1: 1. It significantly enhances the proton conductivity of the phosphorosilicate sol-gel nanocomposite, it being stable in a wide temperature range (20 – 140 °C) at the level of 10^{-3} – 10^{-5} $Ohm^{-1}cm^{-1}$. The addition of a small amount of polyionenes ($8 \cdot 10^{-4}$ mol polyionene I-1/1 mole TEOS) leads to increasing the proton conductivity of nanocomposites by 1-2 orders of magnitude. The maximum value of proton conductivity is stated in nanocomposites most resistant to hydration at temperatures of 200-400 °C.

In the fifth chapter conductivity of inorganic and organic-inorganic protonconducting materials was studied. Such materials show high levels of conductivity (up to 10^{-2} $Ohm^{-1} \cdot cm^{-1}$) in the temperatures range of 293-353 K. They are promising for use as materials for electrically active elements of sensor controls which functionate by the phenomenon of ionic (proton) conductivity. A number of laboratory samples of sensor controls of humidity which impedance characteristics have been investigated at specially developed stand in a range of frequencies from 10 up to 100000 Hz in conditions of controllable temperature and humidity have been made with using materials of described type. The variation of relative humidity from 20 up to 60% at 293 K is stated to result in reversible changes of ionic conductivity from 10 up to 40% depending on nanocomposite membranes structure. Value of conductivity of an alternating current in a studied range of conditions of relative humidity change from 400 up to 2000 depending on frequency of measurement, and the optimal appeared a range of frequencies 0,1-1 kHz.

In the sixth chapter results of study of the composite sorbents ‘silica gel – crystalline hydrate’ synthesized by sol – gel method are presented. The sorption capacity of the composites ‘silica gel – sodium sulphate’ is stated to exceed noticeably the linear superposition of the sorption capacity of mechanical mixture of the silica gel and salt. The composites composition and structure were studied by IR-spectroscopy and wide-angle X-ray scattering. The changing of adsorption properties of crystalline salts at their distribution in silicon oxygen matrix is stated to result from dispersion up to nanoscale. The experimental setup for effectiveness tests of synthesized composite adsorbents is developed. The correlation of their composition, structure, water adsorption kinetic and operating characteristic as heat storage material is revealed. The heat storage capacity is stated to be at least 0.36 kWh/kg (258 kWh/m^3).

In the seventh chapter, results of study of the composite sorbents ‘silica gel – crystalline hydrate’ synthesized by sol – gel method as heat storage media are given. The sorption capacity of the composites ‘silica gel – sodium sulphate’ is stated to exceed noticeably the linear superposition of the sorption capacity of mechanical mixture of the silica gel and salt. The composites composition and structure were studied by IR-spectroscopy and wide-angle X-ray scattering. The changing of adsorption properties of crystalline salts at their distribution in silicon oxygen matrix is stated to result from dispersion up to nanoscale. The experimental setup for effectiveness tests of synthesized composite adsorbents is developed. The correlation of their composition, structure, water adsorption kinetic and operating characteristic as heat storage material is revealed. The heat storage capacity is stated to be at least 0.36 kWh/kg (258 kWh/m^3).

The eighth chapter is devoted to developing the technology of nanocomposites based on polyamide-6 and organo-montmorillonite. The technological parameters of extrusion and injection molding, and their influence on the formation of nanoscale inorganic phase of the polyamide were studied. Morphology and thermal behavior of polyamide-6 (PA-6), composites based on PA-6 and modified montmorillonite are studied by WAXS, XRD, DTA and DSC data. New γ-phase formation is revealed in composite polyamide-6 – 1 – 2% of modified montmorillonite. Optimal concentrations of modified montmorillonite in the composites stated as 1 – 2% are shown to correspond with composites improved properties as compared to initial polyamide-6. It is explained by increasing crystallinity degree, which resulted from acting of modified montmorillonite particles as nucleation heterogenetic agents.

The ninth chapter is focused on the processes of obtaining montmorillonite modified by polyionenes. Intercalation of macromolecular polymer of a quaternary ammonium salt in montmorillonite intercrystalline space is exhibited to be accompanied by increasing interlayer distances from 1.08 nm to 1.67 nm. The technique of synthesis of montmorillonite modified with polyionenes is proposed. Optimal conditions for sorption of polyionenes molecules by montmorillonite are stated as follows: the concentration of aqueous dispersion of montmorillonite equals 1%, the temperature of the reaction medium is 40 °C, the montmorillonite-polyionene ratio is 3: 1, the processing time is 24 hours.

In the tenth chapter the technology of nanocomposites based on polyamide-6 and organo-montmorillonite is developed. The technological parameters of extrusion and injection molding, and their influence on the formation of nanoscale inorganic phase of the polyamide are investigated. Morphology and thermal behavior of polyamide-6, composites based on PA-6 and modified montmorillonite are studied by WAXS, XRD, DTA and DSC data. New γ-phase formation is revealed in composite polyamide-6 – 1 – 2% of modified montmorillonite. Optimal concentrations of modified montmorillonite in the composites stated as 1 – 2% are shown to correspond with composites improved properties as compared to initial polyamide-6. It is explained by increasing crystallinity degree, which resulted from acting of modified montmorillonite particles as nucleation heterogenetic agents.

In the eleventh chapter properties of nanocomposites based on polyamide-6 and organ modified layered layer silicates were analyzed according theories of Halpin-Tsai and Mori-Tanaka. The morphology of nanocomposites under consideration was analyzed by radiographic analysis and transmission electron microscopy. The degree of delamination of the structure of layered silicates in the polymer matrix is shown to be the crucial factors to improve the physical and mechanical characteristics of the nanocomposites. A high level of orientation of elementary particles of layered silicates is stated when shear forces applied in comparison with fibers both due to individual features and due to a more favorable geometric shape, correlation of the simulation and experimental data being shown.

Acknowledgment

The present work was partially supported by Ministry of Education and Science of Ukraine within the frames taxpayer-funded research (grant numbers 0119U002243 and 0122U001727).

Introduction

Global trends of today suggest that 21st century science and technology will be nanoscale, as traditional technologies have exhausted the potential for miniaturizing individual elements, prompting the search for alternative pathways. Nanophase materials science differs from the traditional one not only by the creation of fundamentally new materials, the functional parameters of which are determined by the properties of microregions, but also by processes which take place at the atomic, molecular level, monolayers and nanovolumes.

Design at this level can only be done with the participation of clusters of molecules and nanosized particles. The transition from micro to nanotechnology is not so much quantitative as qualitative, i.e. physical parameters such as melting point, binding energy per atom, change in interatomic distances, ionic or electronic conductivity, optical and magnetic properties show non-monotonic dependence on size. The thermodynamic approach allows to identify the conditions of nanophase formation and to identify the factors of control of these processes on the basis of models of homogeneous and heterogeneous nucleation, study of the dynamics of particle growth and their structural organization.

Nanoparticles with developed interfacial surfaces have excess energy in comparison with homogeneous materials. Energy saturation and, consequently, high chemical activity of nanoparticles is an extremely useful and important property, but it causes huge obstacles in the creation, storage and use. Passivation and involuntary aggregation of particles leads to the formation of powders of ordinary dispersion, and, accordingly, to the loss of all the gains and advantages of nanoscale effects.

In the world of the above, an extremely important task of physical chemistry of nanoparticles is to find effective ways of controlled chemical passivation and stabilization of the latter. Its solution is significantly simplified due to the high adsorption capacity of nanoparticles in relation to surfactants and many, both ionic and nonionic macromolecular compounds. Due to this steric stabilization, formations are formed, which, in the simplest case, are a lyophobic nucleus surrounded by a lyophilic steric barrier formed by solvated polymer chains. In the trivial case, use a minimum amount of polymer stabilizer to overcome the coagulation threshold - that is, the role of the polymer is only in the shielding of nanoparticles. Creation of polymer-inorganic nanostructured hybrid materials with identification of fundamental aspects of physicochemistry of nanocomposites, adsorption of macromolecules on the surface of nanoparticles, formation of interfacial boundaries and modeling of properties is a priority of modern materials science.

Promising polymer stabilizers are polyelectrolytes, which in certain solvents exist in the form of micelles, in the polar nuclei of which, as in a kind of nanoreactors, nanoscale particles are formed, and hydrophobic segments serve as stabilizers.

The most intensively developed sol-gel methods of obtaining hybrid materials and intercalation of polymers in nanostructured layered systems using the methods of chemistry of intracrystalline structures such as guest-host. The main role belongs to the interphase interactions between inorganic and polymeric components in the formation of such hybrids. An important prerequisite for this is the environmental friendliness and relative ease of obtaining such polymer-hybrid materials, as well as the availability of starting materials.

Solving such problems of scientific and practical nature will contribute to the creation of new materials and technologies in the fields of catalysis, adsorption, transfer and conductivity, chemistry and technology of polymeric and composite materials.

The present work is devoted to development of physical and chemical principles of technology for polymer-inorganic nanostructured composites based on amorphous silica, layered silicates and polyionenes to use in the creation of composites for technical purposes.

To achieve this goal, the following research objectives were identified and solved:

- To reveal the influence of polyionenes on the processes of morphology formation, complex of technological, physicochemical and operational properties of polymer-inorganic materials based on amorphous silicas obtained by sol-gel method and template synthesis and intercalation method in layered silicates;

- To state physicochemical factors and technological principles of sol-gel technology of hybrid organo-inorganic composites based on amorphous silicas and polyionenes;
- To develop sol-gel technology to obtain highly efficient and competitive resistive humidity sensors;
- To establish the effectiveness of polyionenes to regulate the surface properties of mesoporous silica;
- To develop new heat-storage materials based on crystal hydrates and organosilicon matrix with heat energy storage density at the level of 1.5-1.7 MJ/kg;
- To state the features of structure formation and development of technology of intercalation of polyionenes in layered silicates for decrease in level of interlayer interactions;
- Theoretical substantiation and experimental determination of physical and mechanical factors of strengthening of polymer composites during delamination of organomodified layered silicates in the process of extrusion with polymer matrix.

Chapter 1
Sol Gel Processes of Obtaining Nanostructered Composites

ABSTRACT

In this chapter, the authors present state of the art and formation of sol-gel technology of obtaining multifunctional hybrid silicate materials with technically improved properties. The crucial steps and new approaches to establish the mechanism of formation of noncrystalline systems (sols and products of sol-gel synthesis) are substantiated to be study of the fractal structure of sols, gels and synthesized silicate nanocomposites. The use of sol-gel technology is proved to be especially relevant for the formation of new silicate materials and coatings required for power engineering. New products of sol-gel synthesis, such as catalytic coatings for gas sensors and next-generation electrochemical devices, are shown to be extremely necessary in transport engineering, as well as to ensure the safety of harmful industries, mining, agriculture, and everyday life.

BACKGROUND

Polymer-inorganic materials are evolved by means of a mechanical mechanism and thermostability in the result of optimal heat transfer powers Wilde A. (2021). The whole class of new materials is the synergy of the powers of the external components. The organic phase can swallow nanoparticles in the middle of a solid "paste" with the optimal parameters of an oxide mesh or a polymeric lanka. As inorganic precursors oxides of silicon and aluminum,

DOI: 10.4018/978-1-7998-9728-6.ch001

vanadium and molybdenum, glass, clay, layered silica and zeolite, phosphates and chalcogenides of metals, oxychloride zaliza, graphite. Particularly, it is a good idea of the method for regulating the size of the pores. These materials are already at once known to exhibit a wide range of components of metal-matrix composites; the stench puffs the taste and hardness of soft metals - such as aluminum, polish, thermochemical, rheological, electrical optical power Koo J. (2006). Such composites can also be enhanced in the quality of chromatographic materials, membrane materials and other materials.

Ecologically optimal are drainless methods of obtaining composite materials, which include the sol-gel method (sol-gel or spin-on-glass process) - such materials as ceramics, phosphors, carriers Rivero-Antúnez P., Zamora-Ledezma C. et al. (2023), catalysts Al Khudhair A., Bouchmella K. et al. (2021), reinforcing agents for plastics and rubber Liu W., Yi S. et al. (2022), adsorbents for the pharmaceutical and cosmetic industry, etc. Gomes A. L.M., Andrade P.H.M. et al. (2021).

To date, the generally accepted classification of sol-gel hybrid organo-inorganic nanocomposites is absent. However, it is possible to list some, often used names of organo-inorganic hybrids differing in a ratio of organic and inorganic phases, structure, initial components of synthesis and functional purpose.

- ORMOSIL - a material synthesized by mixing the calculated amounts of alkoxysilane Si $(OR)_4$ (R = C_2H_5, CH_3) with polyethylene glycol (PEG) and its derivatives Barshilia H.C., Lakshmi R.V. et al (2022);
- ORMOCER - ceramics modified with organic matter Pirmoradian M., Al-Bakhakh B.A.J. et al. (2022);
- CERAMER - glassy or ceramic materials containing comparable amounts of organic and inorganic components in the volume of the hybrid material Vora R. H. & Lau K.S.Y. (2022);
- Air Force (IPN - interpenetrate networks) - a hybrid material formed by interpenetrating networks of inorganic and organic polymers Smith G. N., Brok E. et al. (2021);
- ORMOLYT - a modified organosilicon electrolyte Imai Y. Naka K. & Chujo Y. (1998).

Hybrid organo-inorganic materials and coatings are also divided into two types by type of chemical bonds, namely Huang H.–H. & Wilkes G.L. (1987):

1) Organic and inorganic groups are connected by stable chemical bonds: covalent, coordination, ionic.
2) Organic macromolecules are included in the inorganic networks of the molded gel, or vice versa, inorganic molecules or their aggregates are embedded in organic macrostructures. Thus organic and inorganic fragments can be connected only by weak physical communications.

The nature of the interaction between organic and inorganic fragments is largely determined by the conditions of synthesis of hybrid organo-inorganic composites. The date of birth of sol-gel synthesis is considered to be 1844, when Ebelmen first received tetraethoxysilane Si $(OC_2H_5)_4$ [Ebelmen (1844)], and then in 1846 discovered that in the presence of moisture it easily turns into a gel [Ebelmen (1846).]. A little later, this process was observed by Friedel and Crafts [Friedel C. & Crafts J.M. (1867)]. They clarified the formula of tetraethoxysilane (TEOS) and described the reaction products of its hydrolytic polycondensation [Friedel C. & J.M. Crafts (1870)]. The beginning of practical application of Si (OEt) 4 dates back to 1902, when a method of impregnation of asbestos fiber products was patented in order to provide them with water, acid and fire resistance [Siemens, Halske A.G. (1902)].

Reviews are devoted to the further use of the hydrolytic polycondensation reaction of TEOS to improve the various properties of numerous materials and the application of protective silica coatings on their surface Liu Y., Lin Q. et al. (2022). Concepts such as "sol-gel" processes, "sol-gel" technology appeared much later than the first studies in this area. Initially, terms such as "colloidal" or "semi-colloidal solution" were most often used to describe sol-gel systems. The method of obtaining monolithic materials and coatings is often called "colloidal treatment", "soluble ceramics" Liu J., Li Y. et al. (2023).

The emergence of the terms 'sol-gel system', sol-gel transition, and the very formation of sol-gel processes as an independent scientific field, can be attributed to the release in 1990 of the book by Brinker & Scherrer "Sol-gel science. Physics and Chemistry of Sol-Gel Technology (1990), which for two decades is a monograph that most fully and systematically summarizes the basic ideas of sol-gel processes, modern methods of research of sols, gels and materials, as well as known scientific concepts that used to interpret the observed phenomena.

The sol-gel method can be used to obtain various materials Brinker C.J. &Scherer G.W. (1990) (Fig. 1) - glass, ceramics, fibers, powders, membranes, coatings. According to many experts, further prospects for the use of this

Figure 1. Basic plot to obtain various materials, cores and powders be sol gel tecniques Brinker C.J. &Scherer G.W. (1990)
I - maturation of sol and / or gelation: sol (1)> gel (2);
II - drying under supercritical conditions or washing the gel in solvents: gel (2)> airgel (3);
III - drying under normal conditions: gel (2)> xerogel (4);
IV - deposition of nanoparticles: sol (1)> powder (6);
V - application of sol on substrates> sol (1) -layer of xerogel (7)
VI -burning: xerogel (4) or xerogel layer (7)> monolithic glass and ceramics (5) or coatings (8).

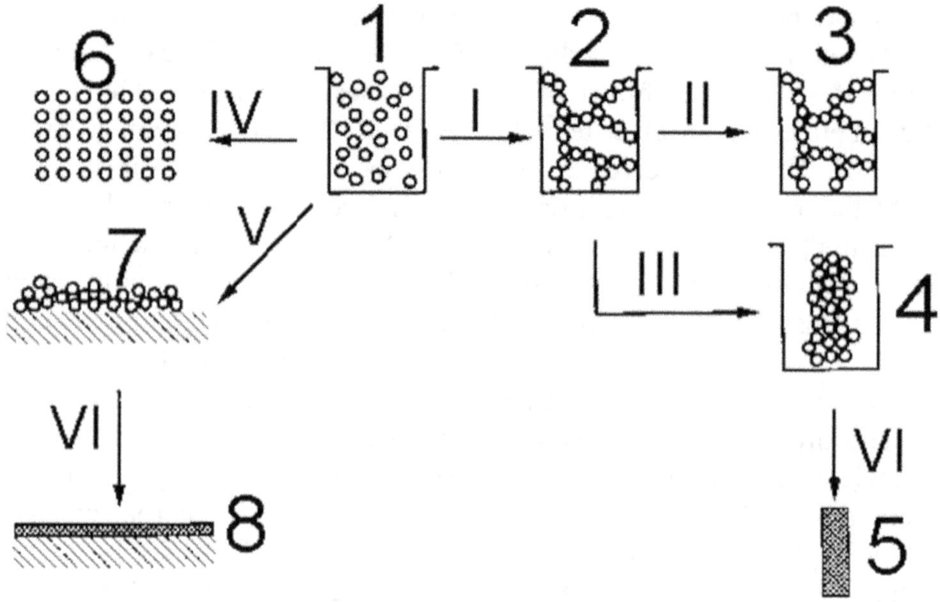

method will be associated primarily with the production of products for specific applications. In addition, more than half of the products of the latest glasslike materials can make up new unpredictable materials that can only be obtained using sol-gel processes.

The prospect of successful further development of sol-gel technology is associated with a number of advantages of this method:

- Processes of sol-gel synthesis are simple, technological, easily fit into standard industrial technologies adopted in microelectronics, in the production of optical devices and a number of other industries, are subject to mechanization and automation Broadstreet S.W. (1955), Liu J., Li Y. et al. (2023);
- The composition of the obtained materials can be reproduced in a wide range Zanurin A., Johari N.A. et al. (2022);

- Synthesis is carried out in milder conditions than the processes of traditional ceramic technology - the synthesis temperature is significantly lower Klein L.C. (1996).
- It is possible to carry out templant synthesis of materials, ie to control the structure and shape (if they are nanoparticles) of the obtained materials by introducing special inorganic or inorganic additives into the ash and varying the gelation conditions Ali S.M. & Emran K.M. (2022);
- Synthesis of a new generation of materials - organo-inorganic hybrids, which in some respects have better properties than traditional ceramics and polymers, as well as the use of organic molecules for molecular engineering and modeling of materials with a given structure I. Gautier–Luneau, A. Denoyelle, J.Y. Sanchez & C. Poinsignon (1992).

The materials obtained in the process of sol-gel synthesis belong to nanomaterials, which is associated with the uniqueness of many of their properties. There are many fundamental problems associated with the development of general concepts of structure formation and gelation in sols based on metal alkoxides and silicon in the presence of inorganic and organic modifiers, as well as identifying mechanisms of processes occurring in multicomponent hybrid organo-inorganic sol-gel systems. Many unsolved technical problems remain in the sol-gel technology for multicomponent multifunctional materials.

FEATURES OF FORMATION OF MATERIALS FROM SOLS

Sol is a colloidal suspension of fine solid particles in a liquid. According to the classical definition: "colloids are suspensions in which the dispersed phase is so small (~ 1-1000 nm) that the gravitational forces are negligibly small, and short-acting forces, such as van der Waals interactions between surface charges, are dominant. The inertia of the dispersed phase is quite small, so there is a Brownian motion of particles, ie random jumps caused by kinetic energy reported due to collisions with molecules of the dispersion medium Brinker C.J. & Scherer G.W. (1990).

The particle size of the sol given in this definition is quite wide, which is opposed by many experts in the field of colloid chemistry, who consider it appropriate to reduce the upper limit of their size to 100 nm. Sometimes ashes are divided into colloidal and polymeric. In colloidal sols, the solid dispersed phase is formed by inorganic particles, and in the polymer sol it is formed from crosslinked macromolecules. The mechanism of gelation for colloidal and polymeric types of sols is different.

The gel of polymer sol is formed during the polymerization of monomers and polymers in the liquid phase. A polymer molecule of three-dimensional structure, a giant cluster, is gradually formed from polymerizable oligomers. In other words, if one molecule reaches the macroscopic size and extends to the entire volume of the sol, we can say that there was a transition of the sol to the gel. The gel consists of a continuous structural mesh impregnated with a continuous liquid phase.

The formation of a colloidal gel occurs by a different mechanism. Inorganic particles of the dispersed phase under the influence of the dispersion forces of attraction stick together (as in the case of the polymer sol), forming a network of inorganic polymer.

Sol-gel systems are not equilibrium systems, on the contrary, these systems are continuously variable systems that are self-developing and self-organizing. Obviously, it is possible to speak about their stability only for individual periods of their existence.

Polymer gels are characterized by the presence of covalent bonds between molecules and fragments of molecules that form a giant cluster. Colloidal gels are formed due to van der Waals forces of attraction between particles. In colloidal systems, these bonds can be reversible, ie destroyed in the process of shaking, and then restored. In polymer gels, covalent bonds are stable.

Based on the above mentioned, the concept of gel can be interpreted as a state of matter when it consists of continuous solid and liquid phases. According to modern notions, the formation of a gel from a sol begins with the formation of a fractal structure, the growth of fractal aggregates, until they begin to collide with each other and connect, as described by the theory of percolation. Near the point of gelation randomly located adjacent clusters (consisting of polymers or aggregates of particles) are connected together, forming a single structural network. The gelation point corresponds to the percolation threshold when a single cluster is formed that has expanded over the entire volume of the sol. After passing the point of gelation, the sol loses mobility and turns into a so-called "wet gel". The wet gel usually takes the form of a vessel in which the sol was. Gel formation does not stop at the point of gelation. Usually in the process of aging the gel "ages". This completely scientific term "aging" of the gel refers to the processes of structural changes that occur after the point of gelation in the wet gel. In the wet gel formed, a single cluster coexists with a sol containing many small clusters. At the same time, the process of continuous joining of the latter to the common skeleton - a continuous giant cluster - takes place. In addition, condensation reactions that have not been completed in the sols can continue in the gels, the processes of redeposition of monomers or oligomers take place, and solid-liquid-liquid phase transitions take place.

In the process of aging gels, shrinkage usually occurs - deformation of the inorganic polymer network and removal of liquid from the pores. When dried, the gel is converted into a xerogel, and decreases in volume by 5-10 times compared to the volume of the wet gel. The wet gel can also be converted into an airgel, which retains a loose structure in which the pores are filled with air. One of the modern methods of obtaining airgel is supercritical drying. To do this, the wet gel in the autoclave is dried under supercritical conditions, ie at high pressure (for example, in a water vapor atmosphere) at a temperature lower than that required for drying at atmospheric pressure. This eliminates the action of capillary forces that prevent the removal of fluid from the pores.

And then xerogels or aerogels for hardening are subjected to heat treatment in the temperature range of 250-1500°C, in the process of which glassy or ceramic materials are formed. At the same time many difficult physical and chemical processes caused by destruction of the organic components connected with an inorganic grid of a xerogel (usually alkoxygroup), and also the adsorbed and chemically bound water and solvents proceed. Upon further heating, the structure of the inorganic polymer is rearranged due to sintering and sometimes crystallization processes.

Of all the variety of materials that can be obtained by the method of sol-gel technology, we focused on the synthesis of sols from tetraethoxysilane in the presence of organic modifiers. The purpose of these studies is to study the structure and properties of nanostructured organo-inorganic materials formed as a result of hydrolytic polycondensation of TEOS, promising for use in energy and sensor technology.

The basis of the processes of gel and film formation of silicons and metals of alkoxides is the reactions of their hydrolytic polycondensation Brinker C.J. & Scherer G.W. (1990), Adnan M.M.; Nylund I.-E. et al (2021). Basically, these are the three most important reactions that take place in sol-gel systems:

$$(RO)_3SiOR + HOH \text{ « } (RO)_3SiOH + HOR \quad (1.3)$$

$$°SiOH + HOSi° \text{ « } °SiOSi° + HOH \quad (1.4)$$

SiOH + ROSi° « °SiOSi° + HOR (1.5)

In sol-gel systems on the basis of alkoxides of silicon and metals similar reactions proceed quickly enough practically to complete disappearance of OH and OR groups. In these cases, the end products of the sol-gel process are silicon dioxide, metal oxides or silicates. Prior to heat treatment, xerogels based on tetraalkoxysilanes contain residues of alkoxy and hydroxy groups, so that their composition can be represented by the formula: $SiO_{2-1/2(m+n)}(OH)_m(OR)_n$.

The content of OH and OR groups in the xerogel depends on the hydrolytic polycondensation reaction conditions. These include: the nature of the alkyl group R in tetraalkoxysilane, solvent and catalyst, the concentration of $Si(OR)_4$, the amount of H_2O and organic solvents, temperature and duration of the process, the method of homogenization of the reaction mixture.

In the synthesis of dispersed sol-gel systems (sol-filler) there are classic problems for ceramic technology of coating formation: uniformity of filler distribution, sedimentation stability of dispersions and their ability to wet (Jalil et al., 2021).

To ensure the above requirements, the obligatory technological technique used in both ceramic and sol-gel technology is homogenization. There are many ways to homogenize the starting components by grinding and mixing. The most modern and effective method of homogenization of sol-gel systems and dispersions based on them is ultrasonic exposure (Greil, 1998). This method refers to mechanochemical methods of initiation (Cheng, L. Cai, J. & Ke, Y., 2020). When sound waves propagate in liquids, a field of variable pressure is created, under the influence of which air bubbles are formed, enlarged and burst. Ultrasonic cavitation in liquid media acts simultaneously due to several effects (Kamila et al., 2022). First of all, this mixing effect of resonating bubbles, which increases the area of contact between the liquid components of the sol-gel system and promotes the dissociation of the starting components into radicals. For example, in aqueous media during adiabatic slamming of cavitation bubbles, water vapor dissociates into radicals (H • and • OH), which initiate redox reactions, polymerizations and other transformations of solutes.

In addition to the above methods of homogenization and activation of dispersed sol-gel systems to increase their sedimentation stability, as well as to form a given microstructure of the obtained ceramic and glass-ceramic materials in some cases, special stabilizing additives are used. Organic

Figure 2.

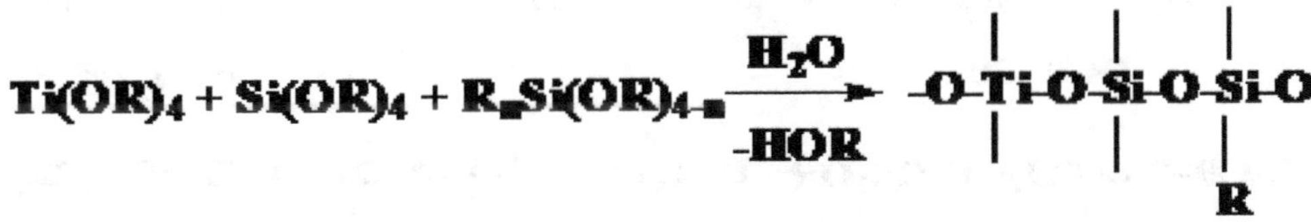

oligomers and polymers, such as polyethylene glycol or polyvinyl alcohol of various molecular weights, are most commonly used (Kamila et al., 2022).

SYNTHESIS OF ORGANO-INORGANIC MATERIALS BY SOL-GEL METHOD

Several ways of sol-gel synthesis of organo-inorganic materials have been developed, namely:

- Historically, the first method of obtaining hybrid organo-inorganic nanocomposites - organically modified silicates (ormosils) is based on the interaction of peralkoxy derivatives Si, Ti, B, Al, Ge, Sn and others. with organyltrialkoxysilanes of General formula $R_{4-m}Si(OR¢)_m$ (m=3,2)). Ormosils are the most numerous class of nanohybrids formed by the reaction of hydrolytic polycondensation of alkoxides (1-3) Abdollahi A., Roghani-Mamaqani H. et al. (2018):

Hybrid nanocomposites also include polyalkylsilsesquioxane structures $(RO)_3Si(CH_2)_nSi(OR)_3$ (n = 2-14) based on α,ω-bis (trialkoxysilyl) alkanes Schubert U. (2003).

Along with the $(CH_2)_n$ groups, other organic fragments (eg, -C6H 4-4-) can be used in ormosilmonomers, including those with heteroatoms in the main or side chain (O, N, S, etc.) (Pinero M., La Rosa–Fox N.D., Erce–Montilla R. & Esquivias L., 2003). The purpose of modifying silicate mesh with organic spacers is to limit its bonding (by including monomers of lower functionality) and to reduce the density of the structure (by increasing porosity or creating free volumes). Changing the nature of organic groups at the silicon atom and the porosity of the forces allows you to effectively regulate the level of gas or liquid permeability.

Inorganic alkoxides and organic monomers, oligomers or polymers containing functional groups (OH, COH, NH_2, etc.) capable of forming organo-inorganic associative systems in which there are no covalent bonds

Figure 3. Structural formula of a telehele polymer - a typical precursor for the synthesis of organo-inorganic hybrids from a high volume of an organic fragment

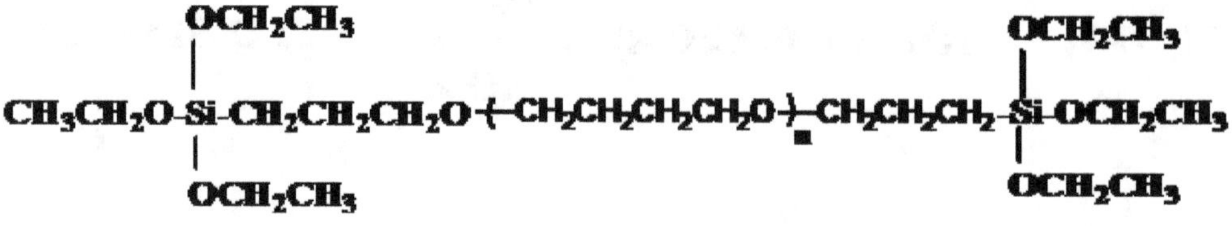

between the organic component and the inorganic network specifically interact with the inorganic matrix. Such systems are able to have properties typical of covalently coupled systems. Associative binding is manifested in organo-inorganic hybrids with specific interactions between organic and inorganic components (hydrogen bonds, ionic, dipole-dipole interactions, vandervaals forces, etc.) Haruvy Y., Ryabov Y. et al.(2002). In most cases, the intensity of such interactions is insufficient for homogenization of the system in the organo-inorganic composite and phase separation can occur with the release of the organic phase. However, in some cases, the mixing of organic and inorganic components occurs at the micro level with the formation of transparent hybrid nanocomposites Fidalgo A. & Ilharco L.M. (2003). However, there is still clearly insufficient experimental data to establish the principles of directed change of the phase structure of hybrid nanocomposites of the associative type. To solve this problem, it is necessary to study systems that include organic polymers of different nature, content, and location of functional groups. The available single studies in this area cannot answer this question.

The method of production includes elements of the two methods described above. It is based on the use of telehelic oligomers and polymers that contain functional groups at the ends of chains (OH, COH, OR, etc.) (Fig. 2) Ou D.L. & Chevalier P.M. (2003).

Condensation of these groups with alkoxy groups of the precursor (for example, $Si(OC_2H_5)4$ provides the formation of a hybrid organo-inorganic network. Such systems are almost impossible to prepare in the first way used for the synthesis of ormosyls. They provide good opportunities for comparing the structure and properties of hybrid nanocomposites with identical molecular structure of organic fragments, but with different degrees of organic and inorganic grids.

Creation of organo-inorganic nanocomposites such as interpenetrating inorganic and organic polymer grids (HPN) R.A. Sá Ferreira, (2003). The phase separation of both covalent and non-covalent air forces usually stops at the nanoscale, resulting in the formation of nanosized regions of different phases.

PROTON-CONDUCTING MATERIALS BASED ON TETRAETHOXYSILANE AND ORTHOPHOSPHORIC ACID

Electrochemical devices for generating and storing electricity, which use as an electrolyte separator polymer-electrolyte membranes (fuel cells, solid electrolyte batteries, accumulators and capacitors of the double electric layer (supercapacitors), are the most promising and environmentally friendly sources of electricity Agrawal R.C., R.K. Gupta (1999).

Capacitors of a double electric layer (supercapacitors) are devices of storage of electric energy with a possibility of its use for small intervals of time (emergency support of computers, food of afterburner of engines of electric cars, etc.) Larminie J., Dicks A. (2003). Low molecular weight inorganic and organic ammonium salts are widely used for the manufacture of electrolyte membranes of supercapacitors M. Ishikawa, M. Ihara, M. Morita & Y. Matsuda (1995). In the presence of tetraalkylammonium salts, proton solid conductors doped with orthophosphoric acid were synthesized by the sol-gel method Pell W.G. & Conway B. E. J. (2001). To improve the mechanical strength, organic elastomer additives are added to the xerogels before pressing them (eg, ethylene-butylene-styrene and sulfonated styrene-isoprene-styrene) Agrawal R.C., R.K. Gupta (1999)., or even cast thin-layer coatings, such as epoxy silicate Shin S.–J. et al. (2002).. However, silicophosphate proton-conducting membranes that combine high proton conductivity and satisfactory mechanical strength and chemical stability have not yet been obtained. Therefore, work in this direction continues.

Fuel cells based on polymeric proton-conducting membranes are devices capable of direct and continuous conversion of chemical energy into electrical energy Gray F.M. (1997). The efficiency of such devices is significantly higher than internal combustion engines *Larminie J., Dicks A. (2003)*. The electrodes of the fuel cell are layered structures consisting of a carbon substrate, a diffusion layer and a catalytic layer Gray F.M. (1997). The electrodes, together with the membrane located between them make up the membrane-electrode assembly.

The source of protons that are catalytically generated at the anode of the fuel cell is hydrogen, or even organic liquids such as methanol, ethanol, gasoline or biomass. Protons penetrate through the membrane that insulates the electrodes by electron current, and, combining in the catalytic layer of the cathode with the oxidant (synthetic or atmospheric oxygen), form pure water as a by-product.

Unfortunately, to date, the efficiency of using catalysts (mostly fine platinum) used for this purpose is extremely low. Despite the need for fuel cells, the problems of their commercialization depend on the reduction of the specific content of platinum in the catalytic layers (without reducing the efficiency of operation), at least by an order of magnitude Gillett S.L (2002). A major obstacle to this is the anodic poisoning of the platinum catalyst by traces of carbon monoxide Marr C. & Li X. (1999). due to chemisorption of traces of carbon monoxide, which impedes the access of hydrogen to the catalyst surface. The most effective possibility of CO desorption is provided by increasing the operating temperature of fuel cells from commonly used (80-90ºC) to 140-160°C. Such an increase in the operating temperature of the fuel cell requires the creation of a new generation of thermostable polymeric proton-conducting membranes. To solve this problem, sol-gel technology can be used, which will create both nanocomposite electrocatalytic layers with a significantly reduced platinum content (due to the high degree of platinum dispersion) and solid proton-conducting thermostable membranes.

FRACTAL STRUCTURE OF ORGANO-INORGANIC HYBRID SYSTEMS

Knowing the molecular structure of matter, its crystal lattice and the spatial arrangement of groups, it is possible, in principle, to determine the coordinates of all atoms in a crystal of any size and with a high probability to predict most of its physical characteristics.

However, such regular ordered systems are a minority. Irregular objects prevail over them. That is why at present the interest in the objects of research (especially micro- and nanoscale), which are irregular systems or phenomena constructed in a completely irreversible way, has grown enormously Urban P.M. et al.(2001). Such objects include sols, gels, some xerogels and some synthesized nanocomposites, various supramolecular aggregates that occur when the stability of colloids and in the processes of sol-gel synthesis.

Since before the concept of fractals appeared there were no sound approaches to their understanding and description, they were all outside the scope of human scientific activity.

The formation of the main provisions of the concept of fractals Baig N., Kammakakam I. & Falath W. (2021) has aroused great interest in its application to describe such irregular structures and phenomena. This purely descriptive stage was successfully completed by establishing the presence of fractal features in most irregular structures. Later it became obvious that, in addition to describing such disordered objects and nondeterministic phenomena, it is necessary to find approaches to establishing internal mechanisms of their formation.

Sol-gel systems at different stages of formation, including xerogel formation, have properties inherent in fractal objects. This was first established by Schaefer D.W. & Keefer K.D. (1984) on the example of the most popular (and studied) products of sol-gel synthesis (silicates) formed under acid- or base-catalyzed hydrolysis of tetraethoxysilane followed by condensation of hydrolysates and generation of fractal clusters. These authors foresaw exceptional prospects for the use of small fractal systems in the technologies of the future, suggesting, in particular, that this kind of structures formed by sol-gel synthesis, in contrast to systems based on dense colloids, play a key role in forming high-quality silicate coatings.

Here are two definitions of the term fractal, which from different positions, but quite fully reflect its essence:

- Fractals are approximate or fragmented images that can be divided into parts, each of which (at least approximately) is a reduced copy of the whole.
- Fractals are objects of complex configuration that cannot be measured by standard methods, such as known measures of length, and are characterized by a fractal dimension.

To date, it has been established that the fractal nature of aggregation is inherent in most silicate and other inorganic sol-gel systems Brinker C.J. et al. (1982)., as well as organo-inorganic sol-gel materials Vollet D.R., Donatti D.A.&Ibanez Ruiz A. (2002). The peculiarities of the evolution of the fractal structure of such systems in the process of sol-gel synthesis were determined. The process of obtaining sol-gel nanocomposites can be represented as a sequence of several characteristic stages Geiculescu A.C. & Rack H.J. (2002).:

generation of sol fragments, formation of the chemical structure of the gel from sol particles, formation of dried xerogel (nanocomposite).

At this time, the processes of structure formation in silicate sol-gel systems have been studied at all stages of nanocomposite formation. However, studies of sol-gel nanocomposites of another nature are limited to single examples. Among the latter is a recent study of the fractal structure of xerogels based on aqueous solutions of zirconium tetraacetate Nogami M. & Moriya Y. (1980). In this case, for xerogels prepared at different temperatures, the values of fractal dimensions were found to be much smaller than in the case of silicate xerogels in similar modes of heat treatment Schaefer D.W., Martin J.E. & Keefer K.D. (1985). These data suggest that the availability of detailed data on the fractal structure of one type of system (silicate) can not be transferred to the sol-gel system of another nature. The latter, in particular, include the well-known silicophosphate sol-gel systems and materials based on them, which have not been studied so far in terms of their fractal structure. The study of such systems is undoubtedly of independent interest, as nanocomposites with an important set of functional properties necessary for use in the latest electrochemical devices for storage and generation of electricity (double electric layer capacitors, fuel cells) have recently been obtained on their basis Shin S.–J. et al. (2002).

Different types of fractal structures are known, both in objects and processes of the surrounding world, and in synthetic materials Matejka L., Plestil J. & Dusek K. (1998). The simplest fractals are mainly divided into mass (volume) and surface Olemskoy A.I. & Flat A.Ya. (1989). Mass fractals are characteristic of materials in which the size of the inclusions of one of the phases varies continuously from infinitesimal to macroscopic. The size of the phases in them is limited by the volume of the fractal aggregate (for example, a porous object with the widest possible distribution of pores in size). Surface fractals appear when the interfacial boundary in the composite is rough at all scale levels of the unit. Fractal aggregates with a structure characteristic of both types of fractals are observed in sol-gel systems. A fundamental characteristic of fractal aggregation is the fractal dimension. The fractal dimension of the mass fractal, D_f, which varies within $1 < D_f < 3$, connects the mass, M, of the fractal object with its radius, R, in accordance with the scaling ratio $M=R^{Df}$. It follows that the density of very large aggregates such as a mass fractal tends to zero. The larger D_f, the more compact the fractal object ($D_f = 3$ corresponds to a three-dimensional system). When $D_f = 3$, the object loses the properties of the mass fractal, because its fractal dimension becomes equal to the dimension of the inclusion space (ie 3, in the case of three-dimensional space). However, the object may exhibit the properties of a surface fractal. In particular, in the case of a porous medium with the number of pores of radius equal to *R* Olemskoy A.I.. Flat A.Ya. (1989), a valid scaling dependence of type $N(\rho) \sim \rho^{-(Ds+1)}$ ($\rho = R / R_{max}$; R_{max} is the maximum pore radius; D_s is the fractal dimension surface fractal). The total surface of all pores with radii from *R* to R_{max} varies in accordance with the exponential law $S(\rho) \sim \rho^{-(Ds-2)}$. Thus, for the value of D_s due to the inequalities $2 < D_s < 3$, in the limit (taking into account infinitesimal pores, ρ ® *0*) the total pore surface becomes infinitely large (S (ρ) ® ¥). The fractal dimension of the surface fractal varies within these limits (from 2 to 3), and the higher the D_s, the larger the rough surface of the fractal object. The value $D_s = 2$ characterizes a fractal object with a smooth surface, and $D_s = 3$ - an object with the roughest (roughest) surface.

To determine the fractal dimension and size of fractal aggregates in sol-gel systems are used mainly methods of low-angle radiography, neutronography and light scattering, as well as electron and atomic force microscopy. Taken together, these methods allow the evaluation of these parameters in the range of characteristic lengths, l, from 0.1 to 1000 nm. The use of scattering methods is associated with the need to select those angular intervals on the scattering curve, which concentrate information on the size of fractal aggregates and the fractal dimension itself. For dilute systems that polymerize, based on the size of the primary particles, a, forming a fractal aggregate, and the radius of rotation, R_g, the aggregate itself, you can use Schaefer D.W., Keefer K.D. (1984) the simplest schematic representation of the scattering curve in the coordinates $log\ (I) - log\ (qR_g)$. Here I is the intensity of ray scattering (for example, in cm-1 when normalizing the cross section of X-ray scattering, expressed in cm^2), $q = 4\pi \sin(\theta) / \lambda$ (q is the value of the wave vector in nm^{-1}, θ is half scattering angle in degrees, λ is the wavelength of radiation in nanometers), associated with the ratio: $q = 2\pi / 1$. Olemskoy A.I.. Flat A.Ya. (1989). A similar scheme for the simplest structure of the mass fractal type and the lack of order in both the location of fractal aggregates and in the primary particles (atoms or molecular fragments) is shown in Fig. 3.

The scattering intensity in the region of extremely small angles (limit mode) allows to estimate the concentration of monomers and the average numerical degree of polymerization; Guinea region - estimate the radius of rotation of the units (Guinea radius); area Porod's (mode) - estimate the fractal dimension of a mass fractal; Bragg region (mode) - to determine the nature of ordering at distances comparable to the length of chemical bonds. Intensity values in the Porod's and Guinea regions are directly used to estimate

Figure 4. Schematic representation of the X-ray scattering curve in the coordinates log (I) - log (qR_g). I - limit mode; II - Guinea's mode; III - Porod's mode; IV - Bragg mode

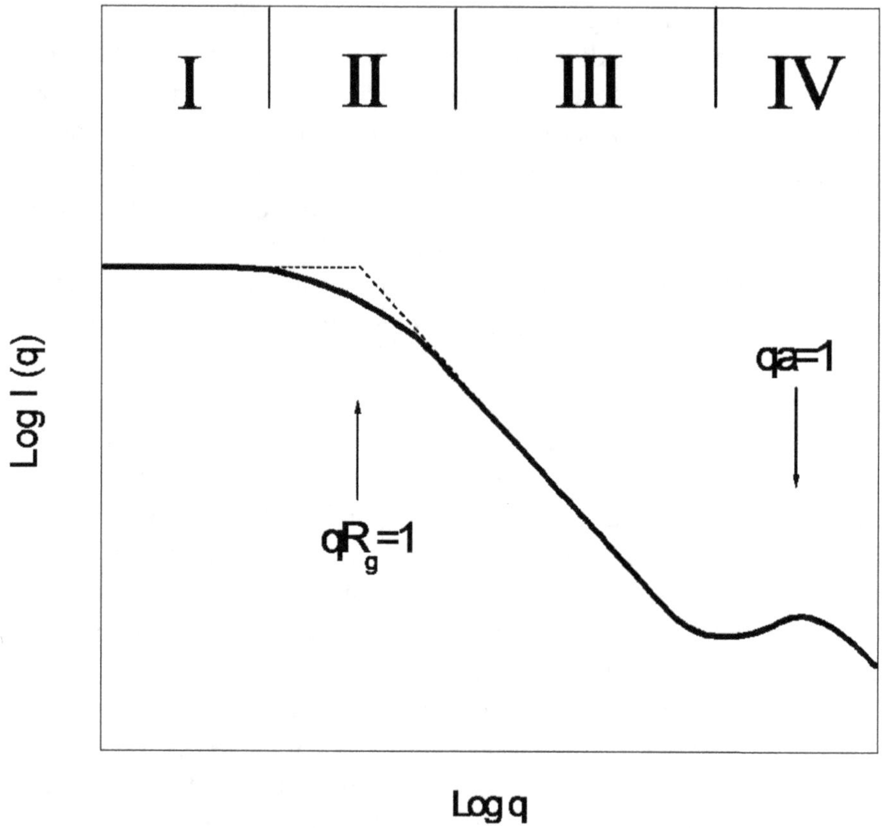

the parameters of the fractal structure. In particular, the fractal dimension is determined from the slope of the intensity curve (in the coordinates *log (I) - log (q)*) in the region of the Porod's, P.

If the absolute value of P is in the range from 1 to 3, fractal aggregates belong to the type of mass fractals and fractal dimension is equal to D_f = ½P½. The absolute value of P in the range from 3 to 4 corresponds to the surface fractal, the fractal dimension of which is determined from the ratio $D_s = 6$ - ½P½. From the slope (α) of the scattering curve in the Guinea's region (in the coordinates log (I) - q^2) is determined, according to the ratio $\alpha = -q^2 R_g^{2/3}$, the average radius of rotation of the fractal aggregate. The position of the maximum of the curve in the Bragg's region allows us to determine the average value of the interplanar distance in the quasicrystalline structure of near order at the level of interatomic distances. Possible complications of the structure of this type can be associated (in the simplest case) with the presence of a two-level fractal structure and a more developed order in the arrangement of atoms of the quasicrystalline lattice. A schematic representation of the scattering curve for ordering of this type, proposed by the authors Roldugin (2003), is presented in Fig. 4.

In this case, it is assumed Roldugin (2003) that the quasicrystalline structure of the atomic environment is formed by particles with average size *a*. The latter form fractal aggregates of the surface fractal type with an average size of. These fractal formations are aggregated into mass fractals with an average radius of rotation R_g.

Many studies of fractal structures in sols are based on the use of sol-gel systems based on TEOS, formed in different conditions (Roldugin, 2003). Changes in the conditions of hydrolysis and condensation significantly affect the structure of silicates synthesized by the sol-gel method at scale levels of the order of 1–20 nm (Schaefer & Keefer, 1984). The following regularities are observed Park Y. (2001).

In two-stage synthesis, both acid-acid and acid-base catalysis, weakly branched structures such as mass fractals with a fractal dimension of 1.9 and 2.0, respectively, are formed. At the same time, one-stage bases catalyzed hydrolysis at equimolar water content and TEOS (1: 1) leads to the formation of fractal particles such as mass fractals with a fractal dimension $D_f = 2.8$.

In one-stage mainly catalyzed hydrolysis with a double excess of water (H_2O: TEOS = 2: 1) surface fractals with a fractal dimension Ds = 2.7 are formed. Further increase in water content leads to the formation of non-fractal smooth colloidal particles. The following trend is observed: acid-catalyzed hydrolysis at low water content is accompanied by the formation of weakly branched polymer sols. At the same time at the basic hydrolysis and the high water content highly condensed sols like nanoparticles are formed.

Determining the size of fractal aggregates in the Guinier's region shows that for systems with a molar ratio of water / TEOS of the order of 4-5, with two-stage acid-acid catalysis, significantly (almost 4 times) increases the size of aggregates (already when diluted with ethanol 10 times) Mandelbrot B. (1982). At the same time, in the two-stage acid-base synthesis, with ten-fold dilution with ethanol, there are no changes in the Guinier radius Mandelbrot B. (1982).

In the two-stage acid-acid hydrolysis of TEOS at the scale level of 1-10 nm at small degrees of conversion at different stages of sol development, the fractal dimension of the mass fractal is slightly reduced Tretiakov Yu.D.

Figure 5. Schematic representation of the X-ray scattering curve in the coordinates log (I) - log (qR_g) for a two-level fractal structure and the developed order of atoms in the quasicrystalline lattice at the level of interatomic distances (Bragg region)

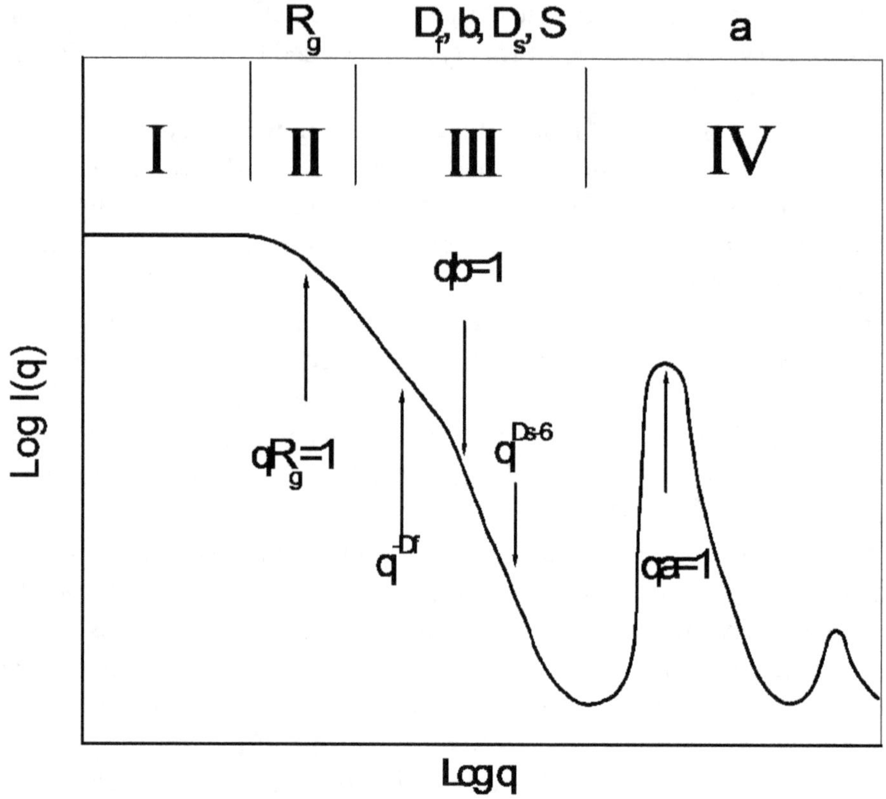

(2003). At later stages of condensation, the size of the scattering particles increases significantly with small changes in their number.

The fractal structure of gels is studied mainly at high dilutions and using methods of static or dynamic light scattering. Comparison of the obtained values of the fractal dimension at higher scale levels (34 - 440 nm) with the fractal dimension at scale levels of the order of 10 nm shows that the factors controlling the structure at both levels are significantly different.

Of particular interest are the processes of conversion of the gel into a xerogel (dried gel). The methods of small-angle scattering of X-rays and neutrons Park Y. (2001) show that xerogels, depending on the conditions of their synthesis, form three types of structures: non-fractal, mass fractals and surface fractals. In particular, the xerogel prepared by one-stage hydrolysis is a surface fractal with dimension $D_s = 2.7$, and its fractal dimension practically does not change at scale levels from 0.3 to 5.0 nm Tretyakov Yu.D. (2003). It is noteworthy that the features of the fractal surface of this gel are preserved during drying, although its volume is reduced by almost 50%. For xerogels obtained by slow drying of the gel obtained by acid hydrolytic polycondensation TEOS (H_2O: TEOS = 2: 1), the absence of a fractal structure on a scale level of the order of 0.1–1.0 nm was observed. At the same time, the source ashes were fractals at this scale. This means that the fractal structure inherent in the original precursor is lost during drying. Thus, the fractal features of the structure of silicates obtained by the sol-gel method are better preserved when drying gels based on systems synthesized under conditions of basic rather than acid catalysis Tretyakov Yu.D. (2003).

Recent X-ray low-angle data indicate that gels formed during single-stage acid hydrolytic polycondensation of TEOS, after 20 days of drying at 60°C, form surface fractals with Ds = 2.5 Schaefer D.W., Keefer K.D. (1984). Their further heating from 120 to 300°C leads at 120°C to the sequential formation of extremely coarse surface fractals with $D_s = 3.0$; at 150 ° C - mass fractals with $D_f = 2.4$; at 200°C - the origin of the combination: surface fractal (D_s = 2,4) + mass fractal (D_f = 2,4); at 300°C - to the formation of surface fractals with a perfectly smooth surface ($D_f = 2.0$) (with the simultaneous disappearance of mass fractals). Further heating to 1100°C does not change the characteristics of fractal dimensions.

TEMPLATE SYNTHESIS OF MATERIALS WITH A STRUCTURE ORDERED AT THE MOLECULAR LEVEL

Template synthesis is one of the most successful methods of sol-gel technology, which allows to obtain nanocomposite materials that meet the following requirements Stein A. (2001):

- A certain size and shape of crystallites or crystals;
- Narrow distribution of pore size in a given range;
- Formation at the molecular level of a specific structure of the nanocomposite, for example, a material with anisotropic organization at the meso level (10-1000 nm).

The use of templates is known in metallurgy. One of the ancient directions is to change the shape of crystallites formed from solutions and melts of

Figure 6. Molecular design of silicate gel when introduced into its structure non-covalently bound template agent

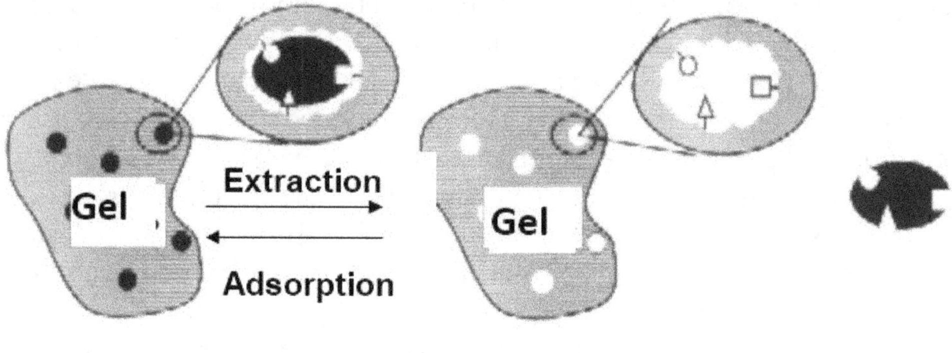

different nature. In sol-gel technology, where the term templates or templates are most often used (from the English template - template, stencil), template synthesis has been consciously used relatively recently.

The addition of surfactant cetyltrimethylammonium chloride to the sol formed in the $Si(OEt)_4$ - H_3PO_4 system promotes the formation of a xerogel with uniformly distributed in volume material and homogeneous in size cylindrical pores Loy Douglas A. (2001).. On the contrary, in the absence of this template, pores are usually formed, unevenly distributed in the material, the size of which varies widely.

The use of techniques of template sol-gel synthesis has become widely used to create silicate materials with adjustable porosity. Silicate nanocomposites obtained by the sol-gel method are chemically inert, hard and have a high degree of crosslinking of the silicate skeleton. Such materials also have the ability to incorporate organic matter, which can then be removed, chemically or thermally.

Srebnik S.& Lev O. (2003)consider three main methods for introducing a template agent into a silicate mesh. The most common of these is based on the formation of a gel in the presence of a non-covalently bound template, which is removed after gelation (Fig. 5).

In this variant of synthesis, the size of the templates can vary widely: from a few angstroms to several microns. Relatively small molecules, such as methyl orange, or larger formations, such as latex beads or bacteria, can be used to generate molecular inclusions or to create controlled porosity in silicate nanocomposites obtained by the sol-gel method.

Another approach is to copolymerize two precursors, followed by selective removal of one of them (Fig. 6). This class of template agents is broader because it generalizes the entire range of monomers that have mutual affinity. In particular, the synthesis of macroporous materials by copolymerization of polyvinyl alcohol or polyethylene oxide with tetraethoxysilane is quite successful. Phase segregation of the introduced polymerized second phase and its removal by heat treatment leads to the formation of macroporous materials with a wide distribution of pores in size.

The third method is based on the hydrolytic polycondensation of tetraethoxysilane with organyl or organoxytrialkoxysilane. The structure of the organyl or organoxy group bound to the silicon atom (after its firing or hydrolysis) determines the size and shape of the pores (Fig. 7) (Schaefer D.W., Keefer K.D., Aubert J.H. & Rand P.B., 1986). As pore-forming agents, in particular, can be used bornyloxytriethoxysilane or phenyloxytriethoxysilane.

Adjacent to this method is the sol-gel synthesis of organo-inorganic materials by molecular engineering (molecular design). An example of directed synthesis of hybrids through the use of such precursors is shown in Fig. 8 (Ben, Boury, Corriu, Delord, & Nobili, 2001). The building blocks are specially synthesized precursors containing an organic fragment of a certain topology associated with hydrolytically unstable trialkoxyl groups. It is possible to obtain materials with anisotropic properties, as shown in Fig. 9 (Semichenko, 1997).

However, the desired result can be achieved only when selecting the conditions for hydrolysis and polycondensation of selected precursors, ie with the optimal selection of starting components of sol-gel systems: not only precursors, but also solvents and catalysts of the hydrolysis process.

CONCLUSION

The section considers the historical milestones of the formation of sol-gel technology for the production of multifunctional hybrid silicate materials with technically valuable properties from silicons and metals. Advances in this field of science and technology are associated with the names of such famous scientists as J.D. Mackenzie, C.J. Brinker, G.W. Scherer, S. Sakka, H. Schmidt, G.L. Wilkes, T. Minami B.E., Yoldas, and others, whose studies have allowed the formation of sol-gel synthesis.

The products of nano- and microtechnological methods of sol-gel synthesis are diverse, multifunctional and, as a rule, are nanocomposites. The advantage of the sol-gel method is the simplicity of technology, low energy consumption,

Figure 7. Template sol gel synthesis that involves-the stage of copolimerisation of two precursores (a) and selective remiving one of them (b)

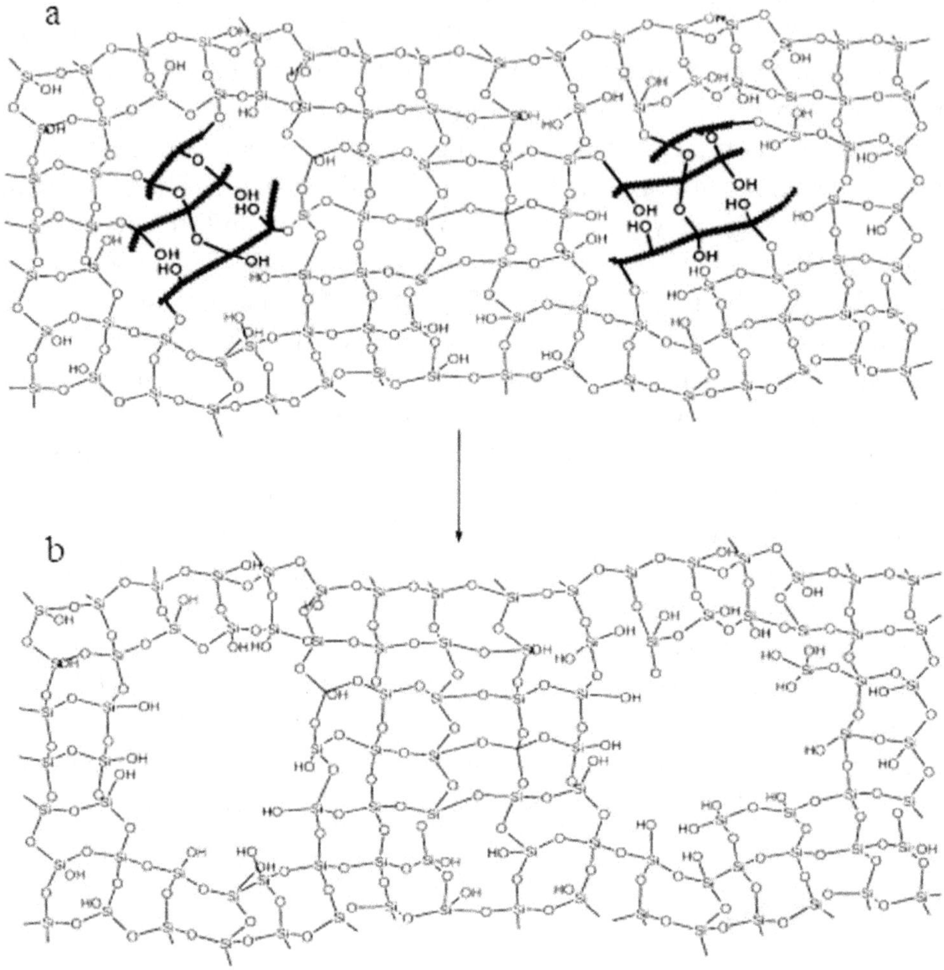

relative environmental safety and the ability to obtain a variety of silicate and hybrid materials with specified properties.

At this time, the use of sol-gel technology has become especially relevant for the formation of new silicate materials and coatings required for energy. Of particular interest are new products of sol-gel synthesis such as catalytic coatings for gas sensors and next-generation electrochemical devices, which are extremely necessary in transport engineering, as well as to ensure the safety of harmful industries, mining, agriculture and everyday life. There is an urgent need to create a sol-gel method of new solid membrane materials

Figure 8. Template sol gel synthesis of silicate porous nanocomposites based on tetraethoxisilane $Si(OCH_3)_4$ and organothree etoxisilane $ROSi(OEt)_3$. Symbol - organyl or organoxy group attached to a silicon atom

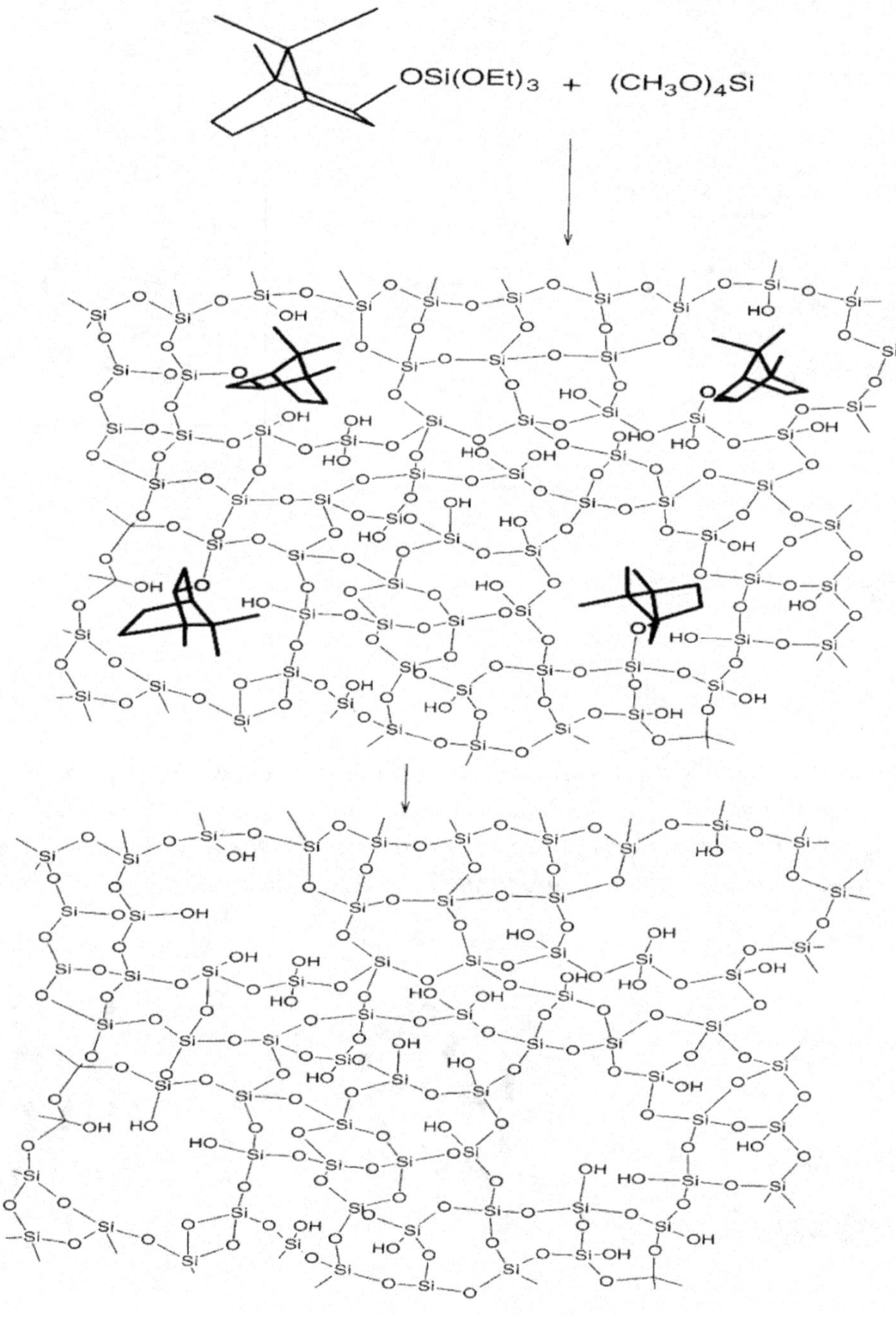

Figure 9. Sol gel synthesis by the method of molecular design

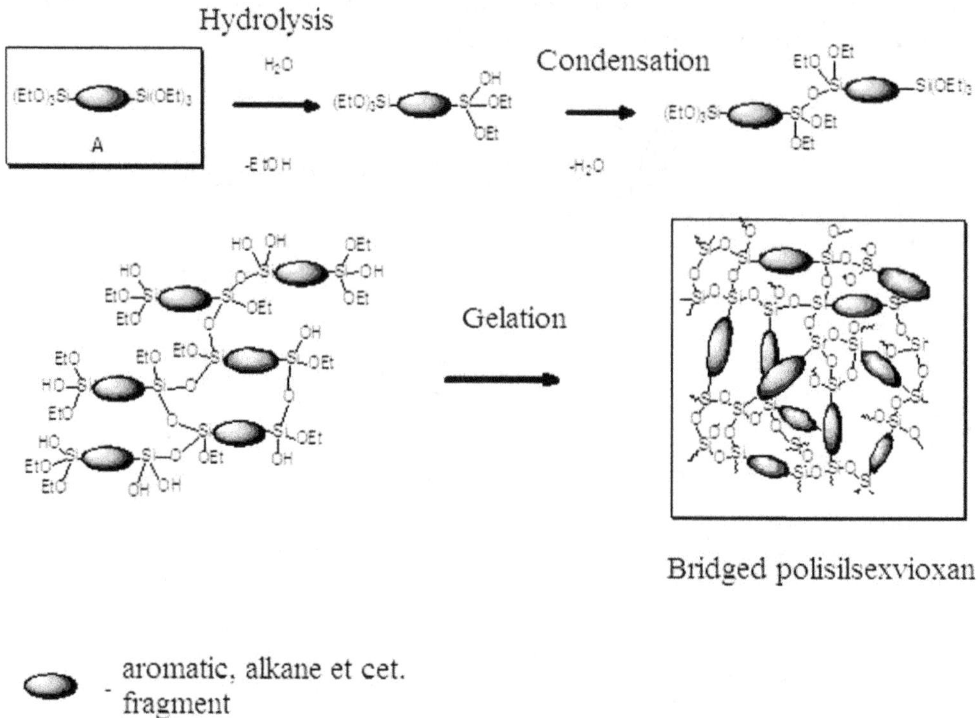

Figure 10. Method of synthesis of materials with anisotropic organization of molecular structure at the mesolevel (10 ... 1000 nm): a - precursor containing functional group R, having anisotropic physical properties (optical, magnetic, etc.); b - isotropic structure; c - anisotropic structure formed by external influence

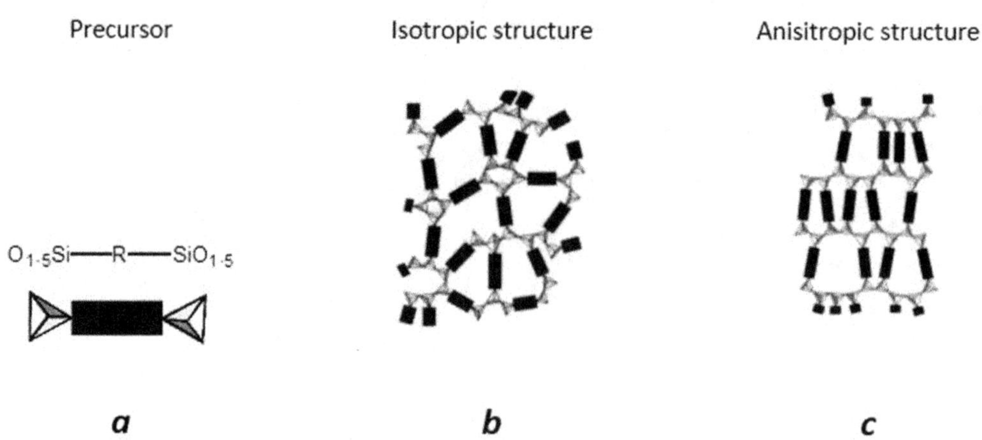

with high proton conductivity in a wide range of temperatures. In the same way, chemically stable coatings that scatter light for use in medicine, space and laser technology are already being created.

Sol-gel technology allowed to synthesize new silicate materials with the necessary technically valuable properties, due to the introduction into the silicate matrix of uniformly distributed heteroatoms of metals or metalloids (dopants). At this time, it was thought that traditional silicate materials have already reached the limits of their technical and operational parameters. Therefore, the development of a new class of silicate hybrid materials - organo-inorganic - has become extremely important. To obtain them, they began to use template sol-gel synthesis, which allows molecular assembly of nanocomposites.

All this indicates that the creation of new materials and coatings required for technical devices of the XXI century, requires further study of the method of sol-gel synthesis on a modern scientific basis. This requires, in particular, systematic studies of the kinetics of hydrolysis and polycondensation processes in sol-gel systems based on TEOS, in the presence of organic modifiers, as well as the use of new alkoxy-, acyloxy- and halogen-derived metals and their complexes in sol-gel synthesis. Particular attention should be paid to the study of physicochemical processes occurring in a thin layer of sols on the surface and in the pores of matrix materials.

Obviously, to establish the mechanism of formation of noncrystalline systems (sols and products of sol-gel synthesis) it is necessary to apply not only a set of modern research methods, but also new approaches, such as studying the fractal structure of sols, gels and synthesized silicate nanocomposites. It is especially promising to use the concept of fractals to describe porous materials for electrochemical purposes, such as, for example, organo-inorganic silicophosphate xerogels for proton-conducting membranes of fuel cells - sources of electrical energy of the future.

Thus, we have to admit that at this time there is a whole set of unsolved scientific and technological problems of sol-gel synthesis of hybrid silicate materials with technically valuable properties and necessary for energy and other industries.

REFERENCES

Abdollahi, A., Roghani-Mamaqani, H., Salami-Kalajahi, M., Mousavi, A., Razavi, B., & Shahi, S. (2018). Preparation of organic-inorganic hybrid nanocomposites from chemically modified epoxy and novolac resins and silica-attached carbon nanotubes by sol-gel process: Investigation of thermal degradation and stability. *Progress in Organic Coatings*, *117*, 154–155. doi:10.1016/j.porgcoat.2018.01.001

Adnan, M. M., Nylund, I.-E., Jaworski, A., Hvidsten, S., Ese, M.-H. G., Glaum, J., & Einarsrud, M.-A. (2021). The Structure, Morphology, and Complex Permittivity of Epoxy Nanodielectrics with In Situ Synthesized Surface-Functionalized SiO_2. *Polymers*, *2021*(13), 1469. doi:10.3390/polym13091469 PMID:34062857

Agrawal, R. C., & Gupta, R. K. (1999). Superionic solid composite phase – an overview. *Journal of Materials Science*, *34*(6), 1131–1162. doi:10.1023/A:1004598902146

Al Khudhair, A., Bouchmella, K., Andrei, R. D., Mehdi, A., Mutin, P. H., & Hulea, V. (2021). One-step non-hydrolytic sol-gel synthesis of mesoporous SiO_2-Al_2O_3-NiO catalysts for ethylene oligomerization. *Microporous and Mesoporous Materials*, *322*, 111165. doi:10.1016/j.micromeso.2021.111165

Ali S.M., Emran K.M. (2022). Surfactant-template sol-gel protective coatings for cast iron corrosion: Effect of the surfactant type and concentration, *Journal of Molecular Liquids*, *367*, 120366.

Baig, N., Kammakakam, I., & Falath, W. (2021). Nanomaterials: A review of synthesis methods, properties, recent progress, and challenges. *Mater. Adv.*, *2021*(2), 1821–1871. doi:10.1039/D0MA00807A

Barshilia, H. C., Lakshmi, R. V., Bera, P., Basu, B. B. J., Srinivas, G., Kumar, V. P., & Manikandanath, N. T. (2022). Transition metal nitride/oxide-based multilayer PVD coating with sol–gel derived ormosil passivation layer as an efficient solar absorber: Studies on high temperature stability and performance evaluation. *Solar Energy*, *239*, 283–293. doi:10.1016/j.solener.2022.05.016

Ben, F., Boury, B., Corriu, R. J. P., Delord, P., & Nobili, M. (2001) Anisotropic silica–based hybrid materials by sol–gel process. *11th International Workshop on Glasses, Ceramics, Hybrids and Nanocomposites from Gels (Sol–Gel 1981–2001): abstract*, (pp. 19–20). University of Padova.

Bokov, D., Jalil, A. T., Supat Chupradit, S., Suksatan, W., Ansari, M. J., Shewael, I. H., Valiev, G. H., & Kianfar, E. H. (2021). Nanomaterial by Sol-Gel Method: Synthesis and Application. *Advances in Materials Science and Engineering, 2021*, 5102014. doi:10.1155/2021/5102014

Brinker, C. J., Keefer, K. D., Schaefer, D. W., & Ashley, C. S. (1982). Sol–gel transition in simple cilicates I. *Journal of Non-Crystalline Solids, 48*(1), 47–64. doi:10.1016/0022-3093(82)90245-9

Brinker, C. J., & Scherer, G. W. (1990). *Sol–Gel Science.The Physics and Chemistry of Sol–Gel Processing*. Academic Press, Inc.

Broadstreet, S. W. (1955). Solution ceramic for enameling *Ceram. Age, 66*(6), 24–27.

Cheng, L., Cai, J., & Ke, Y. (2020). Ultrasonic-Assisted Sol–Gel Synthesis of Core–Shell Silica Particles for High-Performance Liquid Chromatography. *Journal of Inorganic and Organometallic Polymers and Materials, 30*(3), 859–868. doi:10.100710904-019-01239-4

Dahmouche, K., Atik, M., & Mello, N.C. (1998). New Li+ ion–conducting ormolytes. *Solar Energy Materials and Solar Cells, 54.* 1–8.

Ebelmen (1844). Sur les éthers siliciques [On silicic ethers]. *Journ. de Pharm, 5*(1), 262–264.

Ebelmen (1846). Recherches sur les combinaisons des acides borique et silicique avec les éthers [Research on the combinations of boris and silicic acids with ethers]. *Annales de Chemle et de Physique, 16*(3), 129–166.

Fidalgo, A., & Ilharco, L. M. (2003). Thickness, morphology and structure of sol–gel hybrid films: I – the role of the precursor solution's ageing. *Journal of Sol-Gel Science and Technology, 26*(1–3), 363–367. doi:10.1023/A:1020756728315

Friedel C., Crafts J.M. (1867) Research on the ethers of silic acid. *The American Journal of Science and Arts. Ser. 2. 43*(127, 128, 129), 155–171.

Friedel C., J.M. Crafts (1870) The combination of silicon with alcoholic radicals *The American Journal of Science and Arts*. Ser. 2., 49(145, 146, 147), 307–330.

Gautier–Luneau, A., Denoyelle, A., Sanchez, J. Y., & Poinsignon, C. (1615–1618). Denoyelle, J.Y. Sanchez C. Poinsignon (1992) Organic–inorganic protonic polymer electrolytes as membrane for low–temperature fuel cell. *Electrochimica Acta, 37*(9), 1615–1618. doi:10.1016/0013-4686(92)80122-3

Geiculescu, A. C., & Rack, H. J. (2002). X–ray scattering studies of polymeric zirconium species in aqueous xerogels. *Journal of Non-Crystalline Solids, 306*(1), 30–41. doi:10.1016/S0022-3093(02)01056-6

Gillett, S. L. (2002) Foresight Institute. *Nanotechnology: Clean energy and resources for future. White paper for Foresight Institute.* University of Nevada.

Gomes, A. L. M., Andrade, P. H. M., Palhares, H. G., Dumont, M. R., Soares, D. C. F., Volkringer, C., Houmard, M., & Nunes, E. H. M. (2021). Facile sol–gel synthesis of silica sorbents for the removal of organic pollutants from aqueous media. *Journal of Materials Research and Technology, 15*, 4580–4594. doi:10.1016/j.jmrt.2021.10.069

Gray, F. M. (1997). Polymer Electrolytes. RSC Materials Monographs, 175. London: The Royal Society of Chemistry.

Greil, P. (1998). Near net shape manufacturing of polymer derived ceramics. *Journal of the European Ceramic Society, 18*(13), 1905–1914. doi:10.1016/S0955-2219(98)00129-0

Haruvy, Y., Ryabov, Y., Arkhipov, V., Gutinz, A., Axelrod, E., & Feldman, Y. (2002). Fast-sol-gel derived silsequioxane glasses embodying glycerol moieties: Dielectric properties and morphology. *Journal of Non-Crystalline Solids, 305*(1–3), 226–234. doi:10.1016/S0022-3093(02)01096-7

Huang, H.-H., & Wilkes, G. L. (1987). Structure–property of behavior of a TEOS–poly(tetramethylene oxide) hybrid system prepared by the sol–gel process. I. Effect of PTMO molecular weight. *Papers Presented at the ... Meeting. American Chemical Society. Division of Polymer Chemistry, 28*(2), 244–245.

Imai, Y., Naka, K., & Chujo, Y. (1998). Reversible formation of Penetrating Polymer Network Structure in Organic–Inorganic Polymer Hybrid. *Polymer Journal, 30*(12), 90–95. doi:10.1295/polymj.30.990

Ishikawa, M., Ihara, M., Morita, M., & Matsuda, Y. (1995). Electric double layer capacitors with new gel electrolytes. *Electrochimica Acta, 40*(13–14), 2217–2222. doi:10.1016/0013-4686(95)00166-C

Iwamoto, T., & Mackenzie, J. D. (1995). Hard ormosils prepared with ultrasonic irradiation. *Journal of Sol-Gel Science and Technology*, *4*(2), 141–150. doi:10.1007/BF00491680

Kamila, R., Ridwan, M., Akhir, M. P. M., Patriati, A., & Insani, A. (2022). Synthesis of silic a particles through conventional sol-gel and sonochemistry methods and the effect of catalyst, water concentration and sample environment to the particle size. *Journal of Physics: Conference Series*, *2193*(1), 012044. doi:10.1088/1742-6596/2193/1/012044

Kazuki, M., Hu, Y., & Mackenzie, J. D. (1994). The effects of Ultrasonic irradiation on the Preparation and Properties of Ormosils. *Journal of Non-Crystalline Solids*, *3*, 109–116.

Klein, L. C. (1996) Processing of nanostructured sol–gel materials In A. S. Edelstein & R. S. Cammarata (eds), Nanomaterials: synthesis, properties and applications, 147–165. Bristol and Philadelphia: Institute of Physics Publishing. doi:10.1887/0750305789/b712c7

Koo, J. (2006). *Polymer Nanocomposites: Processing, Characterization, And Applications: Processing, Characterization, And Applications*. McGraw Hill Professional.

Larminie, J., & Dicks, A. (2003). *Fuel Cells Systems Explained* (2nd ed.). John Wiley & Sons. doi:10.1002/9781118878330

Liu J., Li Y., Tan N., Hu Z., Zhou Y., Deng Q., Zhang G., Lu Z. (2023) Microstructure and properties of the solid solution ceramic coating by high speed laser cladding, *Optics & Laser Technology*, *158.*

Liu, W., Yi, S., Liu, Z., Mo, Q., Xian, X., Ma, L., & Li, Y. (2022). Sol-gel silica modified recovered lignin particles and its application as blend with HDPE. *Industrial Crops and Products*, *189*, 115842. doi:10.1016/j.indcrop.2022.115842

Liu, Y., Lin, Q., Chen, J., Shao, Y., Wang, Y., & Wang, J. (2022). PDMS-OH and nano-SiO2 Modified KH570-TEOS silica-sol coating and protective effect on concrete. *Colloids and Surfaces. A, Physicochemical and Engineering Aspects*, *648*, 129279. doi:10.1016/j.colsurfa.2022.129279

Loy Douglas, A.(2001). Bridget polysilsesquioxanes. Molecular engineering of hybrid organic–inorganic materials. *11th International Workshop on Glasses, Ceramics, Hybrids and Nanocomposites from Gels (Sol–Gel 1981–2001): abstract*, (pp. 20–21). University of Padova.

Mandelbrot, B. (1982). The Fractal Geometry of Nature. *The Freeman*.

Marr, C., & Li, X. (1999). Composition and performance modelling of catalyst layer in a proton exchange membrane fuel cell. *Journal of Power Sources*, *77*(1), 17–27. doi:10.1016/S0378-7753(98)00161-X

Matejka, L., Plestil, J., & Dusek, K. (1998). Structure evolution in epoxy–silica hybrids–sol–gel process. *Non–Crystal. Solids*, *226*, 114–121.

Nogami, M., & Moriya, Y. (1980). Glass Formation hydrolysis of $Si(OC_2H_5)_4$ with NH_4OH and HCl solution. *Journal of Non-Crystalline Solids*, *37*(2), 191–201. doi:10.1016/0022-3093(80)90150-7

Olemskoy, A.I. Flat A.Ya. (1989). Ispolzovaniye kontseptsii fraktala v fizike kondensirovannoy sredy [Using the fractal concept in condensed matter physics]. *Успехи физических наук [Advances in the physical sciences]*, *163*(12), 1–50.

Ou, D. L., Chevalier, P. M., Mackinnon, I. A., Eguchi, K., Boisvert, R., & Su, K. (2003). Studies on highly porous hybrid prepared by a novel fast gellation process under ambient pressure. *Journal of Sol-Gel Science and Technology*, *26*(1/3), 657–662. doi:10.1023/A:1020737618745

Park, Y. (2001). Fractal geometry of porous materials. *Fractals*, *8*(3), 301–306.

Pell, W. G., & Conway, B. E. J. (2001). Analysis of power limitations at porous supercapacitor electrodes under cyclic voltametry modulation and DC charge. *Power Soureces*, *96*(1), 57–67. doi:10.1016/S0378-7753(00)00682-0

Pinero, M., La Rosa–Fox, N. D., Erce–Montilla, R., & Esquivias, L. (2003). Small angle neutron scattering study of PbS quantum dots synthetic routes via sol–gel. *Journal of Sol-Gel Science and Technology*, *26*(1–3), 527–531. doi:10.1023/A:1020755517353

Pirmoradian M., Al-Bakhakh B.A.J., Behroozibakhsh M., & Pedram P. (2022). Repairability of aged dimethacrylate-free ORMOCER-based dental composite resins with different surface roughening methods and intermediate materials, *The Journal of Prosthetic Dentistry*. doi:. doi:10.1016/j.prosdent.2022.04.001

Rivero-Antúnez, P., Zamora-Ledezma, C., Sánchez-Bajo, F., Moreno-López, J. C., Anglaret, E., & Morales-Flórez, V. (2023). Sol–gel method and reactive SPS for novel alumina–graphene ceramic composites. *Journal of the European Ceramic Society*, *43*(3), 1064–1077. doi:10.1016/j.jeurceramsoc.2022.10.043

Roldugin, V. I. (2003). Svoystva fraktalnykh dispersnykh system [Properties of fractal disperse systems]. [in Russian]. *Uspekhi Khimii*, (11), 1028–1054.

Sá Ferreira, R.A., & Carlos, R.D., V. De Zea Bermude [et al]. (2003). Room temperature visible/infrared emission and energy transfer in Nd^{3+}–based organic/inorganic hybrids. *Sol–Gel. Sciences et Techniques (Paris)*, *26*(1–3), 315–319.

Schaefer, D. W., & Keefer, K. D. (1984). Fractal geometry of silica condensation polymers. *Physical Review Letters*, *53*(14), 1383–1386. doi:10.1103/PhysRevLett.53.1383

Schaefer, D. W., Keefer, K. D., Aubert, J. H., & Rand, P. B. (1986). *Science of Ceramic Chemical Processing* (L. L. Hench & D. R. Ulrich, Eds.). Wiley.

Schaefer, D. W., Martin, J. E., & Keefer, K. D. (1985). *Physics of Finely Divided Matter*. Springer–Verlag.

Schubert, U. (2003). Silica–based and transition metal–based inorganic–organic hybrids materials – a comparison *Sol–Gel. Sciences et Techniques (Paris)*, *26*(1–3), 47–55.

Semichenko G.D. (1997) *Zol–gel protsess v keramicheskoy tekhnologii* [*Zo-gel process in ceramic technology*]. Kharkov: Khark. politekhn.

Shin, S.-J., Lee, J.-K., Ha, H.-Y., Hong, S.-A., Chun, H.-S., & Oh, I.-H. (2002). Effect of the catalytic in preparation method on the performance of polymer electrolyte membrane fuel cells. *Journal of Power Sources*, *106*(1–2), 146–152. doi:10.1016/S0378-7753(01)01045-X

Siemens, H. A. G. (1902) Pat. 128253 Germany C., I. 1902. 448.

Smirnov V.M. (2003) *Sovremennyye vozmozhnosti upravleniya nanostrukturirovaniyem tverdykh veshchestv i materialov Khimiya tverdogo tela [Modern possibilities for controlling solid state chemistry]*. Khimicheskiye problemy sozdaniya novykh materialov. [Chemical problems of creating new materials], 57–87. Sankt–Peterburgskiy Gosudarstvennyy universitet.

Smith, G. N., Brok, E., Schmiele, M., Mortensen, K., Bouwman, W. G., Duif, C. P., Hassenkam, T., Alm, M., Thomsen, P., & Arleth, L. (2021). The microscopic distribution of hydrophilic polymers in interpenetrating polymer networks (IPNs) of medical grade silicone. *Polymer*, *224*, 123671. doi:10.1016/j.polymer.2021.123671

Srebnik, S., & Lev, O. (2003). Theoretical investigation of imprinted crosslinked silicates. *Journal of Sol-Gel Science and Technology*, *26*(1–3), 107–113. doi:10.1023/A:1020701721529

Tretiakov, Yu. D. (2003). Protsessy samoorganizatsii v khimii materialov [Self organization processes in materials chemistry]. [in Russian]. *Uspekhi Khimii*, *72*(8), 731–762.

Urban, P. M., Funke, A., Muller, J. T., Himmen, M., & Docter, A. (2001). Catalytic processes in solid polymer electrolyte fuel cell systems. *Appl. Catalysis. General*, *221*(1–2), 459–470. doi:10.1016/S0926-860X(01)00819-5

Vollet, D. R., Donatti, D. A., & Ibanez Ruiz, A. (2002). A SAXS study of the nanostructural characteristics os TEOS–derived sonogels upon heat treatment up to 1100 °C. *Journal of Non-Crystalline Solids*, *306*(1), 11–16. doi:10.1016/S0022-3093(02)01058-X

Vora, R. H., & Lau, K. S. Y. (2022) Subchapter 10.4 - Research and development of high-performance and thermomechanically stable Ceramer (polymers–ceramic composites materials) and 'CeramImide' nanocomposites material formulations based on fluoro-polyimides for the advanced industrial applications. In H. Dodiuk (ed), Plastics Design Library, Handbook of Thermoset Plastics (Fourth Edition), 517-585. William Andrew Publishing.

Wilde, G. (2021). *Nanostructured Materials*. Elsevier.

Zanurin, A., Johari, N. A., Alias, J., Mas Ayu, H., Redzuan, N., & Izman, S. (2022). Research progress of sol-gel ceramic coating: A review. *Materials Today: Proceedings*, *48*(Part 6), 1849–1854. doi:10.1016/j.matpr.2021.09.203

Chapter 2
Influence of Polyionenes on Rheology of Phospho-Organic Compounds

ABSTRACT

In this chapter authors present basic principles of process control of sol – gel synthesis of new class of organic-inorganic nanostructured materials on basis of tetraethoxysilane and polyionenes doped by orthophosphoric acid with improved functional properties. The chemical compositions improvement results from complex investigation of morphology, physical and chemical properties. Polyionenes are shown to increase the rate of hydrolytical polycondensation reaction of tetraethoxysilane. Raise of the polyionene molecular weight leads to increasing rate of gel formation process. The polyionenes in silicophosphate sols are stated to prevent crystallization processes occurred in silicophosphate xerogels with silica phosphates and pyrophosphates.

BACKGROUND

Sol – gel technology allow incorporating organic molecules into an inorganic network, so that they can be combined virtually at any ratio on the molecular level with the formation of hybrid organic–inorganic nanocomposite materials (Brinker & Scherer, 1990). The mechanical properties of such materials can

DOI: 10.4018/978-1-7998-9728-6.ch002

vary from brittle and hard to rubbery depending on the content of organic modifiers in the inorganic network.

In our earlier works we studied the influence of the introduction of small amounts of organic modifiers into silicophosphate matrix which improved the ionic conductivity and operating characteristics of membranes (Sukhyy et al., 2015). For this purpose, we used oligomeric alkylaromatic quaternary ammonium salts, i.e., polyionenes (Brinker & Scherer, 1990), (Yu et al. 2022). These compounds are appeared to be appropriate for the use in sol–gel processes, because they dissolved in water without breaking the homogeneity of the sol–gel system. These modifiers favor the formation of a xerogel that can be subsequently compacted.

The solution of an important practical problem of introduction into production of scientific developments is promoted by creation of scientific bases of formation of multipurpose hybrid organo-inorganic materials by means of the controlled and directed sol-gel synthesis. To achieve this goal, it was necessary to solve a number of specific problems that arose directly from the practical needs of industry:

- Formation of sol-gel by the method of new effective catalytic coatings for metal oxide humidity sensors,
- Creation of proton-conducting membranes that can be operated at temperatures above 100 °C,
- Creation of highly efficient heat-accumulating materials of sorption type with energy storage density in the range of 1.5-1.7 MJ / kg.

To do this, it was necessary to ensure a favorable course of the hydrolytic polycondensation of tetraethoxysilane in the presence of organic dopants. The introduction of the latter ensures the achievement of the required parameters of the formed materials. It was necessary to optimize the conditions of hydrolysis and polycondensation, both at the stage of sol formation and at the stage of formation of prototypes. An important condition for the preparation of dispersions (sol - metal oxide) are methods of homogenization of the starting components. An obligatory part of the study is the optimization of heat treatment regimes.

To predict the parameters of the created materials it was necessary:

- To study the optimal conditions of hydrolytic polycondensation of Si $(OEt)_4$, H_3PO_4 precursors in the presence of organic modifiers. This allowed to create new organo-inorganic nanostructured coatings on the surface of semiconductor and dispersed ceramic materials, thin-layer glass-ceramic coatings on metal surfaces and solid electrolyte proton-conducting membranes with a given composition, structure and required technical characteristics,

- To formalize the procedure for selecting components, optimizing their ratio in ash and conditions of sol-gel synthesis systems based on tetraethoxysilane using organic dopants,
- To study the influence of anions and cations of organic modifiers as template agents in sol-gel synthesis of nanoscale coatings on their morphology,
- To establish the relationship between the composition of the sol, the structure of the formed xerogel and the proton conductivity of silicophosphate nanocomposites.

EXPERIMENTAL

Silicophosphate sol–gel systems synthesized from tetraethoxysilane $Si(OEt)_4$, orthophosphoric acid H_3PO_4, ethanol EtOH (as a solvent), and hydrochloric acid HCl (as a catalyst of hydrolysis process) were the objects of investigation. The basic reactants were introduced in following mole ratio with variation of concentration of water and orthophosphoric acid: $Si(OC_2H_5)_4$:C_2H_5OH:H_2O:HCl: H_3PO_4= 1: 4: (6-8): 0.01: (0.4-0.8) (Table 1).

A large array of organic-inorganic nanocomposite was studied in present work. Therefore, sufficient rational marking by content of water (W_{mol}) and orthophosphoric acid (PA_{mol}), type of organic modifier and order of component mixing was suggested. Polyionenes (Table 2) were dissolved in water previously, and then introduced into sol comprising TEOS, ethanol, hydrochloric and orthophosphoric acid portionwise. Some sols were exposed to ultrasonic treatment (44 kHz, 20 min.) after that. Then, sols were retained in closed vessels until formation of gel which dried at vacuum during 8 h until coarse-grained powder formation. For investigation this powder was grinded in agathic mortar. Solid films for tests were formed on teflon substrates by watering method with following centrifugation.

The viscosity of the sols was measured on a Höppler rheoviscometer. Thermal analysis (DTA, DTG, and TG) was performed on a Paulik–Paulik–Erdey derivatograph. X-ray powder diffraction analysis was carried out on a Siemens D-500 X-ray diffractometer. IR-spectra were registered on IR-Fourier spectrometer Perkin-Elmer 1760X in transmission regime with resolution of 4 cm^{-1} and averaging of 50 spectra in spectral range of 0 – 4000 cm^{-1}.

Broadband isothermal dielectric relaxation spectroscopy (DRS) measurements were carried out in the frequency range 10^{-1}– 10^{6} Hz and the temperature range from 173 to 293 K, using a frequency response analyzer (Schlumberger SI 1260) supplemented by a buffer amplifier gain (Chelsea Dielectric Interface). The samples were placed between gold-coated brass electrodes.

Colloidal-Chemical Properties of Aqueous Solutions of Polyionenes

Sol-gel technology allows introducing organic molecules into the grid of inorganic polymer. These components can be combined at the molecular level in virtually any ratio, which allows the formation of hybrid organo-inorganic nanocomposite materials with new properties (Brinker & Scherer, 1990). Mechanically, such materials can be both brittle and elastic, depending on the amount of organic component in the mesh of the inorganic polymer. The main requirements for the use of organic modifiers in sol-gel processes are the presence of active functional groups capable of interacting with the organosilicon polymer, good dissolution in aqueous-alcoholic medium without disturbing the homogeneity of the sol-gel system.

Optimal synthetic stabilizers of nanoparticles are amphiphilic polymers - block or grafted copolymers, one of the fragments of which is a polymer insoluble in the dispersion medium, and the other, on the contrary, soluble in it. In this case, the anchor serves to form a bond with the surface (Shilov et al., 2000), and the second component creates a steric barrier required for stabilization. Such modification systems are called surfactants (polymeric surfactants)(Neagu, Neagu, Pissis, Das–Gupta, 1996). In solution, surfactants are microphasically divided into a number of structures, the simplest of which is the micelle formed in dilute solution in a selective solvent (Tadros, 1991).. Self-assembly of surfactants in micelles is one of the most well-known and well-studied phenomena (Tuzar & Kratochvill, 1993).

Polymeric surfactants are able to simultaneously perform several functions: size control, geometric control, and stabilization of nanosized particles. However, they are not only passive participants in the stabilization process: their hydrophobic and electrostatic interactions affect the chemical properties.

Table 1. Reactants ratio and sol – gel system state during silicophosphate and hybrid nanocomposite synthesis Sukhyy K.M., Gomza Yu. P. et al (2015)

<table>
<tr><th colspan="7">Molar concentration of components, mole/mole TEOS</th><th rowspan="2">Gel formation time, day</th><th rowspan="2">Mixing order</th></tr>
<tr><th>marking</th><th>TEOS N° 1</th><th>C_2H_5OH N° 2</th><th>H_2O N° 3</th><th>HCl N° 4</th><th>H_3PO_4 N° 5</th><th>Modifier N° 6</th></tr>
<tr><td>PA_1- W_8</td><td rowspan="15">1</td><td rowspan="15">4</td><td>8</td><td rowspan="15">0,01</td><td>1,0</td><td>-</td><td>6</td><td>1,2,4,5,3</td></tr>
<tr><td>W_8-PA_1</td><td>8</td><td>1,0</td><td>-</td><td>6</td><td>1,2,3,4,5</td></tr>
<tr><td>$PA_{0,5}$- W_8</td><td>8</td><td>0,5</td><td>-</td><td>8</td><td>1,2,4,5,3</td></tr>
<tr><td>W_8-$PA_{0,5}$</td><td>8</td><td>0,5</td><td>-</td><td>8</td><td rowspan="3">1,2,3,4,5</td></tr>
<tr><td>W_2-$PA_{1,5}$</td><td>2</td><td>1,5</td><td>-</td><td>1</td></tr>
<tr><td>W_2-$PA_{2,2}$</td><td>2</td><td>2,2</td><td>-</td><td>0</td></tr>
<tr><td>W_8-PA_1-I-1</td><td>8</td><td>1</td><td>I- 1
$8 \cdot 10^{-4}$</td><td>3</td><td rowspan="9">1,2,3,4,5,6</td></tr>
<tr><td>W_8-PA_1-I-1us</td><td>8</td><td>1</td><td>I- 1
$8 \cdot 10^{-4}$</td><td>3</td></tr>
<tr><td>W_{16}-PA_1-I-1us</td><td>16</td><td>1</td><td>I- 1
$8 \cdot 10^{-4}$</td><td>3</td></tr>
<tr><td>W_2-$PA_{0,1}$-I-2</td><td>2</td><td>0.12</td><td>I - 2
$9 \cdot 10^{-4}$</td><td>14</td></tr>
<tr><td>W_{24}-$PA_{0,1}$-I-2</td><td>24</td><td>0.12</td><td>I - 2
$2 \cdot 10^{-4}$</td><td>14</td></tr>
<tr><td>W_8-$PA_{1,5}$-
I-3us</td><td>8</td><td>0.12</td><td>I - 3
$1.2 \cdot 10^{-4}$</td><td>39</td></tr>
<tr><td>W_8-$PA_{1,5}$-
I-4us</td><td>8</td><td>0.12</td><td>I - 4
$1.2 \cdot 10^{-4}$</td><td>21</td></tr>
<tr><td>W_8-$PA_{1,5}$-
I-5us</td><td>8</td><td>0.12</td><td>I - 5
$1.2 \cdot 10^{-4}$</td><td>33</td></tr>
</table>

In water, 50 to 100 surfactant molecules above the critical micelle concentration (CCM) spontaneously bind to aggregates known as micelles. The formation of water micelles is a cooperative process and is determined by the forces of interaction between the polar main groups and the association between the hydrophobic chains of surfactants. In the general case, the longer the hydrocarbon tail, the lower the CCM.

To determine the adsorption activity of polyonenes, we studied the colloid-chemical properties (at a temperature of 298 K). Concentration dependences of the surface tension of aqueous solutions of salts at the boundaries of the "solution - air" distribution are presented in Fig. 1 - 6. The Rebinder method of measuring surface tension has significant limitations for polymer solutions, because the equilibrium time in the polymer solution is much longer than for solutions of low molecular weight compounds. In view of this, all measurements were performed by sequential dilution and concentration of

Table 2. Polyionenes structural formula Sukhyy K.M., Gomza Yu. P. et al (2015)

Logical number	Structural formula	Molecular weight
I-1	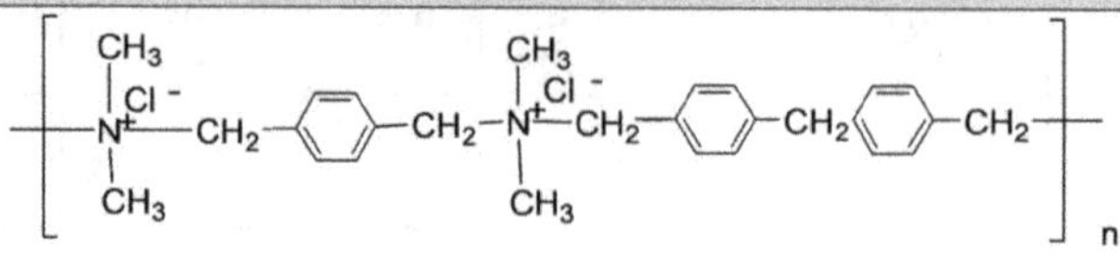	5900
I-2	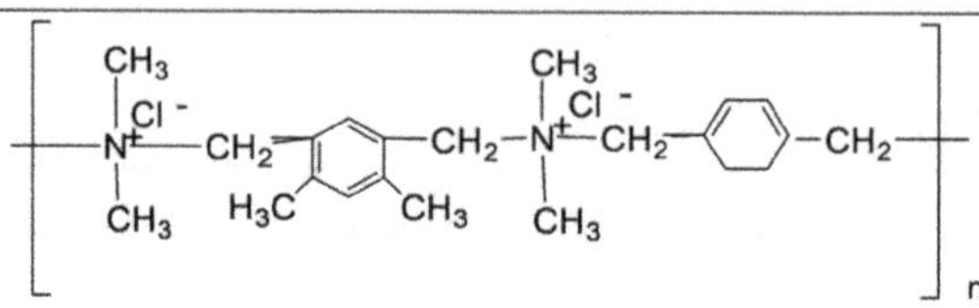	7200
I-3	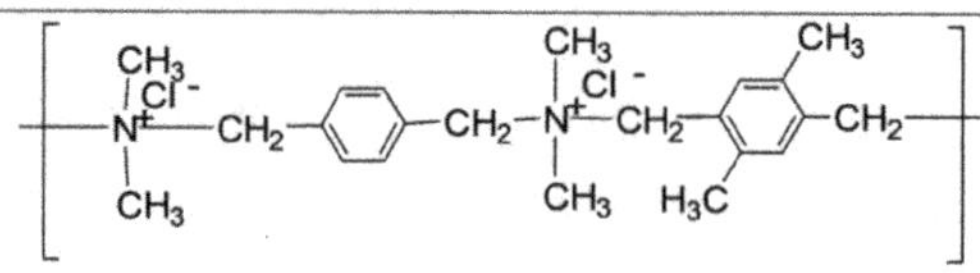	8400
I-4	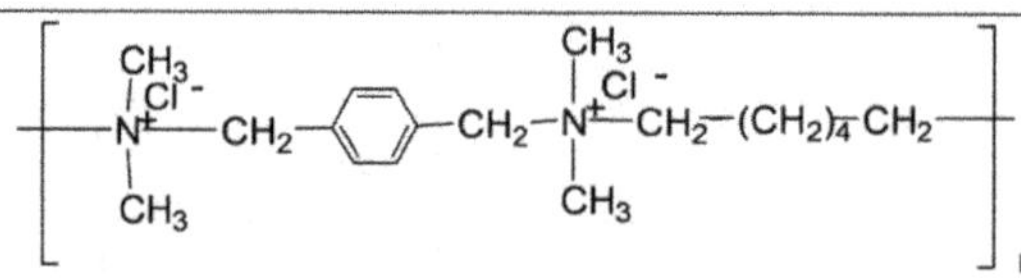	6100
I-5	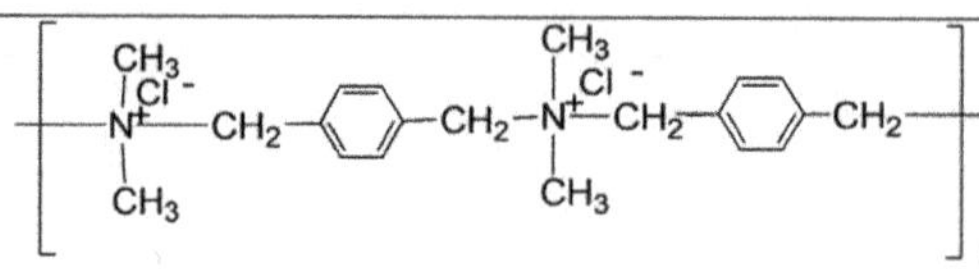	5600

polyionene solutions at least 10 times for each of the samples. The values obtained by both methods form a single curve. Dependencies have an extreme form, i.e. are characterized by the presence of an extremum in the region of low concentrations.

The position and significance of extrema depends on the structure of polyonenes. With the hydrophobization of the structure (replacement of the phenyl radical between the Quaternary nitrogen atoms by a dialkyl-substituted phenyl radical - transition from I-5 to I-1 and by a bionuclear phenyl radical - I-1, I-6) the surface activity increases and increases in the range: I-5 < I-3 < I-1 < I-6. With the introduction of the naphthalene fragment into the structure of polyonenes, the surface activity decreases significantly.

The results of the study of the surface tension of aqueous solutions of polyonenes are consistent with the results of measuring the reduced viscosity of these solutions Price. C. (1982), also presented in Fig. 1-6 - on the concentration dependences there are inflections or extremes at concentrations corresponding to the extremums on the concentration dependences of the reduced viscosity.

The degree of increase in the value of surface tension and reduced viscosity with decreasing concentration reflects the degree of expansion of the polymer ball in solution during dilution and depends not only on the flexibility of the polymer chain, but also on its hydrophobicity.

The obtained effect indicates complex processes that take place in aqueous solutions of polyonenes - micelle formation. The most important point in the study of the adsorption of polymer molecules, in contrast to the adsorption of low molecular weight compounds, is the variety of forms that can take the adsorbed macromolecule on the surface of the adsorbent. Therefore, the question of the conformation of the polymer molecule on the surface of the adsorbent is one of the central in the study of the adsorption of macromolecular compounds. The results of measuring electrical conductivity also testify in favor of micelle formation that occurs in aqueous solutions of polyonenes (Fig. 1 – 6). The results of determining the critical concentration of micelle formation (CCM) by the conductometric method are presented in the form of graphical dependences of the specific conductivity of polyionene solutions on the concentration.

In more concentrated solutions, polyonenes behave like inorganic strong electrolytes, i.e., the dependence $\sigma = f(C)$ is expressed by a straight line with a small slope. However, in contrast to conventional electrolytes, in polyionene solutions, at a concentration corresponding to the CCM, the graphs show an inflection, after which the conductivity decreases more abruptly with decreasing polyionene concentration. This is due to the fact that the formed ion micelles, although conducting current, but due to their large size are less mobile in an electric field than non-aggregated ions. In addition, a strong electrostatic attraction from charged micelles binds part of the counter ions on their surface. This blocking of electric charges reduces the electrical conductivity.

The CMC of the studied polymer quaternary ammonium salts increases with increasing hydrophobicity of macromolecules and increasing the distance between the quaternary nitrogen atoms, i.e., decreasing the concentration of ionic centers (Table 3).

Figure 1. Concentration dependences of specific conductivity, surface tension and reduced viscosity for I-5.

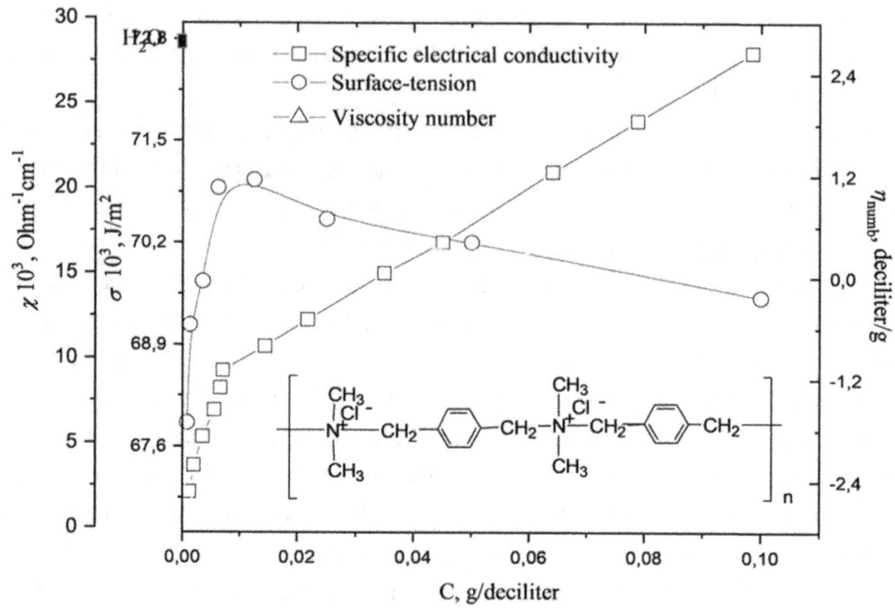

The adsorption of polyonenes at the interface "solution - porous adsorbent" was studied in accordance with the method described in Burmistr et al. (2003). The results of the study of adsorption activity are shown in Fig. 7.

The change in the adsorption values with decreasing concentration for different polyonenes indicates that the adsorption of organic cations depends on both the level of hydrophobicity of the polymer chain and the flexibility of the radicals in the chain. With the replacement of the phenyl radical by a binuclear phenyl radical or by a dialkyl-substituted phenyl radical, the adsorption of organic cations increases. Adsorption proceeds with a significant increase in the free energy of the system with increasing proportion of the hydrophobic component of the surfactant. This indicates an increase in the contribution of van der Waals forces in the total adsorption energy. With respect to polyonenes with naphthalene radicals, it seems that the adsorption activity is lower than the polymers discussed above.

This is due to the decrease in the flexibility of the polymer chain, due to the increase in the size of the radical. Thus, with increasing hydrophobicity of the polymer and decreasing the size of radicals (steric factor) there is

Figure 2. Concentration dependences of specific conductivity, surface tension and reduced viscosity for I-3.

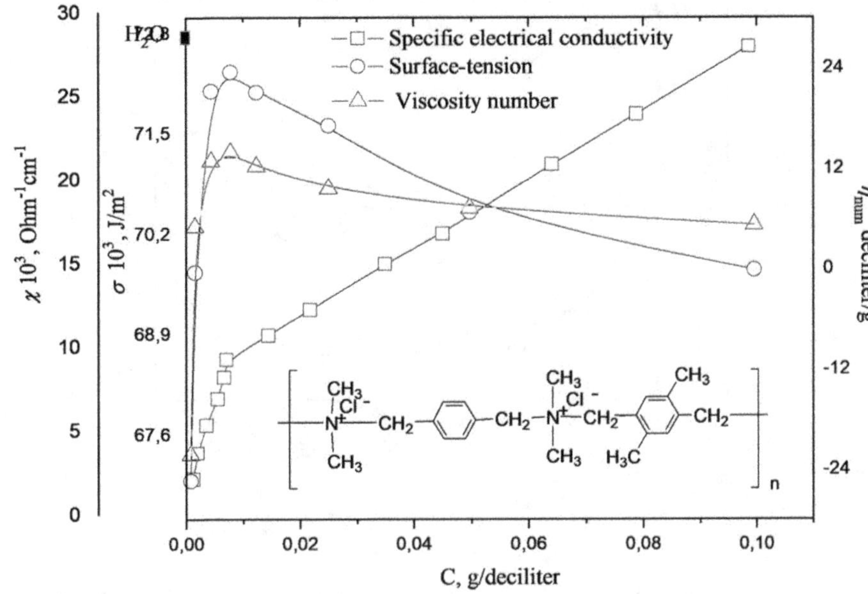

a decrease in entropy, which leads to a loss of mobility of cations during their transition to the adsorbed state Ha C.S. & Park S.S. (2019). Despite the significant adsorption of organic cations from the dispersed phase, the equimolar exchange of inorganic cations for organic ones almost never occurs, and the obtained organomineral derivatives always have a residual exchange capacity Ha C.S. & Park S.S. (2019). Incomplete substitution of inorganic cations for organic ones is explained by a steric factor, low diffusion rate inside the crystal lattice of minerals, as well as partial overlap of the ion exchange centers of the adsorbent with adsorbate molecules.

The adsorption of polyonenes from solutions is characterized by a complex concentration dependence. This is due to conformational structural changes in polymer solutions. The adsorption of polymers is significantly different from the adsorption of low molecular weight substances. These differences are explained not only by the macromolecular nature, but also by the fact that each concentration of the solution from which the adsorption is carried

Figure 3. Concentration dependences of specific conductivity, surface tension and reduced viscosity for I-1.

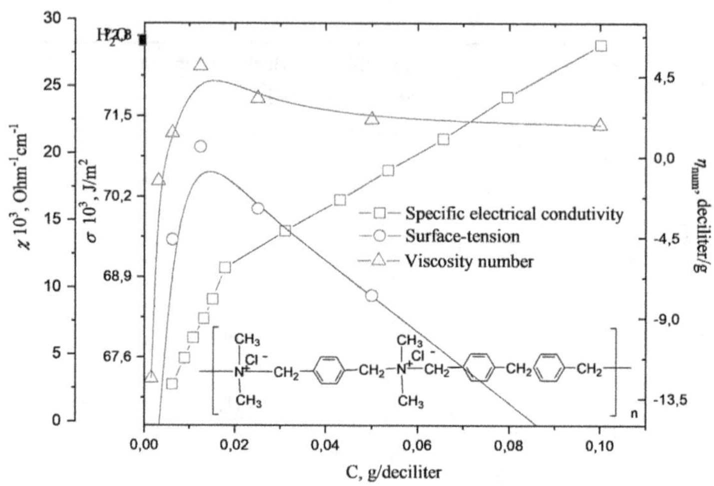

out corresponds to different conformations of the macromolecular ball, the degree of mutual overlap and interpenetration of balls, and the degree of their aggregation.

The values of the critical concentration of micelle formation were used to optimize the concentration of polyonenes in the composition of sol-gel systems. The results of phenomenological observations of changes in the state of phosphorosilicate sol-gel systems are shown in Table 1.

Addition to the sol $PA_{0,1}$- W_8- NH_4OH a small amount of polyonenes I-2, I-4 (TEOS: I = 1: 1.2-10-4), as well as the addition of ammonia, accelerates the processes of structuring sols (Table 1, W_2 -$PA_{0,1}$-I-2) their conversion into a gel is completed faster by 20 and 13 days, respectively.

We found that the presence of polyonenes in sols can contribute to both coagulation processes and stabilization of sols. In this case, the introduction of the sol I-2 in the amount of $6 \cdot 10^{-4}$ mol per 1 mol of TEOS reduces the aggregative stability of the sols. The following are examples where polyionenes increase the stability of sols. From the Table 1, it follows that a significant influence on the state of the sol-gel system has the order of mixing the components. Depending on this, homogeneous or heterogeneous (with sediment) sol-gel systems are formed.

Figure 4. Concentration dependences of specific conductivity, surface tension and reduced viscosity for I-6.

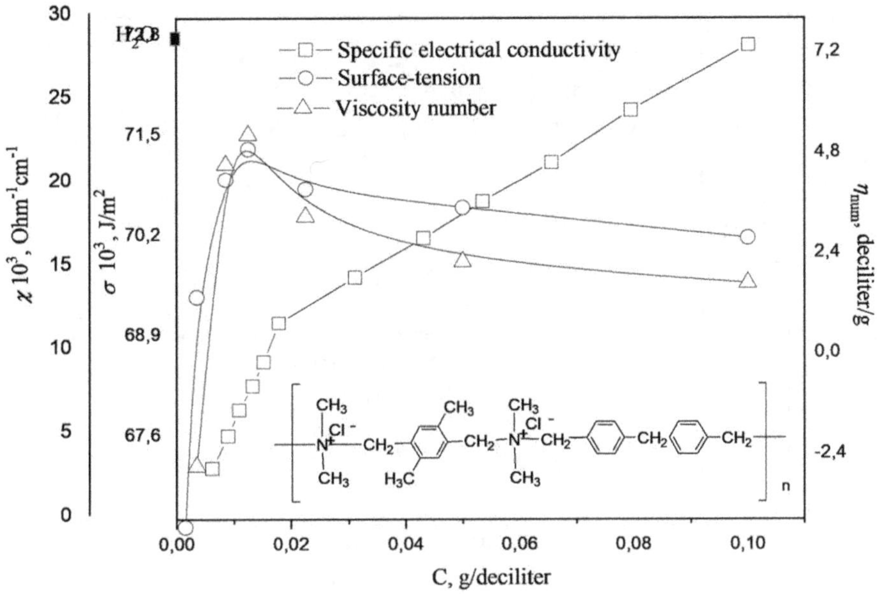

The sequence of water introduction (before or after orthophosphoric acid) also affects the appearance and properties of the resulting xerogel. If water is added at the very end: after mixing all the other components, a transparent sol is formed, after drying which forms a xerogel that crumbles easily (almost dry powder). If water is introduced at the beginning of the sol-gel process, an opalescent solution is formed, and after 6-8 days - a wet gel, which is difficult to dry. Thus, the structure and, as a consequence, the composition of the molded gels significantly depend on the order of mixing of the components (Table 2.1) Burmistr M.V. et al. (2005). Incomplete removal of water during the gel drying process is due to the acceleration of the rate of hydrolytic cleavage of ethoxy groups with the formation of free silanol groups Si-OH, which bind to excess molecular water by hydrogen bonds. As a result, xerogels of different porosity are formed, with the reduction of which it is difficult to remove sorbed water. When using the main catalyst (NH_4OH), as noted by the authors Tsvetkova N. et al. (2008), there is a rapid

Figure 5. Concentration dependences of specific conductivity, surface tension and reduced viscosity for I-7.

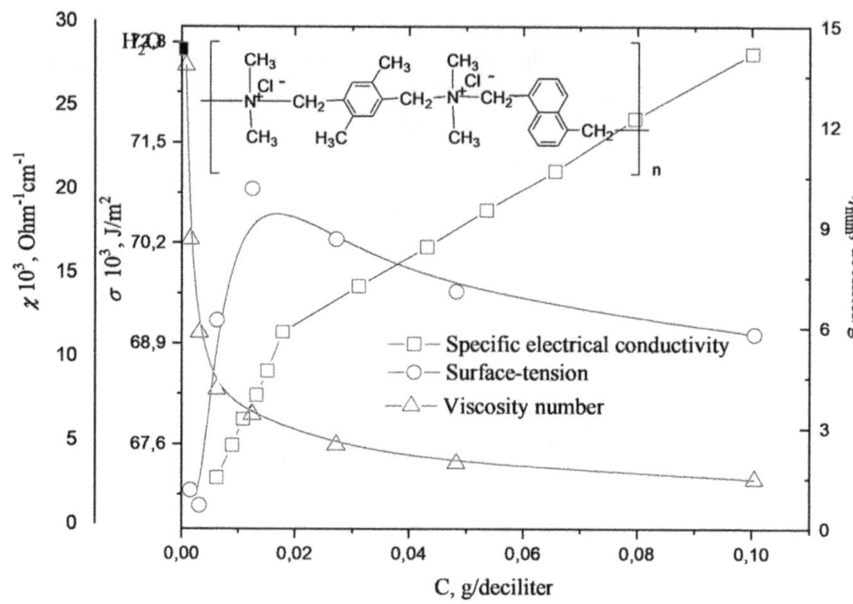

anhydrocondensation of silanol groups with the formation of SiO_2 particles Sviderskyy V. A., Voronkov M.G. & Klimenko S.V. (1996).

The time dependences of the change in the structural viscosity of the base sol without organic modifier (Table 1, W_8-PA_1), sol with the addition of polyionenes (W_8-PA_1-I-1) and sol with the addition of ammonia (W_8-PA_1-NH_4OH) are presented in Fig. 8. It is known that the processes of gel structure formation in sol-gel synthesis are significantly influenced by the amount of water and catalysts, primarily due to the acceleration of the hydrolytic polycondensation reaction of TEOS Tsvetkova I.N. et al. (2007).

As can be seen from Fig. 8, the process of gel formation, which occurs fairly quickly, is preceded by a maturation period in which the structural viscosity of all sols increases monotonically. The gelation of the starting sol W_8-PA_1 is completed in six days. The effect on the sol-gel process of polyonenes and ammonia is different. The introduction of polyonenes accelerates the formation, which ends in 3 days, and ammonia, on the contrary, continues the process of transformation into a gel (11 days).

Figure 6. Concentration dependences of specific conductivity, surface tension and reduced viscosity for I-8.

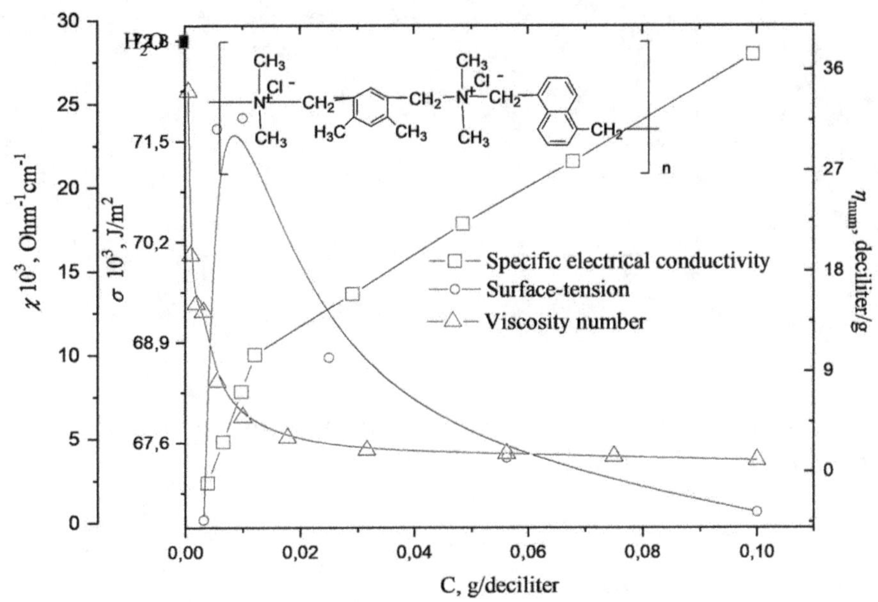

Table 3. CMC values determined from the concentration $\sigma = f(C)$, $\chi = f(C)$, $\eta_{sp} = f(C)$ aqueous solutions of polyonenes

Polymer	Values of CCM, g/deciliter		
	σ = f (C)	χ = f(C)	η_{sp} = f(C)
I-5	0.007	0.006	0.007
I-3	0.005	0.008	0.010
I-1	0.018	0.020	0.017
I-6	0.017	0.019	0.016
I-7	0.015	0.020	-
I-8	0.015	0.016	-

Figure 7. Isotherms of adsorption of polyionenes on bentonite.

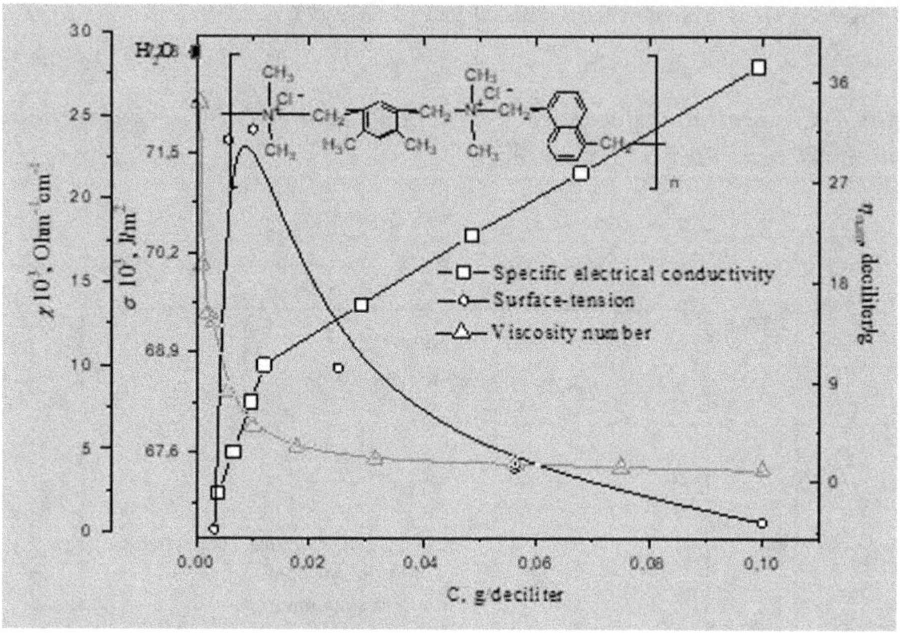

Figure 8. Dependences of the structural viscosity of sols based on TEOS and orthophosphoric acid on time.

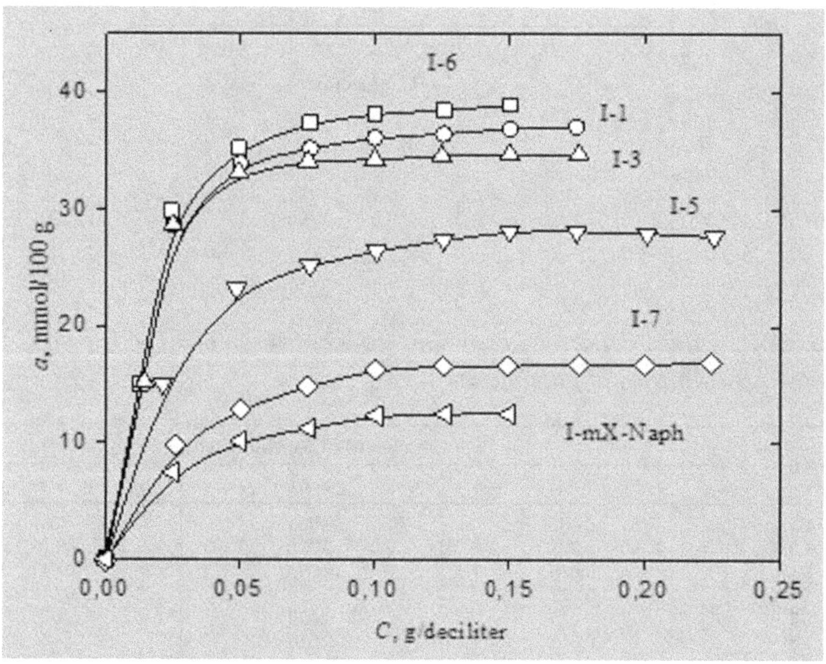

Physicochemical processes occurring in phosphorosilicate nanocomposites

The composition and ratio of the components in the sols varied in such a way as to reduce the gelation time, obtaining homogeneous (without stratification and sediment) gels, to form xerogels from them, and then membranes with high proton conductivity. The results of phenomenological observations of the state of the obtained xerogels, as well as the compositions and description of the conditions of sol gel synthesis are given in Table 1.

It is known that during the hydrolysis of silicon alkoxy derivatives, Si - OR bonds are cleaved with the formation of °Si - OH silanols, which are very unstable and condense, forming a Si-O-Si siloxane bond Burmistr M.V. et al. (2002), which is the main structural element of polysiloxanes. Condensation of alkoxysilanes can occur in the presence and with the participation of polymers having active functional groups (-OH, -Cl, NH_2-, COOH-, etc.) Gresso A.J.et al. (2007), which include the studied polyonenes.

The electron-ion interaction between the nitrogen cation polyonene (polymeric Quaternary ammonium salt) and the silanol oxygen atom in the growing silicate network is shown in Fig. 9. Such an interaction is possible at high concentrations of polyionenes and depends on the presence and concentration of other counter ions in the system Mauritz K.A., Storey K.A., Jones C.K. (1989).

Influence of ultrasonic treating on the course of hydrolysis and polycondensation processes in phosphorosilicate sols

Organo-inorganic composites W_8-PA_1-I-1 and W_8-PA_1-I-1us were studied (Table 1), one of which was subjected to ultrasonic exposure (44 kHz, 10 min.). A sharp acceleration of gelation in this case was not observed. However, studies of the structure of sols using electron microscopy (TEM) showed differences in the kinetics of gelation. As can be seen from Fig. 10, when observing the sol of the same composition applied to the slide, it is clear that gelation (formation of a porous fractal structure) begins if the sol has previously been subjected to ultrasound Sukhyy K.M., Gomza Yu. P. et al (2015).

Figure 9. The mechanism of donor-acceptor interaction between N + and Si-OH groups.

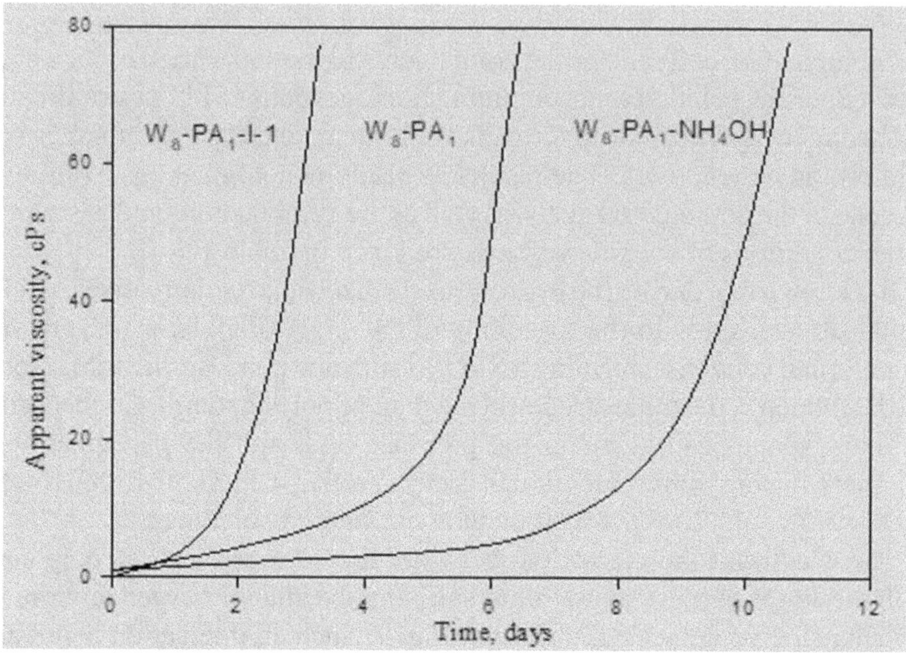

Investigation of the phase composition of phosphorosilicate nanocomposites (according to the results of thermal and X-ray phase analysis)

When drying the obtained gels in vacuum at 80°C, for 5 h, as a rule, coarse-grained white powders are formed. Heat treatment of gels in an air atmosphere at 200°C leads to the formation of coarse-grained brown powders, the color of which deepens in the presence of organic modifiers. Fig. 11 shows some derivatograms of such xerogels.

On all DTA curves of unmodified xerogels there are endothermic maxima at 160 - 280 °C. They are accompanied by weight loss (Table 4), due to the removal of water and ethanol, both adsorbed and sol-gel products formed by thermal destruction, synthesis Mauritz K.A. et al. (1989).

Xerogels PA_1-W_8 and W_8-PA_1, which have the same ratio H_3PO_4: TEOS = 1: 1, but differ in the order of mixing of the components, showed a different nature of the corresponding DTA curves. For the composition PA_1-W_8, obtained by introducing water at the end of the process (in ether-alcohol strongly acidic

Figure 10. TEM image of a drop of phosphorosilicate sol of the same composition W_8-PA_1-I-1 and W_8-PA_1-I-1us (Table 1) without ultrasonic treatment (1) and with the use of ultrasonic treatment (2) (a, b, c - images of different fragments of samples).

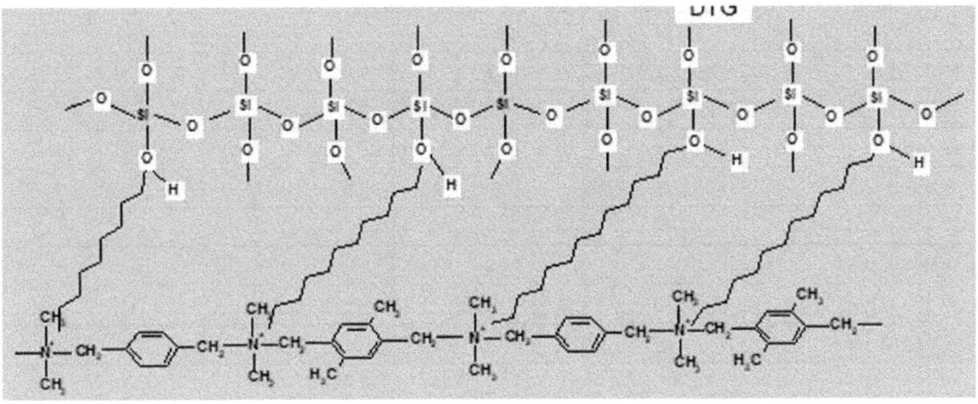

mixture), found two endoeffects at 220 and 320 °C and significantly less weight loss at 400°C than for xerogel W_8-PA_1, where water was introduced in the middle of the synthesis (table 1), with one endoeffect detected at a much lower temperature (150°C). On the curves for the compositions $PA_{0,5}$-W_8, W_8-$PA_{0,5}$ (Table 1), where the acidity of the medium is twice less (H_3PO_4: TEOS = 1: 0.5), the order of mixing of the components has a weaker effect on the course of physical processes, and, as a consequence, the nature of the change in the DTA curves: for xerogel W_8-$PA_{0.5}$ (where water was introduced in the middle of the process) endoeffect is observed at 160 °C, for xerogel $PA_{0.5}$-W_8 (where water was added at the end of the synthesis) - at 180°C.

On the DTA curve $PA_{0,1}$-W_8-NH_4OH, obtained for the sol modified with ammonia, there is a small exoeffect at a temperature of 400 °C, due to the removal of the remaining ethoxy groups. Weight loss in this case, as in the first case, was more for the xerogel, which has an endoeffect at a lower temperature (W_8-$PA_{0,5}$) (Table 4). Based on this, we can conclude that in a more acidic environment, hydrolysis and polycondensation are more complete, forming strong Si-O-Si bonds, provided that water is introduced into the already acidified system. If you add water to the ether-alcohol mixture, then first begins the formation of monodisperse silica particles a few nm in size. When acid is added, electrolyte ions settle on the surface of these particles, which causes an increase in their size, which is manifested in opalescence.

The DTA curve for the W_2-$PA_{1.5}$ xerogel, which contains more orthophosphoric acid, has two compatible endoeffects at 220 and 280°C.

Table 4. The results of thermogravimetric analysis

Marking	Mass loss, % °C			
	0 – 200 °C	200 – 400 °C	400 – 600 °C	600 – 1000 °C
$PA_{0,1}$- W_8- NH_4OH	14	5	2	2
PA_1- W_8	6	8	3	2
W_8-PA_1	15	13	1	1
$PA_{0,5}$- W_8	18	5	1	1
W_8-$PA_{0,5}$	15	17	1	1
W_2-$PA_{1,5}$	6	11	2	2
W_8-PA_1-I-1	11	9	3	2
W_8-PA_1-I-1us	10	10	2	1
W_{16}-PA_1-I-1us	20	9	3	1

These endoeffects, provided that water was not added to the sols, can be attributed to the processes of removal of ethanol and products of sol-gel synthesis, which is formed as a result of thermal destruction.

As well as for unmodified xerogels (Fig. 12), for xerogels with organic modifiers on all curves of DTA the endoeffect at 160-180°C is characteristic. These endoeffects are accompanied by weight loss (Table 4), which is due to the processes of removal of water and ethanol, both adsorbed and products of sol-gel synthesis, which is formed as a result of thermal destruction.

Comparing the DTA curves for samples W_8-PA_1-I-1 and W_8-PA_1-I-1us having the same composition (the sol W_8-PA_1-I-1us was subjected to ultrasonic treatment (44 kHz, 15 minutes)), you can see that the endoeffect at 160°C (W_8-PA_1-I-1) is shifted by 20°C (180°C for W_8-PA_1-I-1us). The nature of the change in mass in these samples is almost the same.

In phosphorosilicate xerogels, heat-treated at 250°C, solid products of variable composition $nSiO_2 \bullet mP_2O_5 \bullet xH_2O$ are formed Sukhyy K.M. et al (2010). Intermediates of hydrolysis of TEOS react in aqueous-alcoholic solution, for example, according to the scheme:

$$(RO)_3SiOH + (HO)_3P{=}O« (RO)_3Si((HO)_2P{=}O + H_2O$$

The formation of phosphorosilicates has been observed previously for similar phosphorosilicate xerogels by Mackenzie J. D. (2003). The thermal transformation of phosphate ceramics Voronkov M.G., Maletina E.A., Roman V.K. (1988) obtained by the sol-gel method has been studied. X-ray phase

Figure 11. DTA and DTG of unmodified xerogels [Sukhyy K.M., Gomza Yu.P., Belyanovskaya E.A. et al. (2015)].

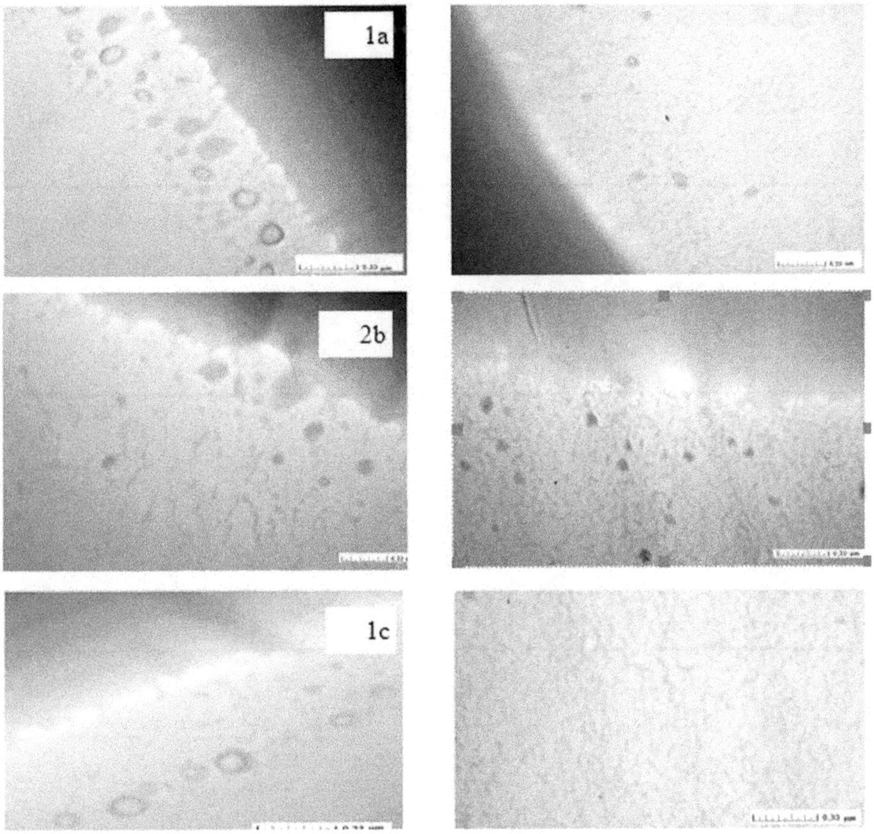

study showed that in the material at a ratio of P: Si = 0.5 and above after heat treatment at 900°C the crystalline phase of $Si_5O\ (PO_4)_6$ is generated. The formation of $Si(PO_4)_4$, which is formed at a temperature above 250°C, is considered undesirable for membrane proton-conducting materials, because the proton conductivity of such xerogels is low. This is explained by the fact that the source of conductivity in phosphorosilicate nanocomposites is the surface groups P - OH.

For all modified xerogels diffraction maxima are absent, i.e. they are X-ray amorphous, which is due to the influence of the organic modifier - polyonene on the processes of structure formation.

Figure 12. DTA and DTG of modified xerogels Sukhyy K.M., Gomza Yu.P., Belyanovskaya E.A. et al. (2015).

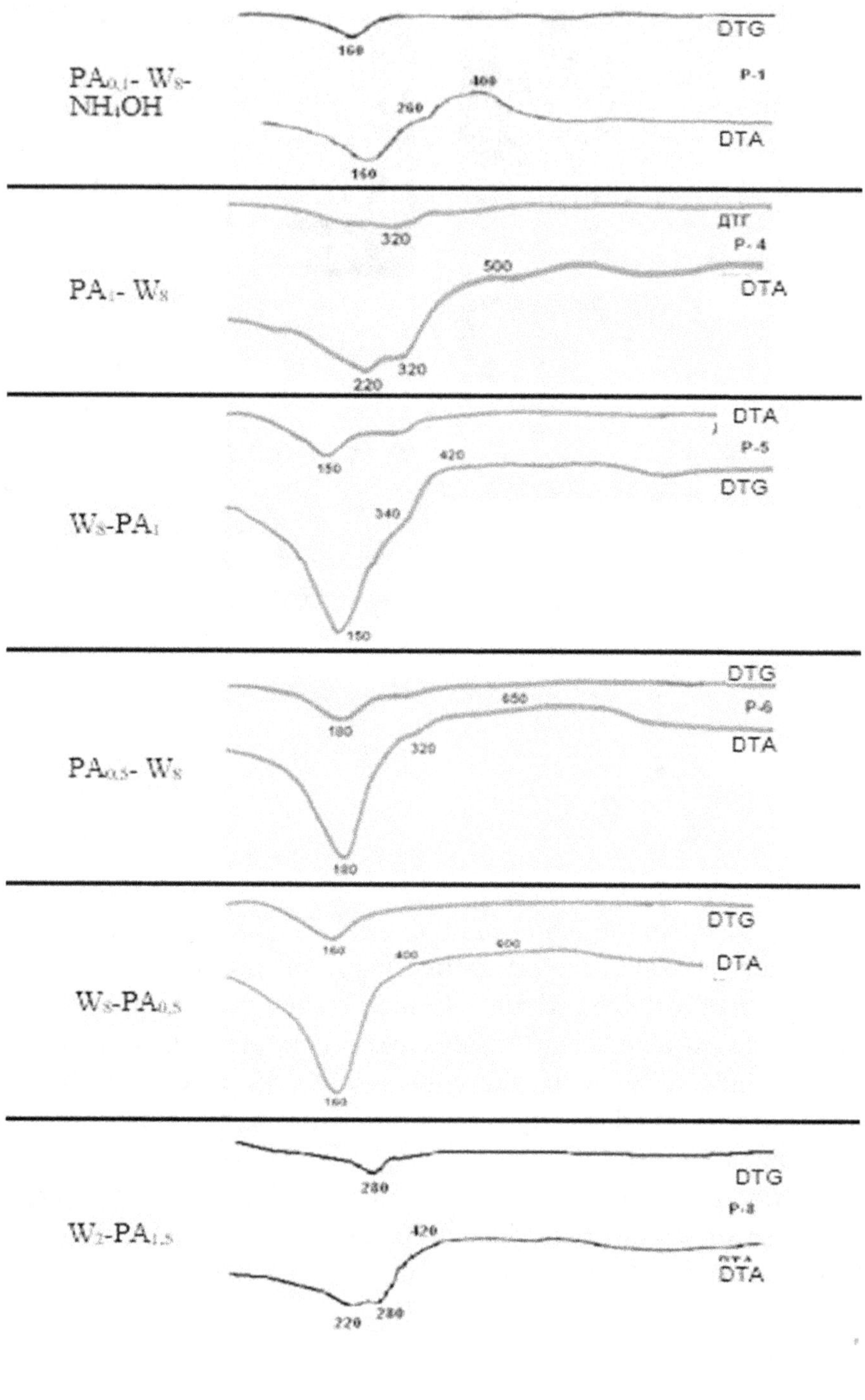

Figure 13. X-ray phase analysis of xerogels, heat-treated at 250°C ÿ- $Si(PO_4)_4$; ¡- SiP_2O_7 Sukhyy K.M., Gomza Yu.P., Belyanovskaya E.A. et al. (2015).

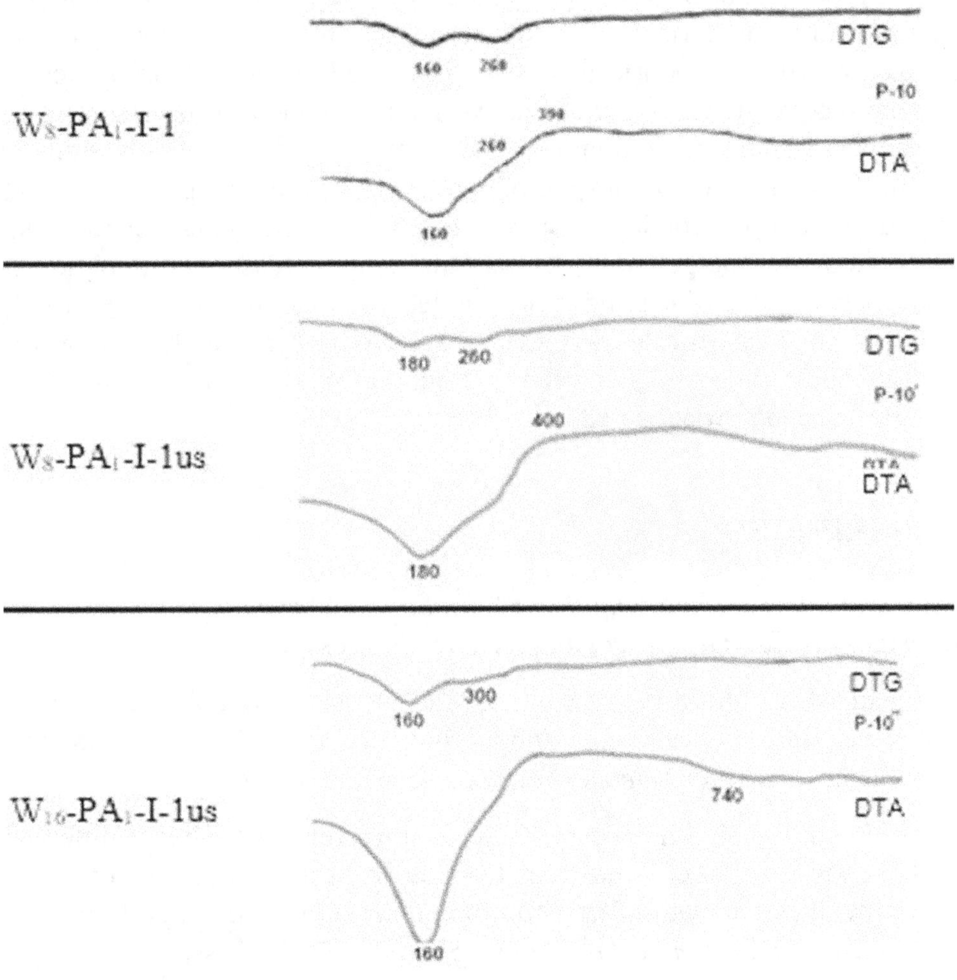

CONCLUSION

Solid inorganic and organo-inorganic phosphorosilicate nanocomposites from sols based on TEOS, orthophosphoric acid and a number of polyonenes were obtained by the sol-gel synthesis method. Morphologically, such materials are nanostructured composites permeated with a large number of mesopores. Functional groups in phosphorosilicate membranes that provide high proton conductivity are P-OH groups, however, it cannot be ruled out that silanol groups Si-OH play a role. It was found that the optimal ratio of H_3PO_4: Si $(OEt)_4$ is 1: 1.

The process control of sol – gel synthesis of new class of organic-inorganic nanostructured materials on basis of tetraethoxysilane and polyionenes doped by orthophosphoric acid with improved functional properties for making electrochemical devices active media technological principles are developed. The chemical compositions improvement is based on results of complex investigation of morphology, physicochemical properties. The rate of hydrolytical polycondensation reaction of tetraethoxysilane is increased by polyionenes, the gelation process rate being increased with polyionene molecular weight. The polyionenes added in silicophosphate sols ($8 \cdot 10^{-4}$ mole polyionenes per mole $Si(OEt)_4$) are stated by X-ray analysis to prevent crystallization processes occurred in silicophosphate xerogels with silicon phosphates and pyrophosphates.

REFERENCES

Brinker, C. J., & Scherer, G. W. (1990). *Sol–Gel Science. The Physics and Chemistry of Sol–Gel Processing*. Academic Press, Inc.

Burmistr, M. V., Shilov, V. V., Sukhoi, K. M., Pissis, P., & Polizos, G. (2003). Dielectric Relaxation and Ionic Conductivity of Oxyethylene–Alkylaromatic Polyionenes. *Polymer Science, Series A*, *45*(8), 785–794.

M. V. Burmistr, K. M. Sukhyy, O.A. Shilova [et al.] (2002). Primenenie poliionenov v zol'–gel' sinteze materialov [Application of polyionenes in sol-gel synthesis of materials.]. *Voprosy khimii i khim. tekhnologii [Questions of chemistry and chem. technologies]. 3*, 153–157. (In Russian)

Burmistr, M. V., Sukhyy, K. M., & Shilov, V. V. [et al]. (. (2005). Structure, thermal and ion–conductivity properties of the polymeric quaternary ammonium salts (polyionenes) containing ethylene oxide and aliphatic fragments in the chain. *Solid State Ionics*, *176*, 1787–1792. doi:10.1016/j.ssi.2005.04.032

Cvetkova I. N., Shilova O. A., Gomza YU. P., Sukhoj K. M. (2007). Zol'–gel' sintez i issledovanie silikofosfatnykh i gibridnykh protonprovodyashchikh nanokompozitov [Sol-gel synthesis and study of silicophosphate and hybrid proton-conducting nanocomposites]. *Al'ternativnaya ehnergetika i ehkologiya [Alternatice energy ecology], 1*, 47–49.

Gresso, A. J., Moore, R. B., Cable, K. M., & Mauritz, K. A. (2007). Chemical modification of a Nafion sulfonil fluoride precursor via in situ sol–gel reactions. *Polymer*, *38*(6), 1345–1356. doi:10.1016/S0032-3861(96)00638-6

Ha, C. S., & Park, S. S. (2019). General Synthesis and Physico-chemical Properties of Mesoporous Materials. In *Periodic Mesoporous Organosilicas. Springer Series in Materials Science* (Vol. 281, pp. 15–85). Springer., doi:10.1007/978-981-13-2959-3_2

Liang, D., Yu, F., Zhu, K., Zhang, Z., Tang, J., Xie, Q., Liu, J., & Xie, F. (2022). Quaternary ammonium salts targeted regulate the surface charge distribution of activated carbon: A study of their binding modes and modification effects. *Environmental Research*, *214*(Pt 4), 114103. doi:10.1016/j.envres.2022.114103 PMID:35987375

Mackenzie, J. D. (2003). Sol–gel research – achievments since 1981 and prospects for the future. J. Sol–gel *Sciences et Techniques (Paris)*, *26*(1), 23–27.

Mauritz, K. A., Storey, K. A., & Jones, C. K. (1989). Multiphase Polimer Materials: Blends, Ionomers, and Interpenetrating. In L.A. Utracki and R.A. Weiss, (eds.) Networks (vol. 395, pp. 401-417). American Chemical Society.

Neagu, E. R., Neagu, R. M., Pissis, P., & Das–Gupta, D. K. (1996) in *Proceedings of the Ninth International Symposium on Electrets*, Shanghai, edited by Z. Xia, & H. Zhang. IEEE: Piscataway, 585.

Price, C. (1982). In: Developments in Block Copolymers, p. 39. London: Applied Science Publ

Price, C. (1982). In: Developments in Block Copolymers. Ed. Goodman I. London: Applied Science Publ, 1, 39.

Shilov, V. V., Shevchenko, V. V., Pissis, P., Kyritsis, A., Georgoussis, G., Gomza, Y. P., Nesin, S. D., & Klimenko, N. S. (2000). Morphology, dielectric relaxation and conductivity of the novel polyurethanes with acid and ionic groups in the polyether segments. *Journal of Non-Crystalline Solids*, *275*(1-2), 116–136. doi:10.1016/S0022-3093(00)00240-4

Sukhyy, K. M., Gomza, Yu. P., Belyanovskaya, E. A., Klepko, V. V., Shilova, O. A., & Sukhyy, M. P. (2015). Resistive humidity sensors based on proton-conducting organic-inorganic silicophosphates doped by polyionenes. *Journal of Sol-Gel Science and Technology*, *74*(2), 472–481. doi:10.100710971-015-3622-7

Sukhyy, K. M., Gomza, Yu. P., Belyanovskaya, E. A., Klepko, V. V., Shilova, O. A., Sukhyy, M. P., & Gavrilko, A. V. (2015). Proton-Conducting Organic-Inorganic Silicophosphates doped by Polyionenes. *Voprosy Khimii i Khimicheskoi Tekhnologii*, *5*(103), 95–102.

Sukhyy K. M., Shilova O. A., Gomza Yu. P. [i dr.] (2010). Vliyanie sostava i uslovij sinteza zol'–gel' silikofosfatnykh ionensoderzhashchikh nanokompozitov na osobennosti ikh struktury i protonnoj provodimosti [Influence of Composition and Synthesis Conditions of Sol–Gel Silicophosphate Ion-Containing Nanocomposites on Features of Their Structure and Proton Conductivity]. *Vopr. khimii i khim. tekhnologii [Question. Chemistry and chem. projects], 4,* 64–68.

Sviderskij, V. A., Voronkov, M. G., & Klimenko, S. V. (1996). Vliyanie prirody rastvoritelya i sootnosheniya reagentov na zol'–gel' process sinteza kremnekislorodnykh kremnezolej [Influence of the nature of the solvent and the ratio of reagents on the sol-gel process of synthesys of silicon-oxeygen silica sols]. [In Russian]. *Russian Journal of Applied Chemistry*, *69*(6), 951–957.

Tadros, T. F. (1991). Experimental investigation of the polymer melt flow during injection post-filling process. *Polymer Journal*, *23*, 683. doi:10.1295/polymj.23.683

Tsvetkova, N., Shilova, O. A., Voronkov, M. G., Gomza, Yu. P., & Sukhyy, K. M. (2008). Sol–Gel Synthesis and Investigation of Proton–Conducting Hybrid Organic–Inorganic Silicophosphate Materials. *Glass Physics and Chemistry*, *34*(1), 68–76. doi:10.1134/S1087659608010100

Tuzar, Z., & Kratochvill, P. (1993). In: Surface and Colloid Science. Plenum Press.

Voronkov M.G., Maletina E.A., Roman V.K. (1988). Kremnekislorodnye soedineniya nemetallov proizvodnye azota i fosfora [Silicon-oxygen compounds of non-metals are derivatives of nitrogen and phosphorus.]. *Novosibirsk: Nauka Sib.otd., 154.*

Chapter 3
Evolution of Organic-Inorganic Nanocomposites During Gel Formation

ABSTRACT

In this chapter, the authors present results of the study of evolution of organic-inorganic nanocomposites during gel formation from sols based on tetraethoxysilane; orthophosphoric acid and a number of polyonenes were studied. Schemes of multilevel fractal aggregation of inorganic nanocomposite and organic-inorganic nanocomposites are suggested according to SAXS data. It was shown that formation of surface fractal aggregates is stated at all structural levels. Mass fractal aggregates are generated on the second structural level (100 – 300 nm). A generation of the initial silicophosphate composite starts from the formation of practically spherical aggregates with order of 12 nm which forms compact lattice of the inorganic polymer on the kind of mass fractal with average particle size of 80 nm on the higher scale level. On the third level this structure occurs to be reorganized with forming of surface fractal clusters with very smooth surface and average particle size of 470 nm. Formation of surface fractal is supposed to strongly affect the protonic conductivity nanocomposites.

BACKGROUND

Sol – gel synthesis is one of the newest and perspective nanotechnologies. A possibility of combination of fragments of inorganic and organic matters

DOI: 10.4018/978-1-7998-9728-6.ch003

on molecular level Matsuda et al. (2003) with realisation of directional synthesis of a new class of compounds, that is hybrid organic-inorganic nanocomposites combined properties of ceramics, as well as polymers, is one of their important advantages. At that, an imperative moment is a possibility of directional change of a structure of nanocomposites synthesized with variation of character and topology not only precursors of sol – gel processes, but organic low- and high-molecular organic compounds and inorganic salts and acids. Thus, nonlinear optical (Yeoman, 2017), heat-sensitive (Yang et al., 2020) and insulating nanocomposite materials (Han, 1999) can be synthesized by sol – gel method. At that, it is succeed to form highly porous monolithic with Nano pores, 1999), materials with pore size controlled (Imai, Yoshida, & Naka. (1999), gas-diffusion electrodes (Deng et al., 1998), membranous materials (Song & Villemure, 2001) including templates (Rabinovich et al,. 2001), thin nano-size films and coatings (Santos et al., 2003), nano-structured thin films Nam CW, Woo SI (1994), water-repellent coatings Tadanaga Katata & Minami (1997), composite and hybrid organic-inorganic materials contained nano-sized inclusions of oxides Armelao L, Barreca D & Moraru B (2003), inorganic salts crystalline hydrate Sukhyy K. M. et al. (2014), metals catalysts Guimaraes AP, Viana APP, Lago RM, Mohalleem L.D.S. (2002), catalyst carriers Biazzotto JC et al. (2002), .Guimarães A.P. et al. (2002), surface active compounds Wu Z, Lee K, Lin Y, Lan S., Huang L. (2003), organic fragments, high-molecular fragments Matsuda et al. (2003), templates Lu Y et al. (1999), catalytically active complexes Biazzotto JC et al. (2002), nontoxic corrosion inhibitors Osborne H, Blohowiak KY, Taylor SR et al (2001) or special additives as pigment Lu X, Manners I, Winnik MA (2001), luminophors Deshpande AV & Kumar U (2002), colorants and chromophores Lu X, Manners I, Winnik MA (2001), biologically active substances Yang P. et al. (2002), biomolecules Pandey PC et al. (2001).

The membranous materials of partially fluorinated sulfonated poly(arylene ether sulfone) copolymers based on imidoaryl biphenol such as 3,8-bis(4-hydroxyphenyl)-N-phenyl-1,2-naphthalimide show protonic conductivity up to $49 \cdot 10^{-3}$ S $\cdot$ cm^{-1} Yang P. et al. (2002), and materials obtained by photo-polymerization of poly(ethylene glycol diacrylate), succinonitrile and lithium bis(trifluoromethane sulfonyl)imide exhibit protonic conductivity up to 10^{-3} S $\cdot$ cm-1 Pandey P.C. et. Al. (2001).

Silicophosphates characterized by high level of proton conductivity are perspective materials for electroactive elements of humidity sensors and electrochemical sensors Mohanty AK et al. (2014). However, their practical use is limited by low-level characteristics of their mechanical properties and hydrolytic stability Echeverri M, Hamad C & Kyu T (2014).Therefore, a determination of optimal compositions and synthesis technologies of membranous materials of this type is expedient.

At the same time, it is obvious that purposeful designing of properties of oxide, as well as organic-inorganic nanocomposites by sol – gel method is impossible without their structure monitoring through all synthesis stages and determination of final products properties.

The aim of present work is to determine of regularities of sol – gel process macroparameters influence, specifically, inorganic and organic polymer components, system composition, synthesis conditions, structure and properties of gel system forming on morphology and final physicochemical characteristics of functional hybrid organic-inorganic nanocomposite systems which are perspective for use as proton-conductive membranes of fuel cells and catalytic electrode materials.

EXPERIMENTAL

The small-angle X-ray scattering curves were obtained in a short-type vacuum chamber.

To model the experimentally obtained intensity curves, we used the global unified Beaucage equation Gkourmpis T. & Mitchell G.R. (2020), which describes the scattering of X-rays by complex systems containing several levels of interrelated structural features. The Beaucage equation describing an arbitrary number of interdependent structural levels can be represented as follows

$$I(q)=\sum_{i=1}^{n}(G_i\exp(-q^2R_{g_i}^2/3)+B_i\exp(-q^2R_{g(i+1)}^2/3)\times\left\{\left[erf(qR_{g_i}/6^{\frac{1}{2}})\right]^3/q\right\}^{P_i})$$

Where G_i is the Guinier preexponential factor for the i-th structural level, B_i is the preexponential factor, which depends on the type of power law of a given section of the scattering curve. B is determined according to the mode, which corresponds to the value of the exponent (for surface fractals, for mass fractals).

The main characteristic of fractal aggregation is the fractal dimension. For a mass fractal D_f that varies within $1 < D_f < 3$, the larger the D_f, the more

Figure 1. Curve of small-angle X-ray scattering intensity for phosphorosilicate nanocomposite PA1-W8; q is the absolute value of the scattering vector. The numbers indicate the slopes of the SAXS curves in the Porod's (Pi) region for each level I (i = 1,2,3)

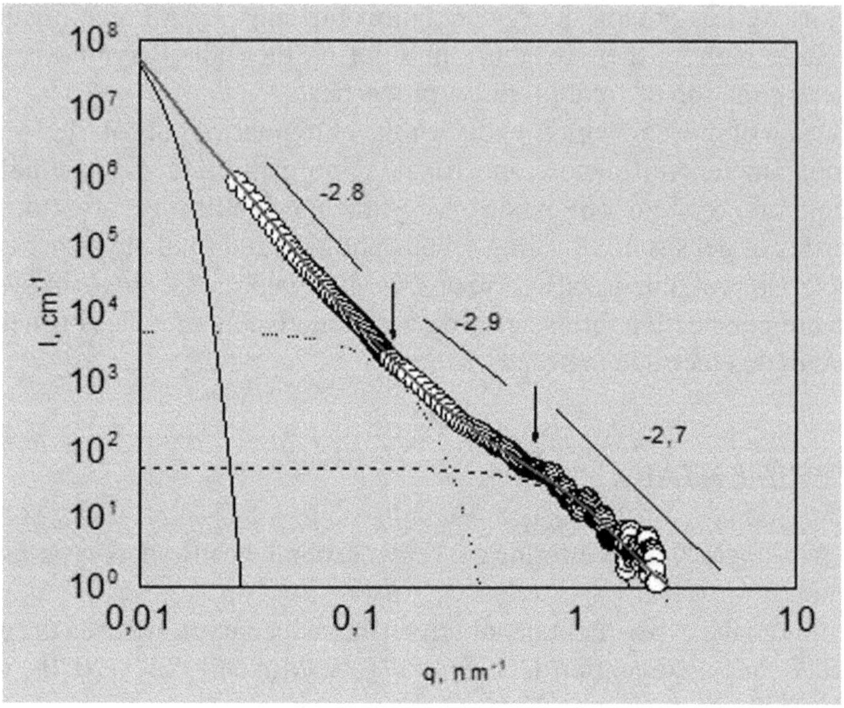

compact the fractal object. The dimension $D_f = 3$ corresponds to a three-dimensional body. However, it may exhibit the properties of a surface fractal. The fractal dimension of the surface fractal varies within $2 < D_s < 3$. The higher is D_s, the rougher is the surface of the fractal object.

In Fig. 1 – 5 shows the curves of low-angle X-ray scattering for unmodified phosphorosilicate sol-gel nanocomposites, the characteristics of the fractal structure are given in Table. 4 .

For phosphorosilicate sample PA_1-W_8 (Fig. 1) which is characterized by the following pattern of structure formation:

(1) the smallest aggregates are mass fractals with a fractal dimension $Df_1 = 2.7$ and an average radius of rotation $Rg_1 = 2.0$ nm;

Figure 2. Curve of the intensity of small-angle X-ray scattering for phosphorosilicate nanocomposite W_8-PA_1; q is the absolute value of the scattering vector. The numbers indicate the slopes of the SAXS curves in the Porod's (P_i) region for each level I (i = 1,2,3)

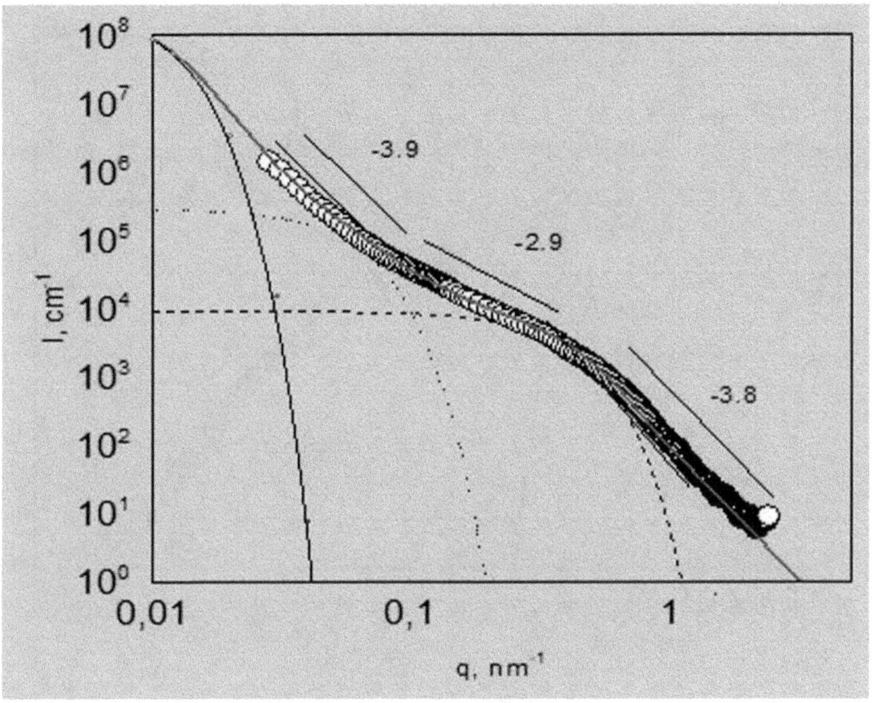

(2) mass fractals with a fractal dimension $Df_2 = 2.9$ and a radius of rotation $Rg_2 = 14$ nm are formed on the basis of the latter, each of these aggregates contains G_2 / G_1 » 10 primary particles;
(3) at the third scale level on the basis of mass fractal aggregates of the second level (as primary particles) mass fractals are also formed (fractal dimension $Df_3 = 2.8$; radius of rotation $Rg_3 = 250$ nm).

Sample W_8-PA_1 is characterized by:

(1) at the first large-scale level of formation of small aggregates such as surface fractals ($Ds_1 = 2.15$, $Rg_1 = 4.7$ nm);
(2) aggregates at the 2nd scale level are mass fractals ($Df_2 = 2.9$, $Rg_2 = 31$ nm);

Figure 3. Curve of intensity of small-angle X-ray scattering for phosphorosilicate nanocomposite $PA_{0,5}$-W_8; q is the absolute value of the scattering vector. The numbers indicate the slopes of the SAXS curves in the Porod's (Pi) region for each level I (i = 1,2,3)

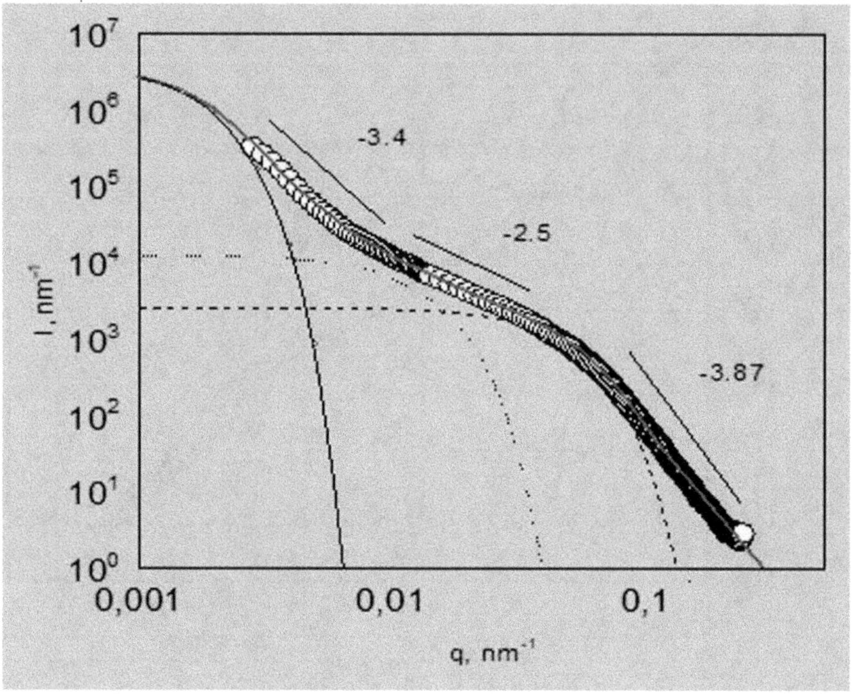

(3) at the third scale level, surface fractals are formed (Ds_3 = 2.1, Rg_3 = 180 nm).

For the sample $PA_{0,5}$- W_8 is characterized by:

(1) the smallest aggregates are surface fractals with a fractal dimension Ds_1 = 2.1 and an average radius of rotation Rg_1 = 3.7 nm;
(2) on the basis of the latter mass fractals are formed with fractal dimension Df_2 = 2.5 and radius of rotation Rg_2 = 14 nm, each of such aggregates contains G_2/G_1 » 250 primary particles;
(3) at the third scale level, surface fractals (D_{f3} = 2.7, Rg_3 = 100 nm) are also formed on the basis of mass fractal aggregates of the second level (as primary particles).

Figure 4. Curve of intensity of small-angle X-ray scattering for phosphorosilicate nanocomposite W_8-$PA_{0.5}$; q is the absolute value of the scattering vector. The numbers indicate the slopes of the SAXS curves in the Porod's (Pi) region for each level I (i = 1,2,3)

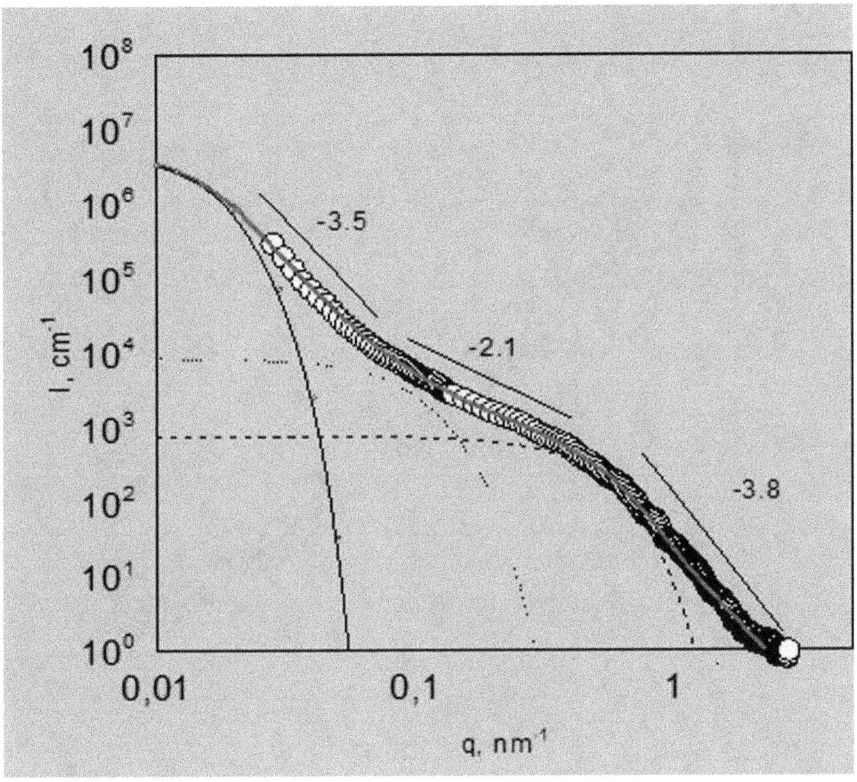

Visual confirmation of the formation of aggregates of the type of mass fractals at the level of 250 nm is presented in Fig. 6.

The initial compositions used for the preparation of two pairs of xerogels, namely: PA_1-W_8, W_8-PA_1 and $PA_{0.5}$-W_8, W_8-$PA_{0.5}$, as already mentioned, are completely identical. Suitable systems differ only in the sequence of introduction of components: for samples W_8-PA_1 and W_8-$PA_{0.5}$ water is introduced before the addition of the catalyst (hydrochloric acid) and phosphoric acid, and for samples PA_1-W_8 and $PA_{0.5}$-W_8 - after that. In the first pair of xerogels (W_8-PA_1 - W_8-$PA_{0,5}$) the introduction of water as the third and not the last component (before the catalyst and phosphoric acid) leads to significant changes in the fractal structure. In this case, first, the surface fractal is significantly loosened: its fractal dimension decreases from

Figure 5. Curve of the intensity of small-angle X-ray scattering for phosphorosilicate nanocomposite W_2-$PA_{1.5}$; q is the absolute value of the scattering vector. The numbers indicate the slopes of the SAXS curves in the Porod's (Pi) region for each level I (i = 1,2,3)

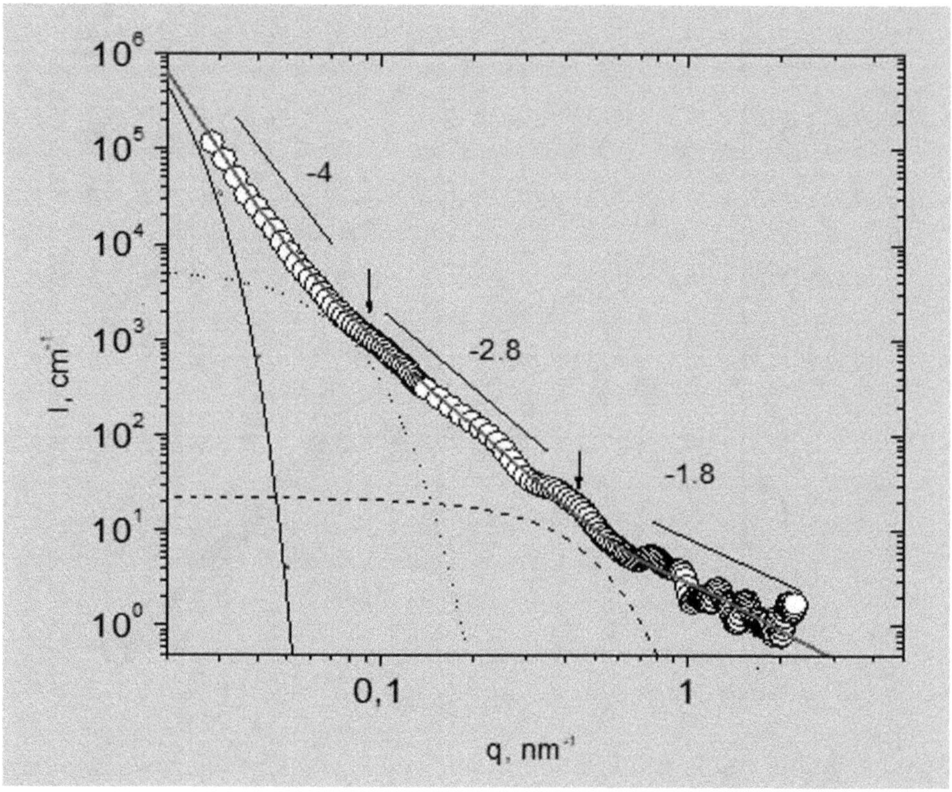

2.9 (PA_1-W_8) to 2.15 (W_8-PA_1). Secondly, the nature of the fractal structure changes dramatically: a more complex structure of a multilevel aggregate (mass + surface fractal) is formed instead of a unit of simple structure (one mass fractal). At the same time, in the case of the second pair ($PA_{0.5}$-W_8 and W_8-$PA_{0.5}$) the case ends with a slight compactification of the mass fractal (fractal dimension of the latter increases from 2.1 ($PA_{0.5}$-W_8) to 2.3 (W_8 -$PA_{0,5}$) while maintaining other parameters of the complex fractal structure are virtually unchanged.

The fractal characteristics of the described samples are given in Table 1, which shows that in a more acidic environment (pH = 2), the order of mixing of the components does not affect the growth of fractals and the alternation of their types: surface > mass> surface.

Figure 6. TEM - image of the formation of aggregation of phosphorosilicate nanocomposites by type of mass fractals for sample W_2-$PA_{1,5}$

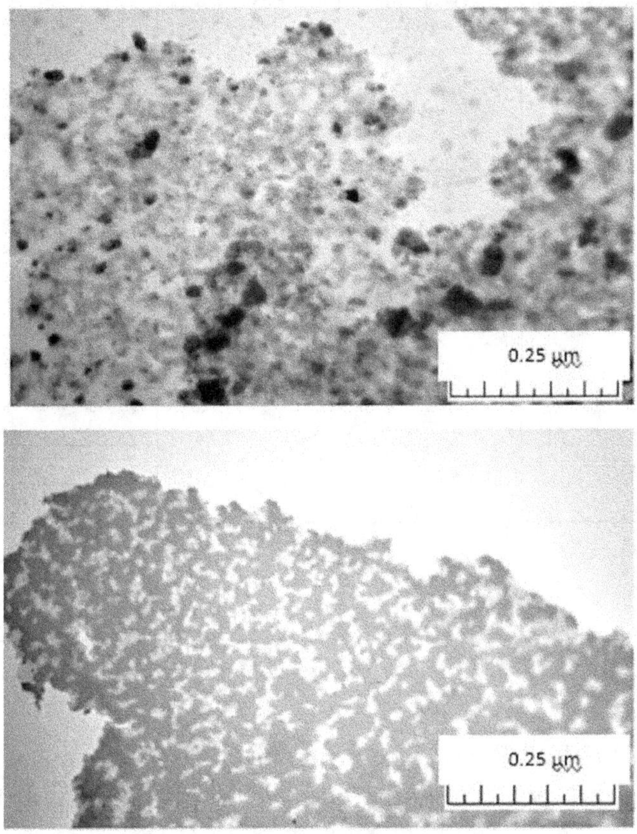

FRACTAL STRUCTURE OF MODIFIED PHOSPHOROSILICATE XEROGELS

Nanocomposites W_8-PA_1-I-1, W_8-PA_1-I-1us, having one composition, differ in that the sol W_8-PA_1-I-1us was subjected to ultrasonic treatment (44 kHz, 10 minutes), and in the sol W_{16}- PA_1-I-1 used more water. Based on the SAXS curves (Fig. 7 - 9), we can conclude that these samples are characterized by the formation of surface fractals with a smooth surface, i.e. approximately spherical particles. The effect of ultrasound is quite noticeable, although it does not lead to qualitative changes in the nature of fractal aggregation: the particles of the first, second and third levels become smaller, varying from 55 nm to 38 nm at the first level, from 160 to 120 nm at the second and from 1500 to 950 - on the third. When more water is used in the initial ash W_8-

Table 1. Fractal characteristics of unmodified phosphorosilicate nanocomposites

Marking	Fractal size Level 1	Fractal size Level 2	Fractal size Level 3	The scale of the aggregate, nm	Type of fractal	pH
PA_1- W_8	2.7	2.9	2.8	2.0	M	4
				14	M	
				250	M	
W_8-PA_1	2.2	2.9	2.1	4.7	S	4
				31	M	
				180	S	
$PA_{0,5}$- W_8	2.1	2.5	2.6	3.7	S	2
				14	M	
				100	S	
W_8-$PA_{0,5}$	2.2	2.1	2.5	3.7	S	2
				18	M	
				120	S	
W_2-$PA_{1,5}$	1.8	2.8	2.0	4.2	M	2
				27	M	
				130	S	

Notes: M – mass fractal, S – surface fractal

PA_1-I-1 (W_{16}-PA_1-I-1), the surface fractions at the first level are also reduced: from 55 to 50 nm, and the system as a whole changes from three-level to two-level Sukhyy K.M., Gomza Yu.P. et al. (2015).

Visual confirmation of the formation of surface aggregates at the level of 250 nm is presented in Fig. 9.

Sample W_8-PA_1-I-1us is characterized by:

(1) the smallest aggregates are surface fractals (Ds_1 = 2.8, Rg_1 = 8.8 nm);
(2) surface fractals are also formed on the basis of the latter (Ds_2 = 2.5, Rg_2 = 12 nm);
(3) at the third scale level, surface fractals (Ds_3 = 2.2, Rg_3 = 95 nm) are formed on the basis of surface fractal aggregates of the second level.

For sample W_{16}-PA_1-I-1 is characterized by:

(1) the smallest aggregates are surface fractals (Ds_1 = 2.3, Rg_1 = 5.0 nm);

Figure 7. Curve of the intensity of small-angle X-ray scattering for phosphorosilicate nanocomposite W_8-PA_1-I-1; q is the absolute value of the scattering vector. The numbers indicate the slopes of the SAXS curves in the Rock (P_i) region for each level I (i = 1,2,3)

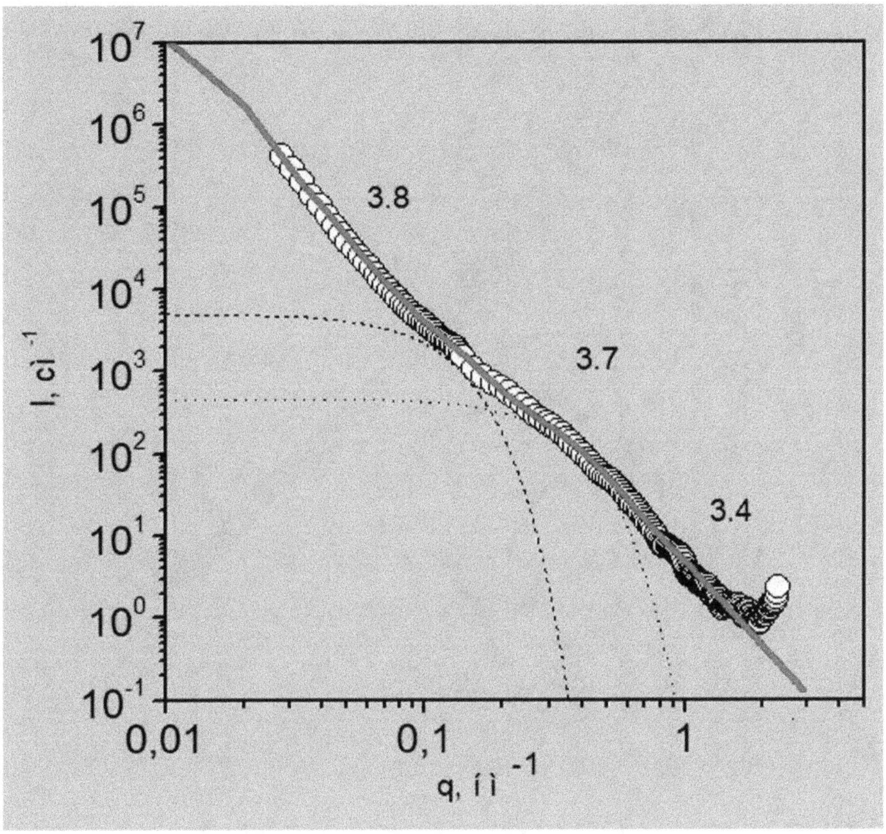

(2) surface fractals are also formed on the basis of the latter ($Ds_2 = 2.5$, $Rg_2 = 14$ nm);
(3) at the third scale level, surface fractals ($Ds_3 = 2.5$, $Rg_3 = 100$ nm) are formed on the basis of surface fractal aggregates of the second level.

It seemed interesting to find out whether there is a structure-forming effect of the organic modifier at extremely small amounts.

For this purpose, xerogels with a low content of orthophosphoric acid W_2-$PA_{0,1}$-I-2 and W_{24}-$PA_{0,1}$-I-2 and their matrix, i.e. xerogels of the same composition, but without polyionene, were studied. As can be seen from the graphs, the slope of the curves (Pi) and hence the fractal dimension (Table

Figure 8. Intensity curve of small-angle X-ray scattering for phosphorosilicate nanocomposite W_8-PA_1-I-1us; q is the absolute value of the scattering vector. The numbers indicate the slopes of the SAXS curves in the Porod's (Pi) region for each level I (i = 1,2,3)

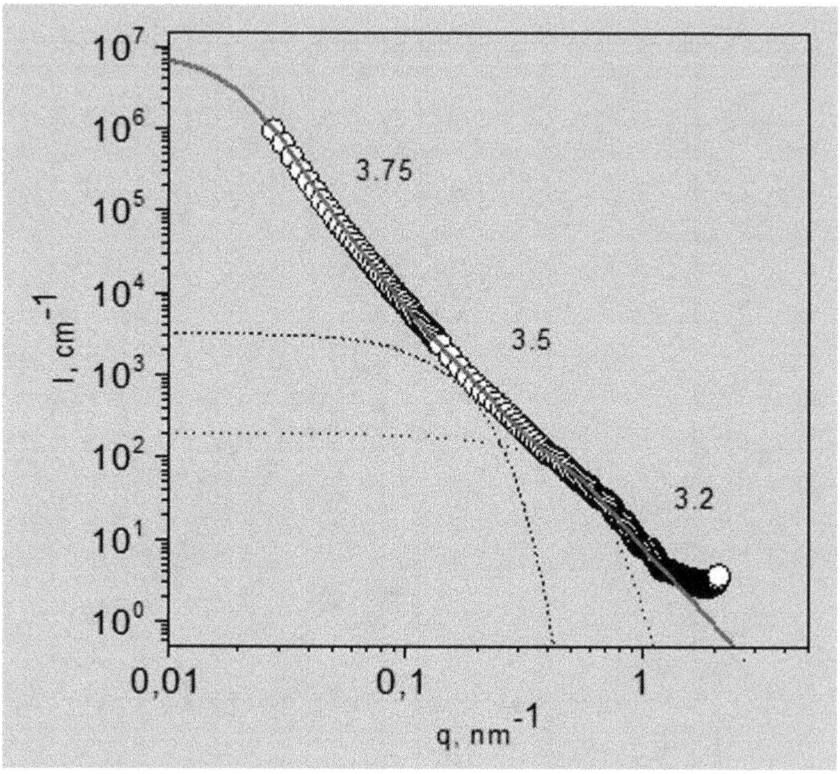

2) have changed significantly: the xerogel matrix W_2-$PA_{0,1}$-I-2, having a non-fractal behavior at the second level, is replaced by a mass fractal when polyionene is added. In the xerogel W_{24}-$PA_{0,1}$-I-2, which contains more water and less polyionene, the non-fractal behavior at the second level during the transition from W_{24}-$PA_{0,1}$ to W_{24}-$PA_{0,1}$-I-2 is preserved (Fig. 10). However, at the first scale level, mass aggregates with a size of 15 nm are formed in the matrix as primary particles instead of surface fractals.

Sample W_2-$PA_{0.1}$ is characterized by (Fig. 11):

(1) at the first large-scale level of formation of small aggregates such as surface fractals (Df_1 = 2.9, Rg_1 = 1.6 nm);
(2) at the second structural level fractal groups are not formed (ß <| 1 |)

Figure 9. Intensity curve of small-angle X-ray scattering for phosphorosilicate nanocomposite W_{16}-PA_{1}-I-1; q is the absolute value of the scattering vector. The numbers indicate the slopes of the SAXS curves in the Porod's (P_i) region for each level I (i = 1,2,3)

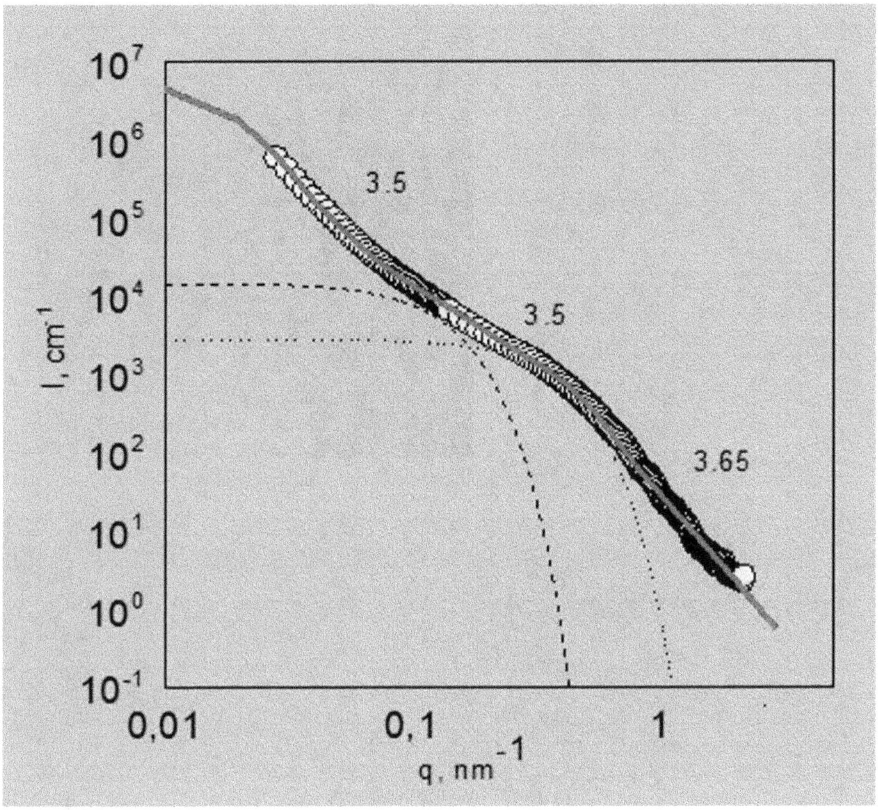

(3) at the third scale level, surface fractals are also formed (fractal dimension $Df_3 = 2.0$, $Rg_3 = 110$ nm).

For sample W_2-$PA_{0,1}$-I-2 is characterized by:

(1) at the first large-scale level of formation of small aggregates such as surface fractals ($Df_1 = 2.7$, $Rg_1 = 2.8$ nm);
(2) at the second scale level of formation of small aggregates of the type of mass fractals ($Ds_1 = 1.5$, $Rg_2 = 17$ nm);
(3) at the third scale level, surface fractals are also formed (fractal dimension $Df_3 = 2.5$, $Rg_3 = 120$ nm).

Figure 10. TEM-images for formation of phosphoro-silicate hybrid aggregates by the type of surface fractals for the sample W_8-PA_1-I-1

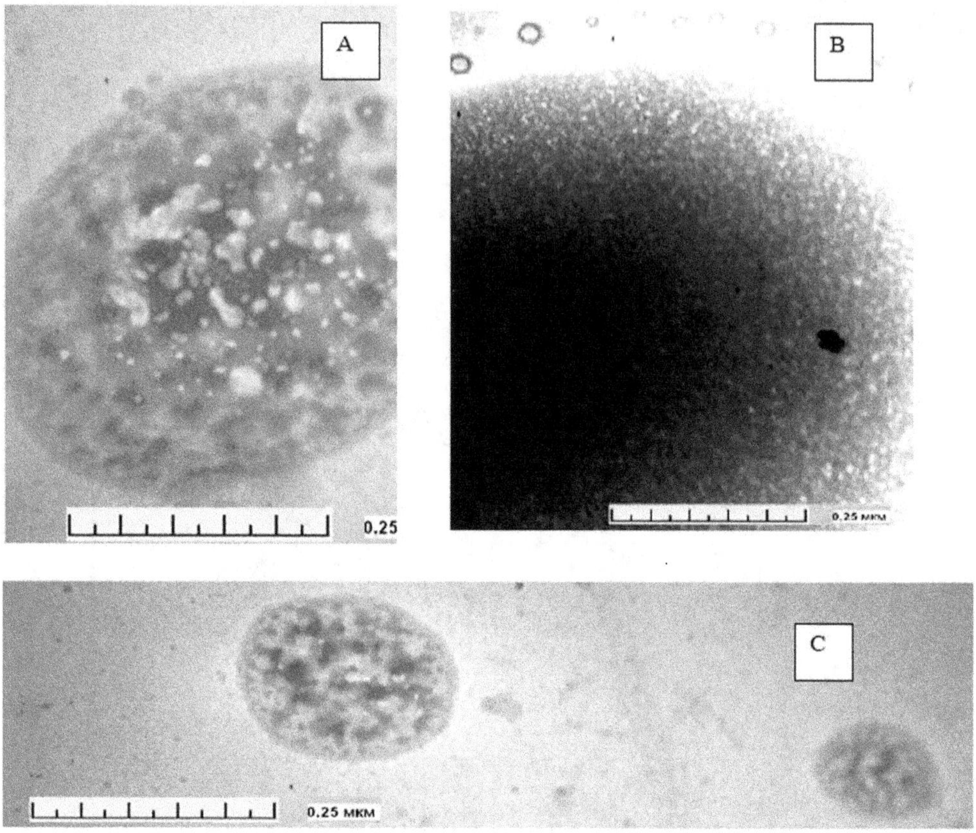

On the different time stages

For sample W_{24}-$PA_{0,1}$ is characterized by (Fig. 12):

(1) at the first large-scale level of formation of small aggregates of the type of mass fractals ($Ds_1 = 2.7$, $Rg_1 = 1.5$ nm);
(2) non-fractal behavior
(3) at the third scale level, surface fractals are formed ($Ds_3 = 2.1$, $Rg_3 = 110$ nm).

For sample W_{24}-$PA_{0,1}$-I-2 is characterized by:

Table 2. Fractal characteristics of organo-inorganic phosphorosilicate nanocomposites

Sample	Level	Fractal*	P	D	R_g, nm	d_{os} Å	G, cm^{-1}	B_o, cm^{-1} $Å^{-P}$
W_8-PA_1-I-1	1	S	3,2	2,8	38	98	195	$5{,}5\cdot10^{-3}$
	2	S	3,5	2,5	120	310	$3{,}3\cdot10^{3}$	$6{,}0\cdot10^{-4}$
	3	S	3,8	2,2	950	2450	$9{,}2\cdot10^{6}$	$2{,}5\cdot10^{-4}$
W_8-PA_1-I-1us	1	S	3,4	2,6	5,5	14,2	460	$1{,}9\cdot10^{-3}$
	2	S	3,7	2,3	16	41	$5{,}0\cdot10^{3}$	$1{,}3\cdot10^{-4}$
	3	S	3,8	2,2	150	390	$2{,}4\cdot10^{7}$	$1{,}2\cdot10^{-4}$
W_{16}-PA_1-I-1	1	S	3,7	2,3	50	130	2830	$6{,}6\cdot10^{-3}$
	2	S	3,5	2,5	140	360	$1{,}5\cdot10^{4}$	$1{,}6\cdot10^{-3}$
	3	S	3,5	2,5	1000	258	$6{,}6\cdot10^{6}$	$7{,}0\cdot10^{-4}$
W_2-$PA_{0,1}$-I-2	1	S	3,3	2,7	28	72	810	$4{,}5\cdot10^{-2}$
	2	M	1,5	1,5	17	440	-	-
	3	S	3,5	2,5	120	3100	$5\cdot10^{6}$	$2{,}5\cdot10^{-4}$
W_2-$PA_{0,1}$	1	S	3,1	2,9	1,6	41	128	$2{,}3\cdot10^{-1}$
	2	-	0.02	6	12	310	-	-
	3	S	4,0	2,0	110	2840	$6\cdot10^{7}$	$8\cdot10^{-5}$
W_{24}-$PA_{0,1}$-I-2	1	S	3,8	2,2	2,4	62	600	$1{,}5\cdot10^{-3}$
	2	-	0,4	-	17	440	-	-
	3	S	3,4	2,6	120	3100	$2{,}3\cdot10^{6}$	$2{,}5\cdot10^{-4}$
W_{24}-$PA_{0,1}$	1	M	2.7	2,7	1,5	39	128	$2{,}3\cdot10^{-1}$
	2	-	0.04	6	13	335	-	-
	3	S	3,9	2,1	110	2838	$2{,}4\cdot10^{7}$	$7\cdot10^{-5}$

(1) at the first large-scale level of formation of small aggregates such as surface fractals ($Ds_1 = 2.2$, $Rg_1 = 2.4$ nm);
(2) non-fractal behavior
(3) at the third scale level, mass fractals are also formed (fractal dimension $Ds_3 = 2.6$, $Rg_3 = 120$ nm).

For the sample W_8-$PA_{1,5}$-I-3us is characterized by:

(1) the smallest aggregates are surface fractals ($Ds_1 = 2.3$, $Rg_1 = 4.2$ nm);
(2) mass fractals are formed on the basis of the latter ($Df_2 = 1.7$, $Rg_2 = 40$ nm);

Figure 11. Intensity curves of small-angle X-ray scattering for organo-inorganic phosphorosilicate nanocomposite W_2-$PA_{0,1}$-I-2 (1) and phosphorosilicate nanocomposite of the same composition without modifier W_2-$PA_{0,1}$ (2); q is the absolute value of the scattering vector. The numbers indicate the slopes of the SAXS curves in the Porod's (Pi) region for each level I (i = 1,2,3)

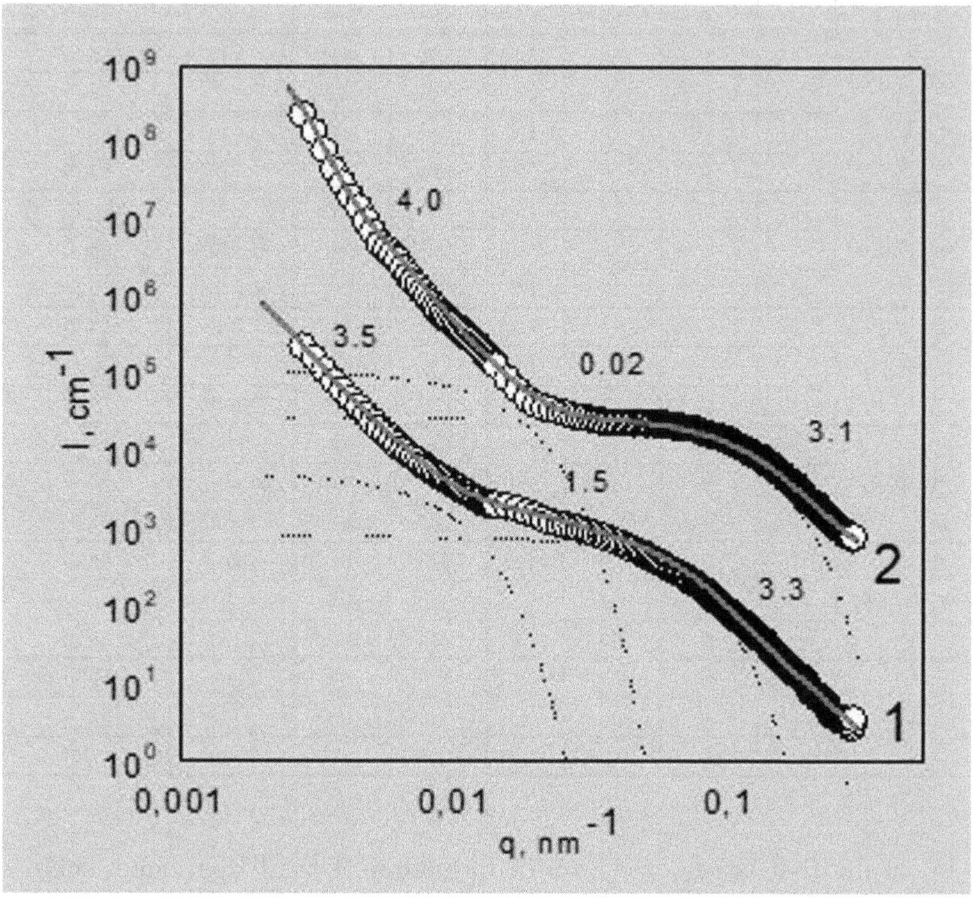

(3) at the third scale level, surface fractals ($Df_3 = 2.2$, $Rg_3 = 160$ nm) are also formed on the basis of surface fractal aggregates of the second level (as primary particles).

For the sample W_8-$PA_{1,5}$-I-4us is characterized by:

(1) the smallest aggregates are surface fractals ($Ds_1 = 2.4$, $Rg_1 = 4.1$ nm);

Figure 12. Curves of intensity of small-angle X-ray scattering for organo-inorganic phosphorosilicate nanocomposite W_{24}-$PA_{0,1}$-I-2 (1) and phosphorosilicate nanocomposite of the same composition without modifier W_{24}-$PA_{0,1}$ (2)

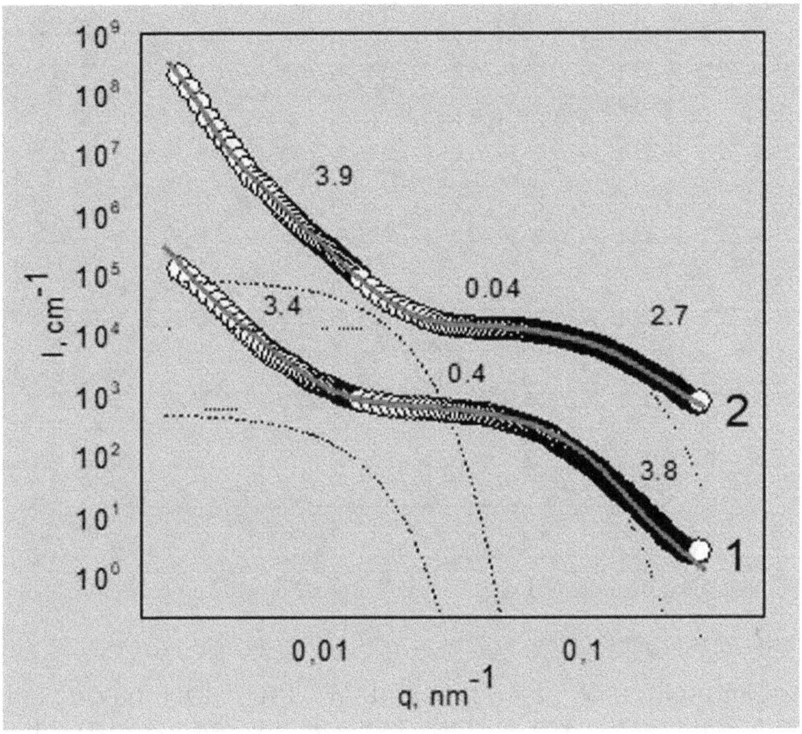

(2) mass fractals are formed on the basis of the latter ($Df_2 = 1.9$, $Rg_2 = 36$ nm);
(3) at the third scale level, surface fractals (Df3 = 2.1, Rg3 = 140 nm) are formed on the basis of surface fractal aggregates of the second level (as primary particles).

For the sample W8-PA1,5-I-5us is characterized by:

(1) the smallest aggregates are surface fractals ($Ds_1 = 2.3$, $Rg_1 = 4.2$ nm);
(2) mass fractals are formed on the basis of the latter ($Df_2 = 2.0$, $Rg_2 = 34$ nm);
(3) at the third scale level, surface fractals ($Df_3 = 2.0$, $Rg_3 = 140$ nm) are formed on the basis of surface fractal aggregates of the second level (as primary particles).

Figure 13. Curves of intensity of small-angle X-ray scattering for organo-inorganic phosphorosilicate nanocomposites W_8-$PA_{1,5}$-I-3us, W_8-$PA_{1,5}$-I-4us, W8-PA1,5-I-5us Sukhyy K.M., Gomza Yu.P. et al. (2015)

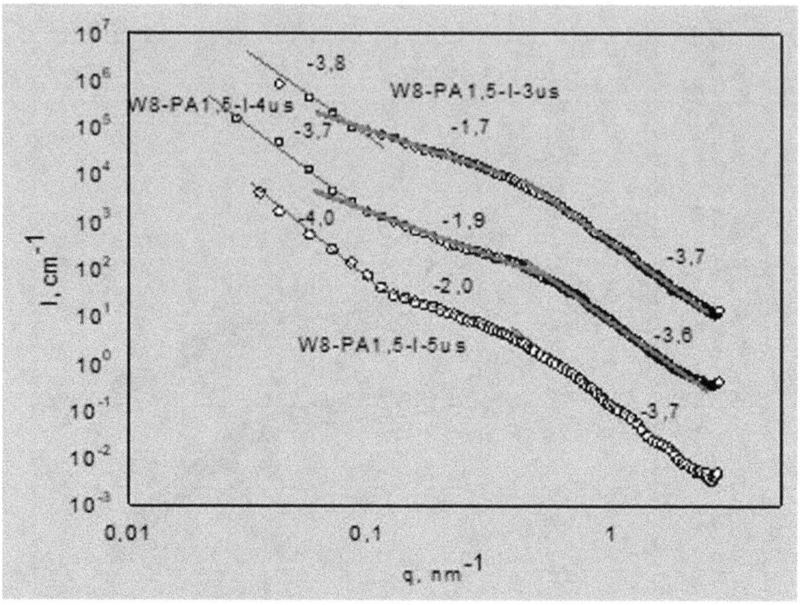

Nanocomposites W_8-PA_1-I-1 and W_8-PA_1-I-1us having identical composition are differed by ultrasonic treatment (44 kHz, 10 min.) of sol W_8-PA_1-I-1us. It could be concluded from SAXS data (Fig. 2 and 3), that forming of surface fractal with smooth surface, that is, close on globules is typical for these samples (Table 3). The ultrasound influence is sufficiently visible, though it does not lead to qualitative changing of characteristics of fractal aggregation, particles of first, second and third levels being decreased changing from 55 to 38 nm on the first level, from 160 to 120 nm on the second, and from 1500 to 950 nm on the third. When a larger quantity of water in initial sol W_8-PA_1-I-1 (W_{16}-PA_1-I-1) is used, surface fractal on the first level decreases too (from 55 to 50 nm), and system as a whole transforms from three-level into two-level.

The scheme of multilevel fractal aggregation of nanocomposites W_8-PA_1, W_8-PA_1-I-1 and W_8-PA_1-I-1us is given on Fig. 13.

When organic-inorganic nanocomposites are formed, a formation of surface fractal aggregates is observed at all structural level, as a whole mass fractal aggregates are generated on the second structural level (100 – 300 nm) when inorganic nanocomposites formed. Systematization and analysis of

Table 3. Fractal characteristics of organic-inorganic silicophosphate nanocomposites.

Sample	Level	Fractal*	P	D	R_g, nm	d_{os} Å	G, cm^{-1}	B, cm^{-1} Å $^{-P}$
W_8-PA_1-I-1	1	S	3,2	2,8	38	98	195	$5{,}5 \cdot 10^{-3}$
	2	S	3,5	2,5	120	310	$3{,}3 \cdot 10^{3}$	$6{,}0 \cdot 10^{-4}$
	3	S	3,8	2,2	950	2450	$9{,}2 \cdot 10^{6}$	$2{,}5 \cdot 10^{-4}$
W_8-PA_1-I-1us	1	S	3,4	2,6	5,5	14,2	460	$1{,}9 \cdot 10^{-3}$
	2	S	3,7	2,3	16	41	$5{,}0 \cdot 10^{3}$	$1{,}3 \cdot 10^{-4}$
	3	S	3,8	2,2	150	390	$2{,}4 \cdot 10^{7}$	$1{,}2 \cdot 10^{-4}$
W_{16}-PA_1-I-1	1	S	3,7	2,3	50	130	2830	$6{,}6 \cdot 10^{-3}$
	2	S	3,5	2,5	140	360	$1{,}5 \cdot 10^{4}$	$1{,}6 \cdot 10^{-3}$
	3	S	3,5	2,5	1000	258	$6{,}6 \cdot 10^{6}$	$7{,}0 \cdot 10^{-4}$

data obtained allow concluding that surface fractal formations predominate in nanocomposites with high protonic conductivity.

These differences are illustrated by schemes of multilevel fractal aggregation of nanocomposites W_8-PA_1 (inorganic nanocomposite), W_8-PA_1-I-1 and W_8-PA_1-I-1us (organic-inorganic nanocomposites) on the plane plotted according to X-ray scattering structural data (Fig. 4). A generation of the initial silicophosphate composite begins from the formation of practically spherical aggregates with order of 12 nm which forms compact lattice of the inorganic polymer on the kind of mass fractal with average particle size of 80 nm on the higher scale level. At last, on the third level occur reorganization of this structure with forming of surface fractal clusters with very smooth surface and average particle size of 470 nm.

A formation of structure of organic-inorganic nanocomposites is substantially differed that could be explained by structure-forming influence of polyionenes. The primary structural element of such nanocomposite is surface fractal or solid particle with average sizes of 14 nm, or 9 nm in the case of ultrasonic homogenizing. On the second structural level in contrast to inorganic composites surface fractal formation with average size of 40 – 50 nm are formed, they being characterized by the occurring of pore system ramified and rough surface. At last, the third structural level is characterized by the formation of surface fractal clusters with average size about of 390 nm with smooth surface. An ultrasonic homogenization have a positive influence on the organic-inorganic nanocomposite structure, the structural elements on each level being characterized by noticeably lesser sizes.

Figure 14. Scheme of peculiarities of multilevel fractal order of nanocomposites investigated: W_8-PA_1–based silicophosphate nanocomposite; W_8-PA_1-I-1 – organic-inorganic nanocomposite; W_8-PA_1-I-1us – organic-inorganic nanocomposite homogenized by ultrasonic during synthesis Sukhyy K.M., Gomza Yu.P. et al. (2015)

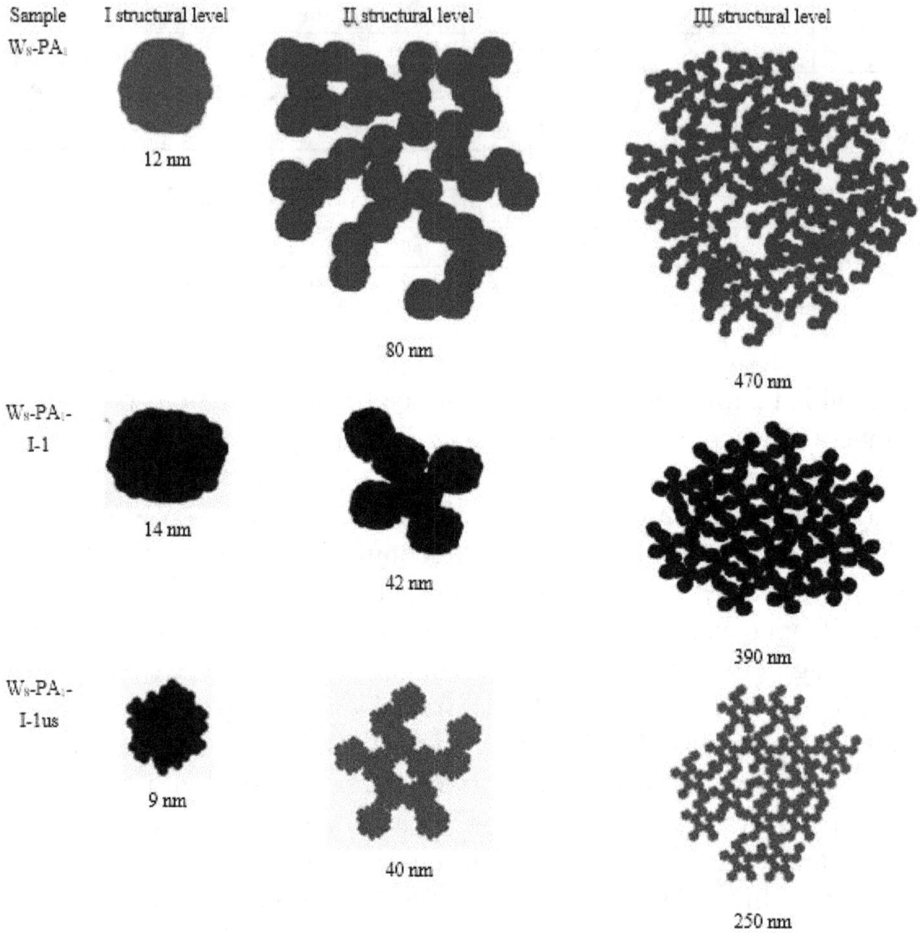

CONCLUSION

Evolution of organic-inorganic nanocomposites during gel formation from sols based on TEOS, orthophosphoric acid and a number of polyonenes was studied. Formation of surface fractal aggregates is stated at all structural level, when mass fractal aggregates are generated on the second structural level (100 – 300 nm). Schemes of multilevel fractal aggregation of nanocomposites W_8-PA_1 (inorganic nanocomposite), W_8-PA_1-I-1 and W_8-PA_1-I-1us (organic-inorganic nanocomposites) are proposed according to SAXS data. A generation of the initial silicophosphate composite is shown to begin from the formation of practically spherical aggregates with order of 12 nm which forms compact lattice of the inorganic polymer on the kind of mass fractal with average particle size of 80 nm on the higher scale level. on the third level this structure is reorganized with forming of surface fractal clusters with very smooth surface and average particle size of 470 nm. Surface fractal formations are supposed to result in high protonic conductivity nanocomposites.

REFERENCES

Armelao, L., Barreca, D., & Moraru, B. (2003). A molecular approach to RuO2-based thin films: Sol–gel synthesis and characterization. *Journal of Non-Crystalline Solids*, *316*(2-3), 364–367. doi:10.1016/S0022-3093(02)01636-8

Biazzotto, J. C., Vidoto, E. A., Nascimento, O. R., Lamamoto, Y., & Serra, O. (2002). Iron(III)-tetra-o-ureaphenylporphyrinosilica obtained by a sol–gel process: A study of EPR, surface area and catalytic activity. *Journal of Non-Crystalline Solids*, *304*(1-3), 101–104. doi:10.1016/S0022-3093(02)01010-4

Deng, Z., Wang, J., Wu, A., Shen, J., & Zhou, B. (1998). High strength SiO2 aerogel insulation. *Journal of Non-Crystalline Solids*, *225*, 101–104. doi:10.1016/S0022-3093(98)00106-9

Deshpande, A. V., & Kumar, U. (2002). Effect of method of preparation on photophysical properties of Rh-B impregnated sol–gel hosts. *Journal of Non-Crystalline Solids*, *306*(2), 149–152. doi:10.1016/S0022-3093(02)01054-2

Echeverri, M., Hamad, C., & Kyu, T. (2014). Highly conductive, completely amorphous polymer electrolyte membranes fabricated through photo-polymerization of poly(ethylene glycol diacrylate) in mixtures of solid plasticizer and lithium salt. *Solid State Ionics*, *254*, 92–100. doi:10.1016/j.ssi.2013.10.050

Gkourmpis, T., & Mitchell, G. R. (2020). The Use of Scattering Data in the Study of the Molecular Organisation of Polymers in the Non-Crystalline State. *Polymers*, *12*(12), 2917. doi:10.3390/polym12122917 PMID:33291414

Guimarães, A. P., Viana, A. P. P., Lago, M. L., & Mohallem, N. D. S. (2002). The effect of thermal treatment on the properties of sol–gel palladium–silica catalysts. *Journal of Non-Crystalline Solids*, *304*(1-3), 70–75. doi:10.1016/S0022-3093(02)01006-2

Guimaraes, A. P., Viana, A. P. P., Lago, R. M., & Mohalleem, L. D. S. (2002). The effect of thermal treatment on the properties of sol–gel palladium–silica catalysts. *Journal of Non-Crystalline Solids*, *304*(1-3), 70–73. doi:10.1016/S0022-3093(02)01006-2

Han, W.-T. (1999). Synthesis and linear and non-linear optical properties of (0.8PPV+0.2DMPPV)/silica glass composites by sol–gel process. *Journal of Non-Crystalline Solids*, *259*(1-3), 107–110. doi:10.1016/S0022-3093(99)00500-1

Imai, Y., Yoshida, N., Naka, K., & Chujo, Y. (1999). Thermoresponsive Organic–Inorganic Polymer Hybrids from Poly(N-isopropylacrylamide). *Polymer Journal*, *31*(3), 258–260. doi:10.1295/polymj.31.258

Li, X., Yang, P., Zhu, Z., You, Z., Zhang, W., Zhang, T., Chen, M., & Zhou, X. (2020). Sol-Gel condensation of temperature sensitive and shape stabilized phase change materials for thermal energy storage. *Thermochimica Acta*, *693*, 178758. doi:10.1016/j.tca.2020.178758

Lu, X., Manners, I., & Winnik, M. A. (2001). Polymer/Silica Composite Films as Luminescent Oxygen Sensors. *Macromolecules*, *34*(6), 1917–1921. doi:10.1021/ma001454j

Lu, Y., Cao, G., Kale, R. P., Prabakar, S., López, G. P., & Brinker, C. J. (1999). Microporous Silica Prepared by Organic Templating: Relationship between the Molecular Template and Pore Structure. *Chemistry of Materials*, *11*(5), 1223–1226. doi:10.1021/cm980517y

Matsuda, A., Nakamoto, N., Tadanaga, K., Minami, T., & Tatsumisago, M. (2003). Proton conductivity at medium temperature range and chemical durability of phosphosilicate gels added with a third component. *Solid State Ionics*, *162-163*, 247–252. doi:10.1016/S0167-2738(03)00261-3

Mohanty, A. K., Banerjee, S., Komber, H., & Voit, B. (2014). Imidoaryl biphenol based new fluorinated sulfonated poly(arylene ether sulfone) copolymers and their proton exchange membrane properties. *Solid State Ionics*, *254*, 82–91. doi:10.1016/j.ssi.2013.11.029

Nam, C. W., & Woo, S. I. (1994). Characterization of spin-coated silicate and phosphosilicate thin films prepared by the sol-gel method. *Thin Solid Films*, *237*(1-2), 314–315. doi:10.1016/0040-6090(94)90278-X

Osborne, H., Blohowiak, K. Y., Taylor, S. R., Hunter, C., Bierwagon, G., Carlson, B., Bernard, D., & Donley, M. S. (2001). Testing and evaluation of nonchromated coating systems for aerospace applications. *Progress in Organic Coatings*, *41*(4), 217–220. doi:10.1016/S0300-9440(01)00132-1

Pandey, P. C., Upadhyay, S., Tiwari, I., & Tripathi, V. S. (2001). An ormosil-based peroxide biosensor— A comparative study on direct electron transport from horseradish peroxidase. *Sensors and Actuators*, *72*(3), 224–226. doi:10.1016/S0925-4005(00)00669-9

Rabinovich, L., Glezer, V., Wu, Z., & Lev, O. (2001). Naphthoquinone–silicate based gas electrodes: Chemical–electrochemical mode of operation. *Journal of Electroanalytical Chemistry (Lausanne, Switzerland)*, *504*(2), 146–149. doi:10.1016/S0022-0728(01)00406-5

Santos, L. R. B., Belin, S., Brios, V., Santilli, C. V., Pulcinelli, S. H., & Larbot, A. (2003). Study of Structural Surface Modified Tin Oxide Membrane Prepared by Sol-Gel Route Sintered at 400°C. *Journal of Sol-Gel Science and Technology*, *26*(1/3), 171–173. doi:10.1023/A:1020778411051

Song, C., & Villemure, G. (2001). Electrode modification with spin-coated films of mesoporous molecular sieve silicas. *Microporous and Mesoporous Materials*, *44*, 679–682. doi:10.1016/S1387-1811(01)00249-9

Sukhyy, K. M., Belyanovskaya, E. A., Kozlov, YaN., Kolomiyets, E. V., & Sukhyy, M. P. (2014). Structure and adsorption properties of the composites 'silica gel–sodium sulphate', obtained by sol–gel method. *Applied Thermal Engineering*, *64*(1-2), 408–412. doi:10.1016/j.applthermaleng.2013.12.013

Sukhyy, K. M., Gomza, Yu. P., Belyanovskaya, E. A., Klepko, V. V., Shilova, O. A., & Sukhyy, M. P. (2015). Resistive humidity sensors based on proton-conducting organic-inorganic silicophosphates doped by polyionenes. *Journal of Sol-Gel Science and Technology*, *74*(2), 472–481. doi:10.100710971-015-3622-7

Tadanaga K, Katata N, Minami T (1997) Super-Water-Repellent Al2O3 Coating Films with High Transparency J Am Ceram Soc 80, 1040-1042.

Valverde, G., Macedo, J. G., & Cruz, D. (2003). Photoconductivity in Mesostructured Thin Films *J Sol-Gel. Sciences et Techniques (Paris)*, *26*, 605–608.

Wu, Z., Lee, K., Lin, Y., Lan, S., & Huang, L. (2003). Effects of surface-active substances on acid–base indicator reactivity in SiO_2 gels. *Journal of Non-Crystalline Solids*, *320*(1-3), 168–170. doi:10.1016/S0022-3093(03)00025-5

Yang, P., Lü, M., Song, C., Gu, F., Liu, S., Xu, D., Yuan, Y., & Cheng, X. (2002). Luminescence of CaS and MnS nanocrystallites co-activated sol–gel derived silica xerogel. *Journal of Non-Crystalline Solids*, *311*(1), 99–103. doi:10.1016/S0022-3093(02)01600-9

Yeatman E. (2017). *Sol-gel fabrication for optical communication components: prospects and progress.* Imperial. doi:. doi:10.1117/12.279834

Chapter 4
Influence Fractal Structure on Proton Conductance of Hybrid Organic-Inorganic Nanocomposites

ABSTRACT

In this chapter, the authors present correlation of fractal structure and protonic conductivity of solid inorganic and organo-inorganic phosphorosilicate nanocomposites which were obtained by the sol-gel method. The optimal ratio of H_3PO_4: $Si(OEt)_4$ is stated to be 1: 1. It significantly enhances the proton conductivity of the phosphorosilicate sol-gel nanocomposite, it being stable in a wide temperature range (20 – 140 °C) at the level of 10-3 – 10-5 Ohm-1•cm-1. The addition of a small amount of polyonenes ($8 \cdot 10^{-4}$ mole polyionene I-1/1 mole tetraethoxysilane) leads to increasing the proton conductivity of nanocomposites by 1-2 orders of magnitude. The maximum value of proton conductivity is stated in nanocomposites most resistant to hydration at temperatures of 200-400 °C.

BACKGROUND

Nanocomposites obtained by the sol-gel method are promising as proton-conducting materials for electrochemical sensors. Their electronic conductivity

DOI: 10.4018/978-1-7998-9728-6.ch004

is negligible, and the ionic conductivity can reach 0.1 S/cm. Increased interest in leading dielectrics is caused mainly by the possibility of their use as membranes of fuel cells, the most promising sources of environmentally friendly energy of the future and electrochemical sensors (Matsuda et al., 1998; Di Noto et al., 2012). Known polymeric proton membranes cannot be used in the temperature range of 100-200 °C, the most optimal for the operation of the fuel cell, which provides maximum efficiency of platinum catalysts. Silicophosphate membranes synthesized by the sol-gel method can function at temperatures above 100 °C (Li et al., 2020).

However, their hydrolytic instability, high aggressiveness, tendency to crystallize at elevated temperatures prevent industrial implementation. At present, the possibility of improving these characteristics by modifying silicophosphate materials with organic and inorganic additives (boric acid and its ethers, aluminum alkoxides, elastomers and epoxy resins) is confirmed Matsuda A et al. (2003). According to the previous chapter, formation of sols in presence of polyionenes leads to formation on surface fractals on all of structure levels which results in high protonic conductivity.

RESULTS AND DISCUSSION

The objects of the study are silicophosphate sol-gel systems, for the synthesis of which tetraethoxysilane and orthophosphoric acid were taken as precursors. The solvent was ethanol. The catalyst for the hydrolysis process is hydrochloric acid. The main components were taken in the following molar ratios with variation of water and orthophosphoric acid: $Si\ (OC_2H_5)_4$: C_2H_5OH: H_2O: HCl: H_3PO_4 = 1: 4: (6-8): 0.01: (0.4-0, 8).

To study the proton conductivity, the obtained xerogels after drying in air at 100°C were ground into a powder with a grain size of 0.02 – 0.1 mm, which was pressed at a pressure of 5000 kg/cm^2 into electrolyte films with a thickness of 0.2 mm. These films were introduced into the three-layer electrode structure of the model supercapacitor type porous electrode/electrolyte/porous electrode, manufactured according to the previously described method (Catauro, Dell'Era, & Ciprioti, 2016).

The dependences of the real parts of the complex specific electrical conductivity for the composite $PA_{0,5}$-W_8 on the frequency in the temperature range from 20 to 140°C in the logarithmic scale are presented in Fig. 1. For other studied nanocomposites, the dependences of the real parts of the complex specific electrical conductivity are similar.

For all considered nanocomposites, the real part of the conductivity exhibits the behavior typical of solid electrolytes. In the region of low frequencies on the curves σ'(f) there is a plateau corresponding to the conductivity at direct current $\sigma_{d.c}$ (Sukhyy et al., 2015). From the plateau on the dependences σ'(f) it is possible to determine the conductivity at direct current $\sigma_{d.c}$ by extrapolation to the y-axis. The values of $\sigma_{d.c}$ found in this way are given in table 9.1, and the dependences of $\sigma_{d.c}$, found from the plateau on the frequency dependences σ, on the inverse temperature for the studied nanocomposites are presented in Fig. 1 - 3. As you can see, the curves for all samples can be described by the Arrhenius equation.

Table 1. The value of proton conductivity at constant current of the studied nanocomposites, ($Ohm^{-1} \cdot cm^{-1}$)

Marking of nanocomposite	Mole H_3PO_4 per 1mole of TEOS	20 °C	40 °C	60 °C	80 °C	100 °C	120 °C	140 °C
Proton conductivity of membranes obtained without organic modifiers								
W_2-$PA_{1.5}$	1.5	$1.3\ 10^{-4}$	$2.5\ 10^{-4}$	$4.0\ 10^{-4}$	$6.2\ 10^{-4}$	$1.0\ 10^{-3}$	$1.3\ 10^{-3}$	$2.1\ 10^{-3}$
W_8-PA_1	1.0	$1.0\ 10^{-4}$	$1.6\ 10^{-4}$	$2.5\ 10^{-4}$	$3.6\ 10^{-4}$	$4.7\ 10^{-4}$	$7.0\ 10^{-4}$	$9.7\ 10^{-4}$
W_8-$PA_{0.5}$	0.5	$2.0\ 10^{-5}$	$4.1\ 10^{-5}$	$6.4\ 10^{-5}$	$1.1\ 10^{-4}$	$1.6\ 10^{-4}$	$2.2\ 10^{-4}$	$3.4\ 10^{-4}$
W_2-$PA_{2.2}$	2.2	$1.5\ 10^{-5}$	$4.7\ 10^{-5}$	$1.3\ 10^{-4}$	$3.7\ 10^{-4}$	$9.1\ 10^{-4}$	$2.1\ 10^{-3}$	$5.1\ 10^{-3}$
PA_1- W_8	1.0	$2.9\ 10^{-7}$	$9.9\ 10^{-7}$	$2.9\ 10^{-6}$	$8.0\ 10^{-6}$	$2.1\ 10^{-5}$	$4.8\ 10^{-5}$	$9.6\ 10^{-5}$
$PA_{0.5}$- W_8	0.5	$2.5\ 10^{-8}$	$1.3\ 10^{-7}$	$5.3\ 10^{-7}$	$1.8\ 10^{-6}$	$5.5\ 10^{-6}$	$1.5\ 10^{-5}$	$3.7\ 10^{-5}$
Proton conductivity of membranes obtained without organic modifiers used								
W_8-PA_1-I-1us	1	$1.3\ 10^{-3}$	$2.3\ 10^{-3}$	$4.1\ 10^{-3}$	$5.9\ 10^{-3}$	$9.8\ 10^{-3}$	$1.3\ 10^{-2}$	$2.0\ 10^{-2}$
W_8-PA_1-I-1	1	$1.8\ 10^{-4}$	$4.1\ 10^{-4}$	$8.6\ 10^{-4}$	$1.5\ 10^{-3}$	$2.3\ 10^{-3}$	$3.5\ 10^{-3}$	$5.4\ 10^{-3}$
W_{16}-PA_1-I-1us	1	$1.1\ 10^{-6}$	$5.0\ 10^{-6}$	$1.9\ 10^{-5}$	$6.3\ 10^{-5}$	$1.6\ 10^{-4}$	$4.5\ 10^{-4}$	$1.1\ 10^{-3}$

For ease of interpretation of the data, the studied nanocomposites are divided into three groups with some differences. In the first group of inorganic silicophosphate nanocomposites (Fig. 9.1) the variable factor is the amount of phosphoric acid, which is added during sol-gel synthesis. With a gradual increase in the amount of H_3PO_4 from 0.5 to 1.5 mol/mol TEOS, the proton conductivity of the phosphorosilicate membrane gradually increases and at 1.5 mol H_3PO_4/mol TEOS reaches a maximum of $2 \cdot 10^{-3}$ $Ohm^{-1} \cdot cm^{-1}$. With a further increase in the molar ratio of H_3PO_4: TEOS to 2.2 mol/mol TEOS, the resulting membrane functions effectively only at temperatures above 100°C. This significantly changes the activation energy of conductivity, and, accordingly, its mechanism. In the synthesis of the described objects, phosphoric acid was added last to the already diluted sol.

Figure 1. Isotherms of frequency dependences of the real part of the complex electrical conductivity for $PA_{0,5}$*-*W_8*.*

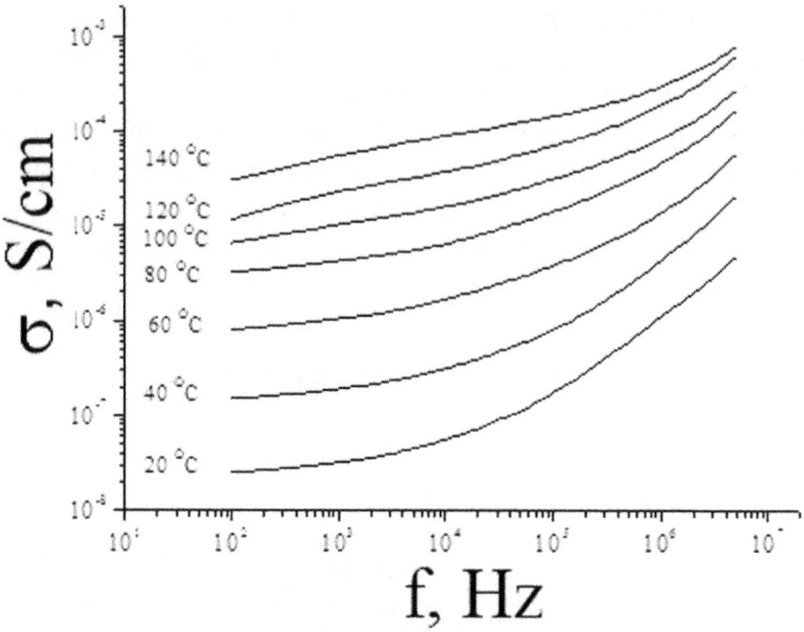

Fig. 2 presents data on the conductivity of sol-gel nanocomposites in which phosphoric acid was added before adding water to the sol. It is seen that the conductivity of such nanocomposites is much lower than discussed above. In addition, the conductivity is much more dependent on temperature.

For hybrid organo-inorganic nanocomposites with low water content with the ratio H_3PO_4: TEOS = 1: 1, in the temperature range of 20 – 140°C conductivity is characterized by high values at the level of $10^{-2} – 10^{-3}$ $Ohm^{-1} \cdot cm^{-1}$ at low energy activation. Ultrasonic treatment of the sol is accompanied by an increase in proton conductivity by an order of magnitude. Dilution of the sol with water adversely affects the conductivity. Thus, the best proton conductivity have phosphorosilicate membranes obtained at a molar ratio of H_3PO_4: TEOS = 1.5: 1 or P: Si = 3: 2. When hybridizing sol-gel systems with polyonenes at a molar ratio of reagents H_3PO_4: TEOS: PI-1 = 1: 1: 8 10^{-4} (ie at a lower content of H_3PO_4), the proton conductivity of the obtained membranes in the temperature range 20-140 °C is maximum and constant . It has also been observed that nanocomposites capable of storing the largest number of water molecules in their structure at temperatures of 200-400°C have the highest proton conductivity. For hybrid organo-inorganic nanocomposites, the conduction process is thermoactivational.

Figure 2. Arrhenius dependences of direct current conductivity $\sigma_{d.c}$ for nanocomposites W2-PA1.5, W_8-PA_1, W8-$PA_{0.5}$, W_2-$PA_{2.2}$ in the temperature range from 20 to 140 °C. Solid lines are approximation curves.

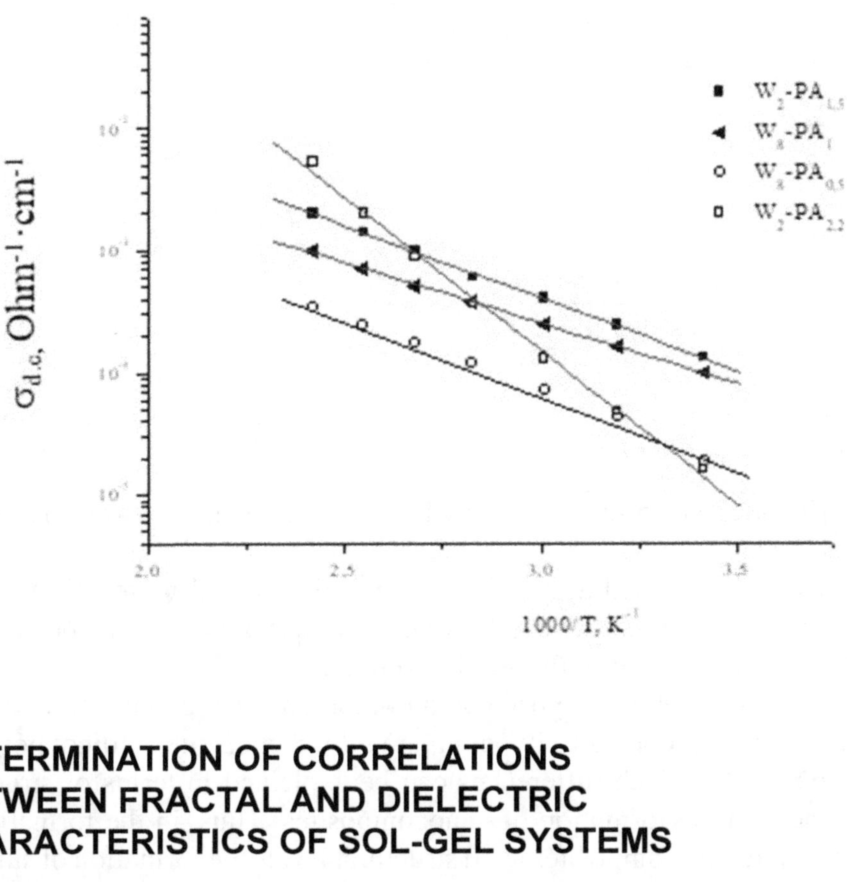

DETERMINATION OF CORRELATIONS BETWEEN FRACTAL AND DIELECTRIC CHARACTERISTICS OF SOL-GEL SYSTEMS

As noted above, the fractal type and fractal aggregation significantly depend on the initial compositions of the sols and the conditions of the synthesis. The use of the Buccage technique to construct an approximation curve from experimental data allowed us to evaluate the structural features of the considered sol-gel systems. Figure 5 shows the characteristic variants of the

Figure 3. Arrhenius dependences of direct current conductivity $\sigma_{d.c}$ for nanocomposites PA_1-W_8, $PA_{0.5}$-W_8 in the temperature range from 20 to 140 °C. Solid lines are approximation curves.

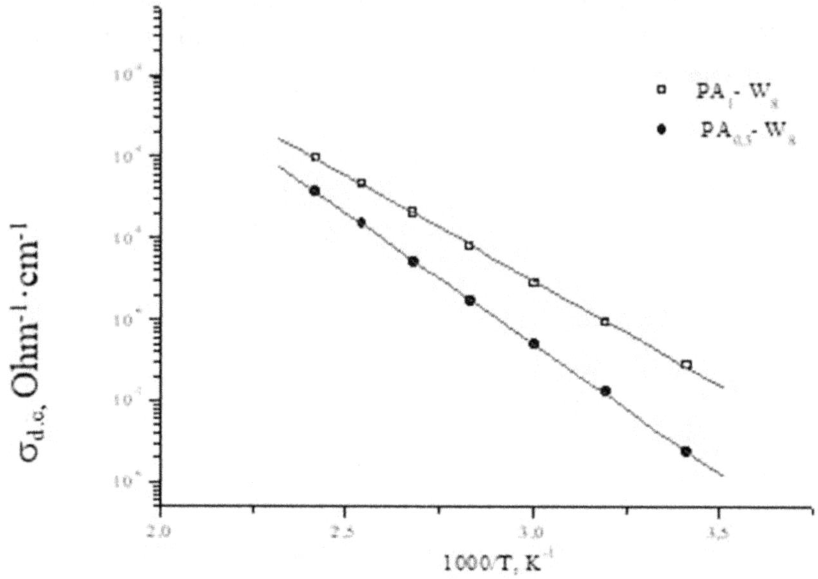

fractal behavior of sol-gel systems, which were found in three dozen xerogels studied by us.

As can be seen from Fig. 5, different variants of fractal aggregation (simple one-level, two-level, three-level), as well as a case of non-fractal behavior (P = -0.45) were revealed in the studied xerogels.

As discussed above, hybrid organo-inorganic nanocomposites exhibit a higher proton conductivity compared to the original silicophosphate nanocomposites. Such differences can be explained in terms of structural differences in the formation of nanocomposites. Thus, in the formation of organo-inorganic composites at all structural levels the formation of surface-fractal aggregates is observed, while in the formation of silicophosphate composites at the second structural level (100-300 nm) mass-fractal aggregates are formed. However, during formation of the hybrid organic-inorganic composites surface fractals aggregates are formed on all structural levels.

Figure 4. Arrhenius dependences of direct current conductivity $\sigma_{d.s}$ for organo-inorganic nanocomposites W_8-PA1-I-1us, W_8-PA1-I-1, W_{16}-PA1-I-1us in the temperature range from 20 to 140 °C. Solid lines are approximation curves.

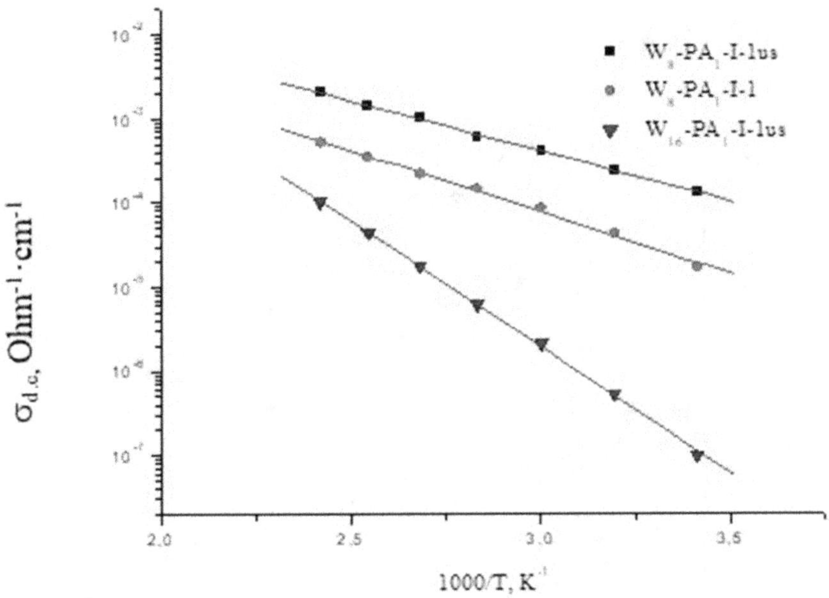

THE MECHANISM OF ACTION OF ORGANIC MODIFIERS TO INCREASE THE PROTON CONDUCTIVITY OF SOL-GEL SYSTEMS

According with results of our research, to improve the proton-conducting properties of modified nanocomposites, it is necessary to introduce into phosphorosilicate sols small additives of polyonenes ($8 \cdot 10^{-4}$ mole polyionene per mole of TEOS).

The effect of modifiers on proton conductivity can be explained as follows. Polyonenes are cationic substances that are ionized in aqueous solution with the formation of positively charged organic ions. Polyionenes contain hydrophilic polar quaternary ammonium groups and a hydrophobic hydrocarbon radical. At high concentrations of polyonenes between the polar ammonium groups and the silanol oxygen atom in the growing silicate network, a donor-acceptor interaction is possible (Fig. 9.6), which depends on the presence of other counterions in the system (Jonscher, 1983).

Figure 5. SAXS curves in log-log coordinates, illustrating the features of the fractal behavior of sol-gel materials.

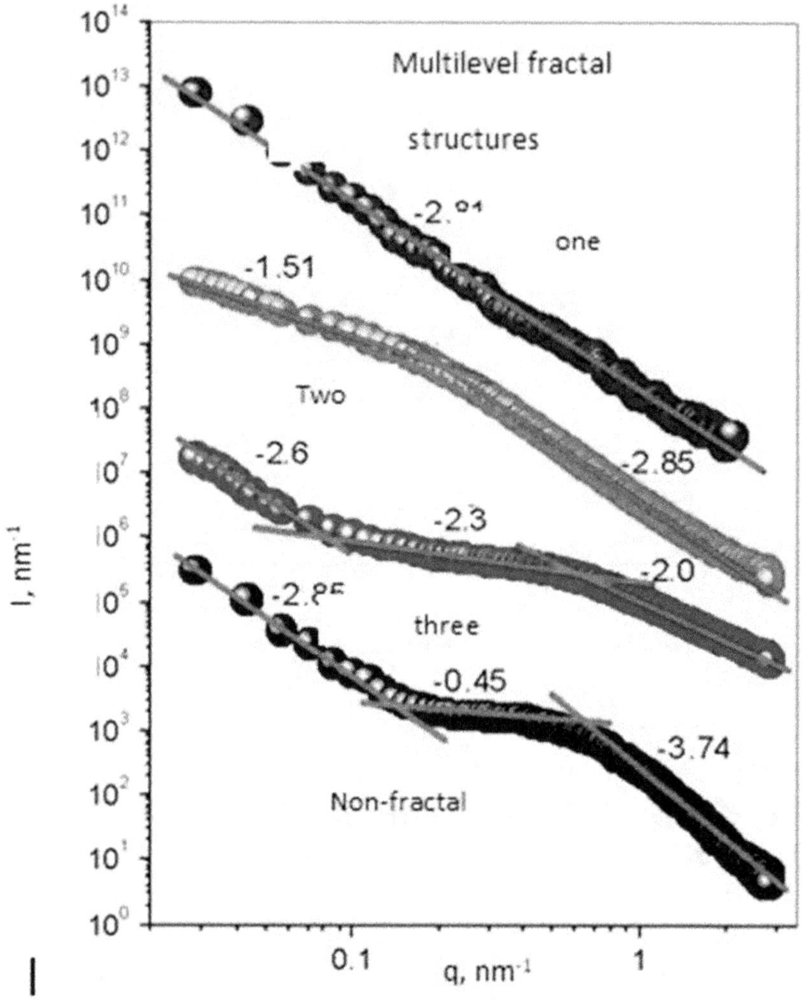

At high content of orthophosphoric acid, the nitrogen cation by analogy can interact with the oxygen atom of orthophosphoric acid. Since the ionization constant of orthophosphoric acid is quite large (K_1 = 7.52 10^{-3}) it can be concluded that the available anions (mainly $H_2PO_4^-$) will be present in the ash, attracted to N^+ and serve as counterions located along the polymer, thereby facilitating the relay the mechanism of proton motion (Fig. 7).

Figure 6. Possible electron-ion interaction between N + and Si-OH groups.

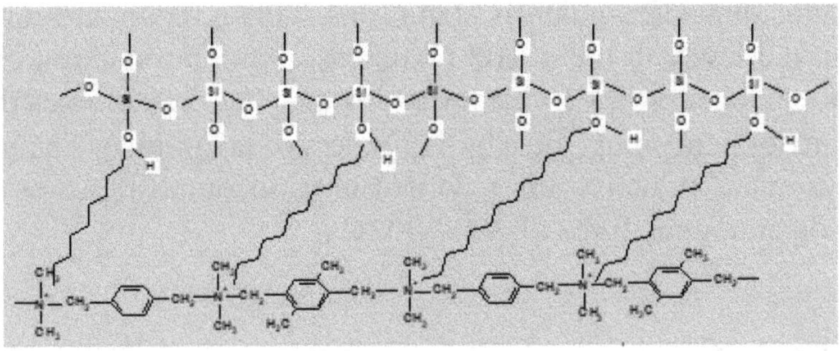

Figure 7. Possible electron-ion interaction between N⁺ and P-OH groups.

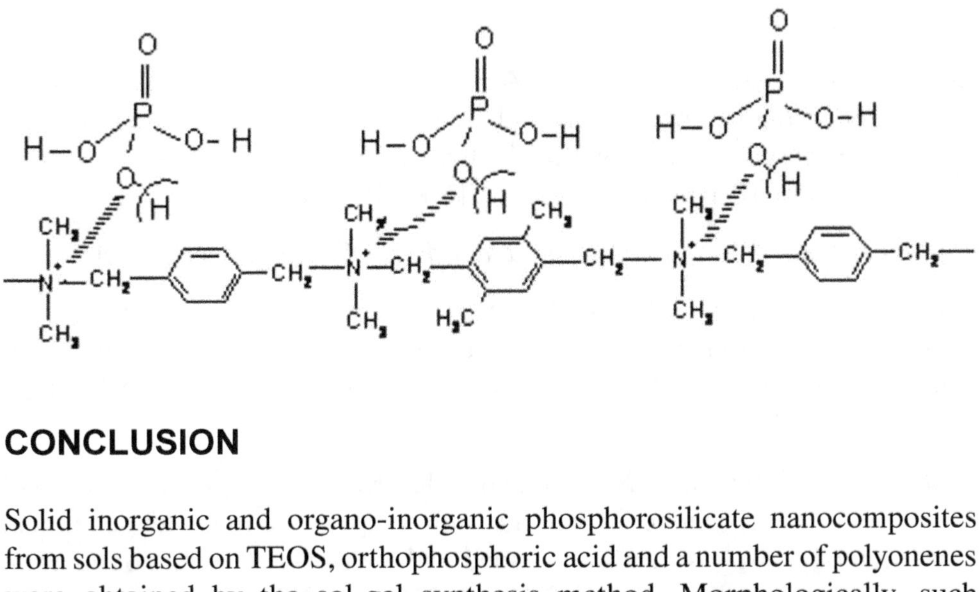

CONCLUSION

Solid inorganic and organo-inorganic phosphorosilicate nanocomposites from sols based on TEOS, orthophosphoric acid and a number of polyonenes were obtained by the sol-gel synthesis method. Morphologically, such materials are nanostructured composites permeated with a large number of mesopores. Functional groups in phosphorosilicate membranes that provide high proton conductivity are P-OH groups, however, it cannot be ruled out that silanol groups Si-OH play a role. It was found that the optimal ratio of H_3PO_4: $Si(OEt)_4$ is 1: 1. Morphologically, such materials are nanostructured composites permeated with a large number of mesopores. Functional groups in phosphorosilicate membranes that provide high proton conductivity are P-OH groups, however, it cannot be ruled out that silanol groups Si-OH play a role. It was found that the optimal ratio of H_3PO_4: Si $(OEt)_4$ is 1: 1.

This significantly improves the proton conductivity of the phosphorosilicate sol-gel nanocomposite, which is stable in a wide temperature range (20 – 140 °C) at the level of 10^{-3} – 10^{-5} $Ohm^{-1}cm^{-1}$. The introduction of a small amount of polyonenes (8 • 10^{-4} mol PI-1/1 mole TEOS) increases the proton conductivity of nanocomposites by 1-2 orders of magnitude. The maximum value of proton conductivity is recorded in nanocomposites most resistant to hydration at temperatures of 200-400 °C.

REFERENCES

Catauro, M., Dell'Era, A., & Ciprioti, S. V. (2016). Synthesis, structural, spectroscopic and thermoanalytical study of sol–gel derived SiO_2–CaO–P_2O_5 gel and ceramic materials. *Thermochimica Acta, 625*, 20–27. doi:10.1016/j.tca.2015.12.004

Di Noto, V., Piga, M., Giffin, G. A., Negro, E., Furlan, C., & Vezzù, K. (2012). New nanocomposite hybrid inorganic-organic proton-conducting membranes based on functionalized silica and PTFE. *ChemSusChem, 5*(9), 1758–1766. doi:10.1002/cssc.201200118 PMID:22807005

Jonscher, A. K. (1983). *Dielectric Relaxation in Solids*. Chelsea Dielectrics.

Liu, Q., Li, Z., Wang, D., Li, Z., Peng, X., Liu, C., & Zheng, P. (2020). Metal Organic Frameworks Modified Proton Exchange Membranes for Fuel Cells. *Frontiers in Chemistry, 8*, 694. doi:10.3389/fchem.2020.00694 PMID:32850683

Matsuda, A., Nakamoto, N., Tadanaga, K., Minami, T., & Tatsumisago, M. (2003). Proton conductivity at medium temperature range and chemical durability of phosphosilicate gels added with a third component. *Solid State Ionics, 162-163*, 247–252. doi:10.1016/S0167-2738(03)00261-3

Matsuda, H., Honjo, K., Hirata, M., Tatsuminago, T., & Minami, T. (1998). Proton conductive silica gels doped with several acids and their application to electric double–layer capacitor. *Chemistry Letters, 27*(11), 1189–1190. doi:10.1246/cl.1998.1189

Sukhyy, K. M., Gomza, Yu. P., Belyanovskaya, E. A., Klepko, V. V., Shilova, O. A., & Sukhyy, M. P. (2015). Resistive humidity sensors based on proton-conducting organic-inorganic silicophosphates doped by polyionenes. *Journal of Sol-Gel Science and Technology*, *74*(2), 472–481. doi:10.100710971-015-3622-7

Chapter 5
Humidity Sensors Based on Hybrid Organic-Inorganic Nanostructured Films

ABSTRACT

In this chapter, the authors present study of properties of inorganic and organic-inorganic proton-conducting materials' conductivity, where it is realized, due to protons that are formed of mobile hydrogen of phosphoric acid groups, which contains in structure of hybrid organic-inorganic nanocomposites and mobile protons of acid groups, which contain in its organic components. Such materials show high levels of conductivity (up to 10^{-2} Ohm^{-1}•cm^{-1}) in the temperatures range of 293-353 K. They are promising for use as materials for electrically active elements of sensor controls which are characterized by the phenomenon of ionic (proton) conductivity.

BACKGROUND

Silicophosphates, which are characterized by high levels of proton conductivity, are promising materials for the electroactive elements of humidity sensors and electrochemical sensors (Pell & Conway, 2001). However, their practical use is limited by low indicators of mechanical properties and hydrolytic stability (Larminie & Dicks, 2003), so it is advisable to determine the optimal composition of membrane materials of this type. Suitable membranes should

DOI: 10.4018/978-1-7998-9728-6.ch005

be characterized by high levels of proton conductivity (up to 0.01 S/cm) and optimized values of strength and elasticity.

Silicophosphate sol-gel materials are nanocomposites, which include a silicate matrix and its saturating ion-conducting component - orthophosphoric acid and may contain modifying organic macromolecular additives of various natures. Organo-inorganic silicophosphate nanocomposite materials using polymeric quaternary ammonium salts (polyonenes) show high levels of proton conductivity (10^{-2} - 10^{-3} $Ohm^{-1}cm^{-1}$ at room temperature) (Gomza, et al., 2006), which makes them promising for use as matrices for manufacture of humidity sensors (Roling, Happe, Funke & Ingram, 1977). Materials of this type are usually structured in the form of fractally aggregated nanostructures. The ionic conductivity levels of the latter depend on the type of fractal aggregation and quantitative parameters of the fractal structure (Sakai, Sadaoka, Matsuguchi, & Kanakura, 1991). Therefore, to control the process of creating membrane materials, it is necessary to monitor their structural and morphological features and properties of ionic (proton) conductivity using the methods of low-angle X-ray scattering (Gomza, et al., 2006) and impedance spectroscopy.

The studies indicate the fractal nature of the morphology of the formed nanocomposites, i.e. the presence of both mass and surface fractal aggregates. Fractal structures are multilevel. Quantitative characteristics of fractal aggregation, such as fractal dimension values, fractal aggregate sizes, and the degree of aggregation of lower-level particles in higher-level fractal aggregates depend on the ratio of inorganic components (Si_2/H_3PO_4), molecular structure, organic modifier content, and anion type.

The direct functional characteristic of the synthesized systems (the level of ionic conductivity) is determined from the data of impedance spectroscopy, on the basis of which isotherms of proton conductivity, dielectric constant, dielectric losses and Nyquist diagrams are calculated.

The obtained results of impedance spectroscopy demonstrated the possibility of achieving high levels of conductivity of inorganic silicophosphate sol-gel materials (up to 10^{-2} $Ohm^{-1}cm^{-1}$).

Solid polymer electrolytes are intensively studied as materials for humidity sensors, because they are characterized by fairly high conductivity values, which vary depending on the humidity of the medium (Roling, Happe, Funke, & Ingram, 1977).

The aim of this study is to create humidity sensors, as the material of the active layer which used organo-inorganic silicophosphate nanocomposite materials and study their proton-conducting characteristics in the frequency range from 0.01 to 100 kHz under controlled temperature and humidity.

The design of humidity sensors based on a corundum substrate, which houses two rows of flat counter-pin electrodes, is described in detail in Gomza et al. (2006). A thin sensitive layer is applied to the surface of such electrodes, as which interpenetrating polymer grids based on hydrophilic and hydrophobic polymers were used. Such sensors showed inverse changes in the impedance studied at a frequency of 1 kHz, which in the range of values of relative humidity from 30 to 90% at a temperature of 303 K varied by 3-4 orders of magnitude.

The compositions of the initial solutions and the characteristics of silicophosphate nanocomposites are given in Chapter 2. The use of polyionene in the synthesis of silicophosphate nanocomposites leads to the formation of materials with higher mechanical strength and elasticity compared to the original silicophosphate nanocomposites. As a result of hydrolysis and polycondensation reactions, a nonporous composite material was formed, which consisted of a SiO_2 framework, forming a system of micro- and mesopores filled with a proton-conducting electrolyte - orthophosphoric acid.

C-50-1-06 sieve plates with dimensions of 15×6 mm^2 were used for the production of laboratory samples of humidity sensors. Counter-pin electrodes with a distance of 400 μm were applied to their surface by spraying. Metallization of two-layer electrodes: lower layer ~ 0.1 μm Cr, upper - ~ 0.3 μm Ni.

Layers of organo-inorganic silicophosphate nanocomposite materials with a thickness of ~ 50 μm were applied to the entire surface of the plate using sols based on tetraethoxysilane.

An experimental stand was made to measure the impedance characteristics of humidity sensors. The sensor was fixed in a clamp with electrical terminals that connect the sensor to the input of the dielectric spectrometer. The clamp was constructively used as a minibus cover, in which due to the presence of supersaturated salt solutions the specified levels of relative humidity were maintained in the range from 23.5% to 96.5% at 303 K (23.5% RH - NH_4NO_3 + $Ca(NO_3)_2 \bullet 4H_2O$, 59.4% RH - $Ca(NO_3)_2 \bullet 4H_2O$, 84.0% RH - KCl, 96.5% RH - K_2SO_4). In fig.10.1 shows the view of the clamp with the sensor under study and the design of the measuring cell assembly.

Dielectric characteristics and proton conductivity levels of humidity sensors with nanocomposite silicophosphate sensitive layer were investigated in the

Figure 1. Humidity sensor in the clamp.

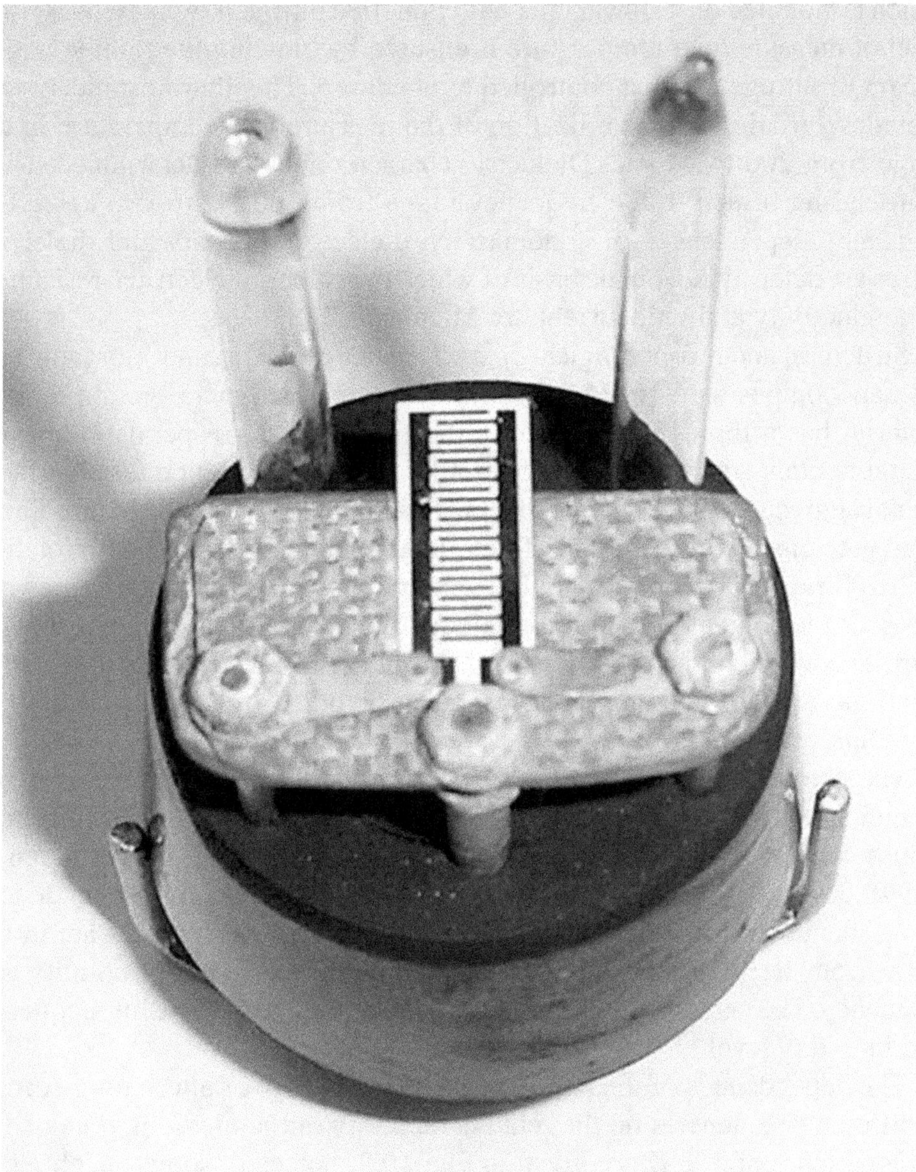

range from 297 to 353 K using the method of impedance spectroscopy. For such measurements, a dielectric spectrometer operating in the frequency range from 0.1 to 100 kHz was used.

The measuring cell with the investigated sensor is placed in a thermostat, which is mounted on a Dewar tank with liquefied nitrogen. Achieving a given level of measurement temperature is ensured by blowing the sample with a stream of nitrogen with a controlled temperature. This thermostatic system provides variation and stabilization of the measurement temperature in the range from 200 to 413 K. Dielectric characteristics are determined using a measuring bridge in the frequency range from 100 Hz to 100 kHz. The frequency dependences of conductivity, dielectric constant and dielectric losses are determined, on the basis of which the values of electrical resistance or conductivity at direct current are calculated.

Structural and morphological characteristics of the organo-inorganic sol-gel nanocomposite W16-PA1-I-1us were studied earlier in section 3. This material has a three-level structure: the first level corresponds to surface fractal fractals with a radius of rotation of about 5 nm, which form surface-fractal aggregates of the second level with a radius of rotation ~ 14 nm. These aggregates, in turn, form surface-fractal agglomerates with a radius of rotation of ~ 100 nm. This complex multilevel structure of the silicate framework provides the content of the proton-conducting electrolyte - orthophosphoric acid. The ionic conductivity levels of this material vary in the widest range of all the organo-inorganic nanocomposites studied by us and reach 10^{-3} - 10^{-7} $Ohm^{-1}\cdot cm^{-1}$.

The frequency dependences of the AC conductivity obtained for the manufactured laboratory sample of the humidity sensor at 303 K and different values of relative humidity in the range from 23.5% to 96.5% are given in Fig 10.2. According to Fig. 10.2, all isotherms of frequency dependences of AC conductivity are characterized by conductivity levels that are in the range from 10^{6} to 10^{1} $Ohm^{-1}\cdot cm^{-1}$ depending on the level of humidity and frequency. The values of AC conductivity increase with increasing frequency and humidity level.

The dependences of the values of the conductivity of alternating current at different frequencies on the relative humidity of the medium, obtained at a temperature of 303 K are given in Fig. 10.3. Selected isotherms obtained at frequencies of 100 Hz, 1 kHz and 100 kHz. As can be seen from Fig. 10.3, the isotherms show a nonlinear variation of the values of the conductivity of alternating current in the studied range of conditions of relative humidity.

According to Fig. 3, the values of the conductivity of alternating current obtained at a frequency of 1 kHz show variations of this characteristic (σ_{max} / σ_{min}) for the humidity sensor in the studied range of relative humidity 405 times, at a measurement frequency of 100 Hz - 1717 times and a frequency of

Figure 2. Isotherms of frequency dependences of AC conductivity obtained at a temperature of 303 K and shown in Fig. relative humidity values.

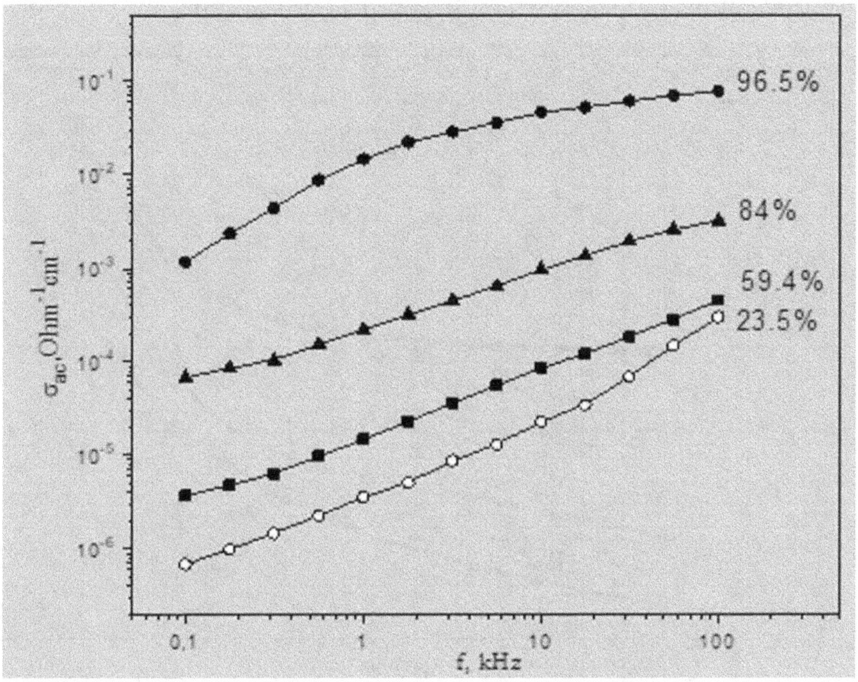

100 kHz - 262 times. Thus, in this case, (the results are consistent with those obtained in Y. Sakai et al. (1991)), the most optimal were the measurements of relative humidity using frequencies not exceeding 1 kHz.

It should be noted that such levels of proton conductivity values and the range of changes in this characteristic of the studied laboratory samples of humidity sensors when varying the humidity are sufficient for the use of the studied nanocomposite materials as the active medium of resistive humidity sensors. It should be noted that the studied organo-inorganic nanocomposites have satisfactory mechanical properties.

Thus, the use of a hybrid organo-inorganic nanocomposite as an active medium for humidity sensors showed that changes in the composition of systems and synthesis conditions allow to vary the value of the main functional characteristic of the material - proton conductivity - in the range of 10^{-6} - 10^{-3} $Ohm^{-1}\cdot cm^{-1}$. Such levels are acceptable for the use of synthesized nanocomposites as the material of the active medium of resistive and electrochemical sensors.

Figure 3. Isfrequency dependences of the specific conductivity of alternating current on relative humidity. The dependences obtained at frequencies of 0.1, 1 and 100 kHz are selected.

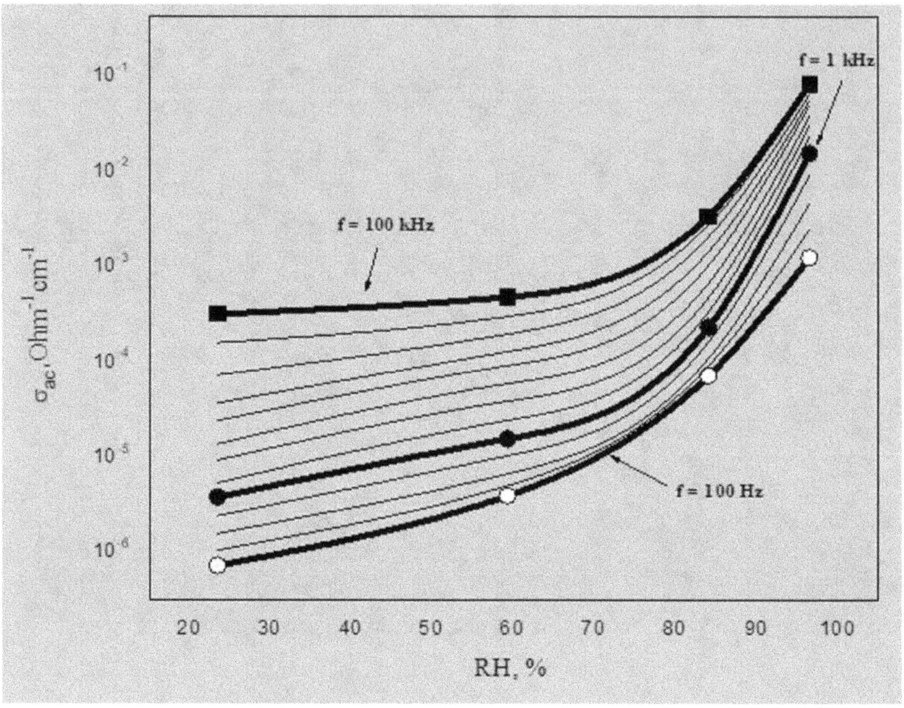

Variation in the nature of the polymer component, as well as its content provides an opportunity to obtain satisfactory mechanical properties and conductivity levels of nanocomposites in order to optimize their performance (Gomza et al., 2015).

Laboratory samples of humidity sensors based on substrates containing interpenetrating electrodes with sensitive layers applied by the sol-gel method are made. It is established that when the values of relative humidity of the medium vary from 23.5 to 96.5%, the values of the conductivity of alternating current increase by 2-3 orders of magnitude.

CONCLUSION

Thus, the use of inorganic silicophosphate nanocomposite as the active medium of humidity sensors showed that changes in the composition of systems and synthesis conditions allow to vary the value of the main functional characteristic of the material - proton conductivity - in the range of 10^{-5} - 10^{-2} $Ohm^{-1} \cdot cm^{-1}$. Such levels are acceptable for the use of synthesized nanocomposites as the material of the active medium of resistive and electrochemical sensors.

Variation in the nature of the polymer component, as well as its content provides an opportunity to obtain satisfactory mechanical properties and conductivity levels of nanocomposites in order to optimize their performance.

Laboratory samples of humidity sensors were made on the basis of substrates, which contain interpenetrating electrodes with sensitive layers of silicophosphate materials obtained by the sol-gel method. It is established that when the values of relative humidity of the medium vary from 23.5 to 96.5%, the values of the conductivity of alternating current increase by 2-3 orders of magnitude.

REFERENCES

Gomza Yu.P., Klepko V.V., Nesin S.D., Sukhyy K.M., Burmistr M.V. (2006). Ion–conducting organic–inorganic sol–gel materials as an active medium for resistive sensors, *Investigation on sensor system and technologies*, 1, 290–303.

Larine, J., & Dicks, A. (2003). Fuel Cells Systems Explained. Second Edition. John Wiley & Sons. doi:10.1002/9781118878330

Pell, W. G., & Conway, B. E. J. (2001). Analysis of power limitations at porous super capacitor electrodes under cyclic voltammetry modulation and DC charge. *Power Sources*, *96*(1), 57–67. doi:10.1016/S0378-7753(00)00682-0

Rolling, B., Happen, A., Funk, K., & Ingram, M. D. (1977). Dielectric studies of same conducting systems. *Physical Review Letters*, *78*, 2160–2164.

Sakai, Y., Sadaoka, Y., Matsuguchi, M., & Kanakura, Y. (1991). Kinetic analysis of Cathodic Deposition. *Journal of the Electrochemical Society*, *138*, 2474. doi:10.1149/1.2085997

Sukhyy, K. M., Gomza, Yu. P., Belyanovskaya, E. A., Klepko, V. V., Shilova, O. A., & Sukhyy, M. P. (2015). Resistive humidity sensors based on proton-conducting organic-inorganic silicophosphates doped by polyionenes. *Journal of Sol-Gel Science and Technology*, *74*(2), 472–481. doi:10.100710971-015-3622-7

Chapter 6
Heat Storage Media Based on Composites "'Silica Gel–Crystalline Hydrate"

ABSTRACT

In this chapter, the authors present results of a study of the composite sorbents 'silica gel – crystalline hydrate' synthesized by sol – gel method. The sorption capacity of the composites 'silica gel – sodium sulphate' is stated to exceed noticeably the linear superposition of the sorption capacity of mechanical mixture of the silica gel and salt. The composites composition and structure were studied by IR-spectroscopy and wide-angle X-ray scattering. The changing of adsorption properties of crystalline salts at their distribution in silicon oxygen matrix is stated to result from dispersion up to nanoscale. The experimental setup for effectiveness tests of synthesized composite adsorbents is developed. The correlation of their composition, structure, water adsorption kinetic, and operating characteristic as heat storage material is revealed. The heat storage capacity is stated to be at least 0.36 kWh/kg (258 kWh/m^3).

BACKGROUND

Adsorptive heat energy storage and conversion is promising technology for systems of heating the residual premises, ventilation and conditioning. However, wide commercialization is strongly complicated with properties of convenient adsorbents such as adsorption, heat of adsorption and regeneration

DOI: 10.4018/978-1-7998-9728-6.ch006

temperature (Sukhyy, Belyanovskaya, & Sukhyy, 2020). Salts which form crystalline hydrate can be considered as alternative. For example, the working pairs $CaCl_2$–H_2O, $SrBr_2$–H_2O, Na_2S–H_2O, and $MgCl_2$–H_2O allow realizing the heat energy storage density of 1.9–2.7 GJ per 1 m^3 of crystalline hydrate (Jong et al., 2014). Nevertheless, their application is restricted by the corrosive activity of these salts along physical and chemical instability at high temperatures, instability in repeated cycling, degradation in time, and also an underdeveloped heat exchange surface. Such problems are typical for every crystalline hydrate.

The most promising attempts to improve adsorptive materials are mainly concerned with synthesis and investigation of composite sorbents 'porous matrix-salt' due to increased adsorptive capacity to convenient working fluids such as water, ammonia, methanol, ethanol Cabeza, Solé, & Barreneche (2017).. Moreover, the distribution in the porous matrix results in stabilization of the salt when hydration cycles-dehydration repeated (Jong et al., 2014). As porous matrix, silica gel (Gordeeva, Grekova, Krieger, & Yu, 2013)., zeolites (Cabeza, Solé, & Barreneche, 2017)., vermiculite (Scapino et al., 2017), graphite (Zamengo & Yu, 2017), and salts are used. Magnesium chloride and lithium bromide (Yu et al., 2013), lithium chloride (Grekova, Gordeeva, & Aristov, 2017), magnesium oxide (Grekova, Gordeeva, & Yu, 2017), calcium chloride (Hiremath & Kadoli, 2013). are used as a salt.

Obviously, the heat energy storage by composite sorbents is based on three mechanisms: capacitive, phase-transitional, and adsorptive. The salt reacts with water vapor, which results in formation of crystalline hydrate, further adsorption leads to complex dissolution and transformation into a 'salt-sorbate' solution. Nevertheless, the presence of a crystalline salt in the matrix results not only in the growth of the adsorption capacity, but also leads to a qualitative change in the sorption kinetics. Composite water adsorbents such as 'salt in a porous matrix' in devices for heat energy storage, adsorption refrigeration devices and air conditioning units are known to be widely spread. However, their production is based on rather complex impregnation technologies of dry and wet impregnation of porous media with solutions of crystalline hydrates, often expensive, which impedes commercialization in energy storage devices (Cabeza, Solé, & Barreneche, 2017).

Alternative to impregnation methods is synthesizing composites such as silica gel-crystalline hydrate by sol-gel method (Sukhyy et al., (2014). Nevertheless, the correlation of the properties of heat-storage media (composition, structure, kinetics of adsorption of water vapor) of the composite adsorbents 'silica gel-salt' synthesized by the sol-gel method and the performance characteristics of heat storage devices still require further studies.

The present chapter is focused on the properties (composition, structure, kinetics of adsorption of water vapor) of the composite adsorbents synthesized by the sol-gel method and their performance characteristics as a heat-storage material.

EXPERIMENTAL

Porous composite materials were prepared by deposition of liquid sodium silicate glasses with various acids, i.e. sulfuric and acetic. Composite materials are obtained in a three-dimensional silicate matrix, which contain 20–40 wt. % of sodium sulfate or acetate Serhiienko Ya. O., Sukhyy K.M. et al. (2019). The composite material 'silica gel – Na_2SO_4' was obtained by the neutralization reaction of sulfuric acid H_2SO_4:

$$Na_2Si_3O_7+H_2SO_4=Na_2SO_4+(SiO_2)_n+H_2O \quad (1)$$

To synthesize the composite material 'silica gel–Na_2SO_4', it is necessary to take the components (98% sulfuric acid and 54% liquid sodium silicate glass) in such a ratio (mass. %): the solution of liquid sodium silicate glass is 81.76 and sulfuric acid – 18.24.

The neutralization reaction of 98% acetic acid solution with a 54% solution of liquid sodium silicate glass is described by the following reaction:

$$Na_2Si_3O_7+ 2CH_3COOH=2CH_3COONa+(SiO_2)_n+H_2O. \quad (2)$$

At such concentrations of the reagents, for the synthesis of the 'silica gel-CH_3COONa' composite material the following ratio of components (wt. %) was used: a liquid sodium silicate glass solution of 78.54 and acetic acid of 21.46.

The synthesis of composite sorbents 'silica gel - crystalline hydrate' involves two stages. In the first stage, a polymeric quaternary ammonium salt (PQAS) was added to the 50% aqueous solution of silicate glass (with a silicate module of 3.2) at a rate of 1 g per 100 g of a dry residue of SiO_2 and heated to 35 °C with stirring, after which the appropriate acid (sulfuric or acetic) was added dropwise to the desired pH. Then, the resulting solution was heated for an hour to a temperature of 95 °C. At this stage, the silicon phase nuclei were formed, which was stated by the turbidity of the reaction medium. In the second stage, with intensive stirring, solutions of the corresponding acid and silicate glass of a certain concentration were simultaneously added dropwise. At the same time, the particles coarsened and the highly porous silica gel in the pores of which the corresponding crystalline hydrates (sodium sulfate in the case of sulfuric acid and sodium acetate using acetic acid) introduced is formed. Composite sorbents were precipitated as a fine precipitate, filtered from the liquid phase and dried to constant weight at a temperature of 40-50 °C.

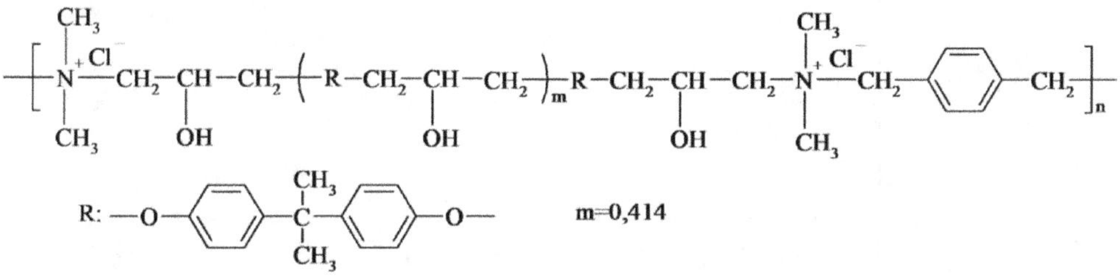

The sorption equilibrium between water vapor and dry sorbent was studied by the weight method. Chemical balance 'AXIS-ANG200C', class II of accuracy were used at temperatures of 20, 40 and 60 °C and pressures of 20 and 760 mm Hg. Prior to the tests, the test substances were kept in a thermal cabinet at a temperature of 140 °C for 4–6 hours. After drying, samples (in an amount of 1–2 g) were placed in a desiccator, at the bottom of which distilled water was contained.

Filled desiccators were loaded into the thermal cabinets with a predetermined temperature. The temperature in the thermal cabinets was maintained with an accuracy of ± 0.1 °C. The necessary pressure was maintained by means of a vacuum pump, which was attached to a desiccator. The pressure was recorded with a sensor "MKS Baratron type 626A". Every 20 minutes, the test samples were removed and weighed on analytical scales. The samples were held at predetermined temperatures until equilibrium was established, until the sorbent absorbed the theoretically required amount of water.

Figure 1. Curves of weight loss of the composite "silica gel-sodium sulfate" in dehydrated (1) and hydrated (2) states

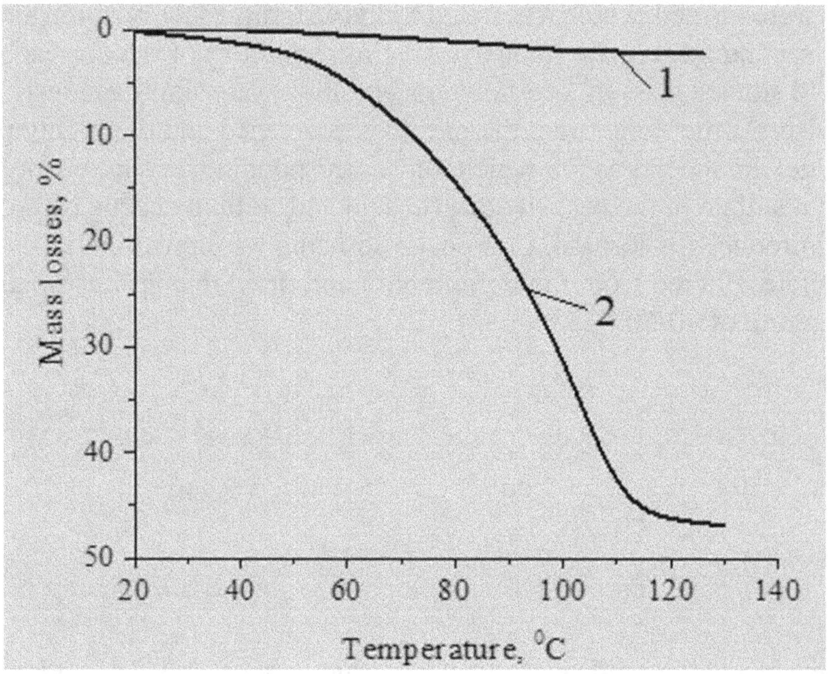

RESULTS AND DISCUSSION

The curves of mass loss for composite sorbents "silica gel-sodium sulphate", "silica gel-sodium acetate" in hydrated and dehydrated states are given in Fig. 1 and 2. Analyzing the curves of weight loss of the composite "silica gel-sodium sulphate", which are given in Fig. 1, it can be noted that the composite after desorption (curve 1) when heated to 1000°C, loses only 2% of the mass. Mass loss for the composite after adsorption of moisture is observed in the temperature range from 32°C to 100°C and equals to 46%. For the composite "silica gel – sodium acetate" (Fig. 2) in the fully hydrated state, the mass loss is 42% in the range of 38–125°C (Belyanovskaya et al., 2014).

To clarify the differences between the thermophysical properties of the studied composite sorbents from the properties of massive crystal hydrates, a differential thermal analysis was performed for 10-aqueous ($Na_2SO_4 \cdot 10H_2O$) and anhydrous sodium sulfate, as well as for dried to constant weight at 100°C silica gel KSM, which can considered as an analogue of the silicon-oxygen matrix of the studied composites Fig. 3.

Figure 2. Mass loss curves of the composite "silica gel-sodium acetate" in dehydrated (1) and hydrated (2) states

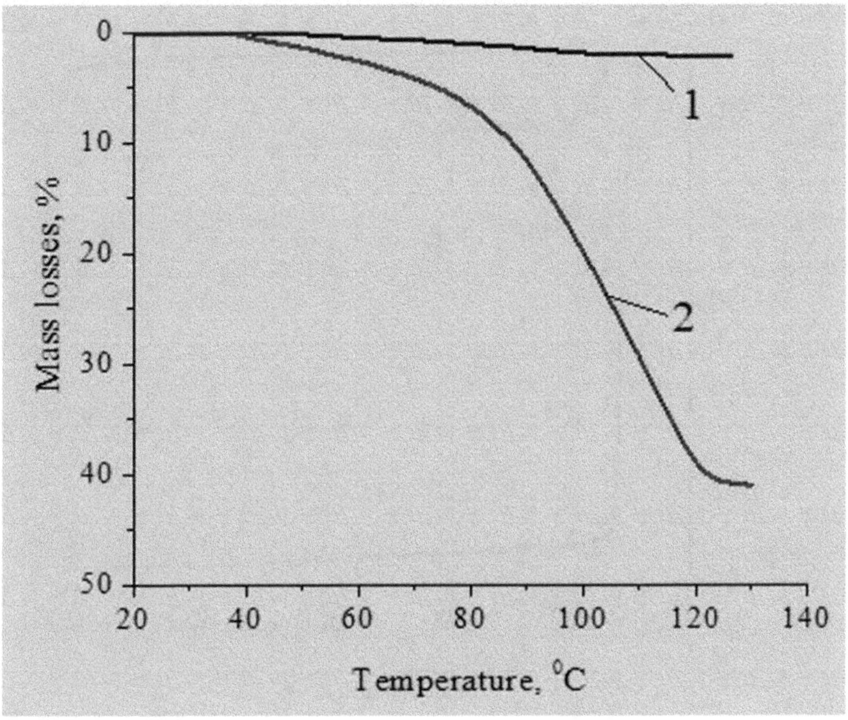

Anhydrous sodium sulphate (curve 1) loses only 5% of its mass in the temperature range of 175–200 ° C when heated to 1000 ° C. For 10-aqueous sodium sulfate (curve 2), the weight loss is 54% in the range of 75–140 °C. The mass loss of KSM silica gel when heated to 1000 ° C is 15% and is associated with the removal of bound water (Sukhyy et al., 2014)..

The sorption properties of the composite "silica gel-sodium sulfate" differ slightly from the properties of solid salt and silica gel due to the peculiarities of this dispersed system. This can be seen by comparing the sorption capacities of these substances. According to the study, the sorption capacity of "pure" silica gel is 0.18 g / g, massive sodium sulphate 1.27 g / g, and the composite "silica gel-sodium sulphate", respectively, 0.85 g / g. The stoichiometric composition of the composite sorbent, according to the synthesis conditions is Na_2SO_4: SiO_2 = 1: 1.27. The linear superposition of the sorption capacity of "pure" silica gel and sodium sulfate was calculated by the following formula:

Figure 3. Curves of weight loss of sodium sulfate after desorption (1), after adsorption (2) and silica gel KSM (3)

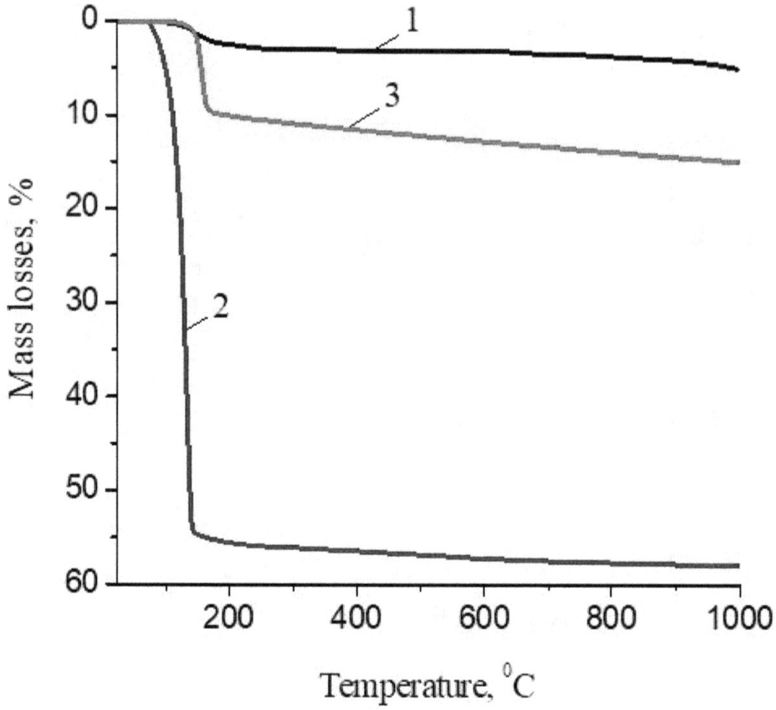

$$m_\Sigma = a \cdot m_{Na_2SO_4} + (1-a) \cdot m_{SiO_2}, \quad (3)$$

where a is mass content of sodium sulfate in the composite a=0,44; $m_{Na_2SO_4}$, m_{SiO_2} are mass of water (g) which absorbed by 1 g of massive salt Na_2SO_4 and 1 g of silica gel, $m_{Na_2SO_4}$=1,27, m_{SiO_2}=0,18.

Substituting the corresponding data in Eq.(3), the linear superposition of the adsorption capacity of the mechanical mixture of silica gel and sodium sulphate, which is 0.66 g/g, is calculated. This indicates that the sorption properties of the synthesized composite sorbents "silica gel-crystal hydrate" are not a linear combination of the properties of silica gel and massive salt, i.e. the synthesized composite sorbents have a unique structure that increases the rate of reactions between crystal hydrates and water vapor in developed pores.

To determine the composition and structure of the obtained composites, IR spectroscopic studies of the synthesized composite sorbents "silica gel-crystal hydrate" in comparison with massive salts and starting compounds were

Figure 4. IR – absorption spectra: 1 – sodium sulphate; 2 – composite 'silica gel – sodium sulphate' after desorption; 3 – silica gel KSM

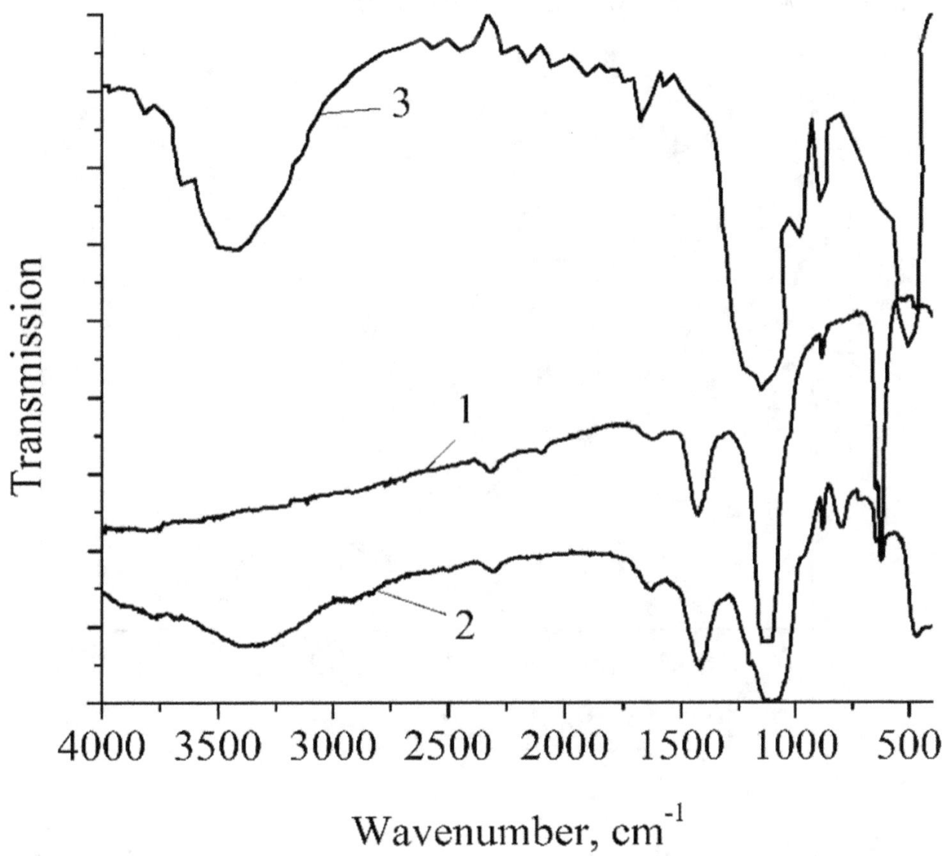

performed. The IR spectra of $Na_2SO_4 \cdot 10H_2O$ (1) crystal hydrate composite sorbents "sodium silica gel-sulfate" (2) and KSM silica gel (3) are given in Fig. 4.

When comparing the spectra of sodium sulfate and KSM brand silica gel, which can be considered as an analogue of the silicon-oxygen matrix of the studied composites with the synthesized composite sorbent "sodium silica gel-sulfate", it can be noted that in the spectra of composite "silica gel-sodium sulfate" and sodium sulfate.

The synthesized "silica gel – sodium sulfate" composite has virtually no intense absorption band in the region of 3749 cm^{-1}, which is characteristic of the IR absorption spectra of substances containing Si – OH groups (Zhong, Zhu, & Wang, 2005). (OH oscillations of the dehydrated silanol surface). In

Figure 5. IR absorption spectrum of fully dehydrated silica gel surface (1), silica gel at the initial stage of water adsorption (2) and highly hydrated silica gel surface (3) Sukhyy K. M., Belyanovskaya E. A. et al (2014)

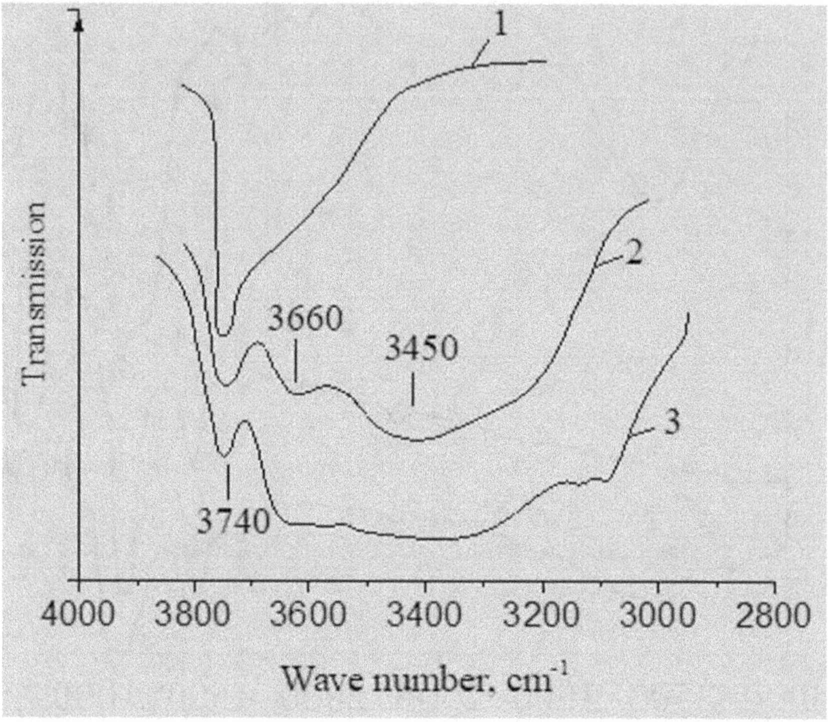

this case, for the composite "silica gel – sodium sulfate" there is an offset wide band 3680–3200 cm^{-1} (Fig. 4), which is characteristic of silica gel at the initial stage of water adsorption (Fig. 4, curve 2) and highly hydrated silica gel surface. (Fig. 4, curve 3). According to Zhong, Zhu, and Wang (2005), it consists of two components: bands with frequencies of 3665 and 3450 cm^{-1} and characterizes the fluctuations of adsorbed water. The low intensity of these bands, as well as the distribution of water molecules between the adsorption centers, for the synthesized composite "sodium silica gel-sulfate" (3665 and 3450 cm^{-1}), depends on the relative vapor pressure of water in the sample cell and the degree of hydration of its surface.

The absorption bands in the region of 3000–3200 cm^{-1}, which are characteristic of the highly hydrated surface of silica gel (Fig. 5, curve 3), which can be attributed to the oscillations of the hydrogen bond of OH groups of silica gel with oxygen atoms of water molecules, are also characteristic of the synthesized composite "silica gel-sodium sulphate" (Fig. 5). It is obvious that such interaction is not carried out at the first moment of water adsorption, as indicated by the physical state of the surface of curve 3 (Fig. 5).

The synthesized "silica gel – sodium sulfate" composite has a number of absorption bands in the region of 1100–900 cm $^{-1}$, which are characteristic of silica gel (Figs. 4 and 5) and are due to valence oscillations of Si – O bonds in silicon-oxygen tetrahedra $[Si_4]^{4-}$.

Intense bands in the region of 470–490 cm $^{-1}$, characteristic of silica gel (Fig. 5), and composite "silica gel – sodium sulphate" (Fig. 6) characterize the deformation oscillations in silicon-oxygen tetrahedra.

Considering the bands associated with the valence and strain oscillations of hydroxils in the range of 1500–4000 cm^{-1}, it can be seen that they coincide for the composite "sodium silica gel-sulfate" and silica gel. Thus, the study of the IR spectra of the studied composite "sodium silica gel-sulphate" allows us to conclude that it has a specific structural and mineralogical characteristics and high adsorption properties.

The main attention in the study of synthesized composite sorbents is paid to their sorption properties, which are the basis of thermochemical transformation and accumulation of low-potential heat. For these applications, it is first necessary to investigate the ability of composite sorbents to reversibly bind and deliver water vapor, to determine the temperature ranges of sorption and desorption processes, and to determine the heat of sorption and desorption of water vapor.

The kinetics of equilibrium in the system of composite sorbent - water provides important information about the mechanism of sorption and time characteristics of this process. In order to determine these parameters, the kinetics of water sorption with composites "silica gel-sodium sulfate" and "silica gel-sodium acetate" at temperatures of 20, 40 and 60 ° C and a pressure of 1010.8 mbar (atmospheric pressure) was studied.

The kinetic curves of water sorption by the investigated substances in comparison with massive salts are given in Fig. 6. The kinetic curves of sorption show that the mechanism of water vapor sorption processes for composite sorbents "silica gel-crystal hydrate" differs significantly from sorption by massive salts. For massive salts, hydration occurs in the kinetic mode, as evidenced by the S-shaped shape of the kinetic curve. For composite sorbents, the initial section of the kinetic curve is almost linear in the coordinates "$A\sqrt{T}$", which indicates the diffusion nature of the process - i.e. the chemical reaction of water with dispersed salt is fast, and the limiting factor is water transport through silica gel pores. Equilibrium for massive salts is reached much later than for composites.

Thus, we can say that the synthesized composite sorbents are a highly porous three-dimensional silicon-oxygen backbone, in the pores of which are placed molecules of crystal hydrate. For massive crystal hydrates, water sorption is complicated by kinetic interference, and for composite sorbents, water transport through the pore system is much easier. As the amount of sorbed water increases, the heat of sorption decreases monotonically, which indicates a decrease in the binding energy between water and salt molecules? This heat is higher than the heat of evaporation of water and the heat of sorption of water by massive salts.

The authors observed a significant change in the sorption properties of crystal hydrates when they are placed in the pores of the silicon-oxygen skeleton. This change may be due to the interaction of the salt with the matrix or to its dispersion to the nanoscale (dimensional effect). In order to assess the contribution of these effects, X-ray diffraction analysis was used. WAXS profiles of KSM silica gel, composite sorbent "silica gel-sodium sulfate" and the original sodium sulfate are given in Fig. 7.

Analyzing the results of the study, we can see that on the WAXS profile of the composite "silica gel-sodium sulfate" (Fig. 7, curve 2) there are reflexes that correspond to crystalline sodium sulfate. The composite "silica gel – sodium sulfate" in the dehydrated state has Na2SO4 microcrystals with the size of the coherent scattering region at the level of 12–17 nm. This indicates that the salt of sodium sulfate is in the pores formed by the silicon-oxygen matrix, and not on its surface. Crystalline phases of the interaction of the salt with the silicon-oxygen matrix are not detected.

Comparing the data of the WAXS profiles, it can be seen that on the radiograph of the hydrated composite 'silica gel modified with sodium sulphate' only dimensional reflexes from SiO_2 are observed, except for the dimensional peak of sodium sulfate decahydrate. These peaks are narrower, and their intensity is less than for the original salt. This may indicate that immediately after the formation of the hydrate is its dissolution, or the crystal hydrate is in the X-ray amorphous state. The obtained data indirectly indicate that salt nanocrystallites have a rather narrow size distribution. In turn, this may be the reason for the rapid transition to sodium sulphate decahydrate, and also confirms the kinetic features of hydration of composite sorbents discussed above.

For a more detailed study of the hydration of the studied objects, diffraction patterns of hydrated samples were taken, shown in Fig. 8.

Thermal energy storage systems can be made in two basic configurations: combined, in which the sorbent and ad (ab) sorbate are located together and

Figure 6. Kinetic curves of water sorption with: a – massive salt sodium sulphate; b massive salt sodium acetate; c – composite sorbent 'silica gel – sodium sulphate'; d – composite sorbent 'silica gel – sodium acetate'. Water vapor pressure 1010.8 mbar. Temperature, °C: 1 – 20; 2 – 40; 3 – 60 Sukhyy, K.M., Belyanovskaya E.A. & Sukhyy M.P. (2020)

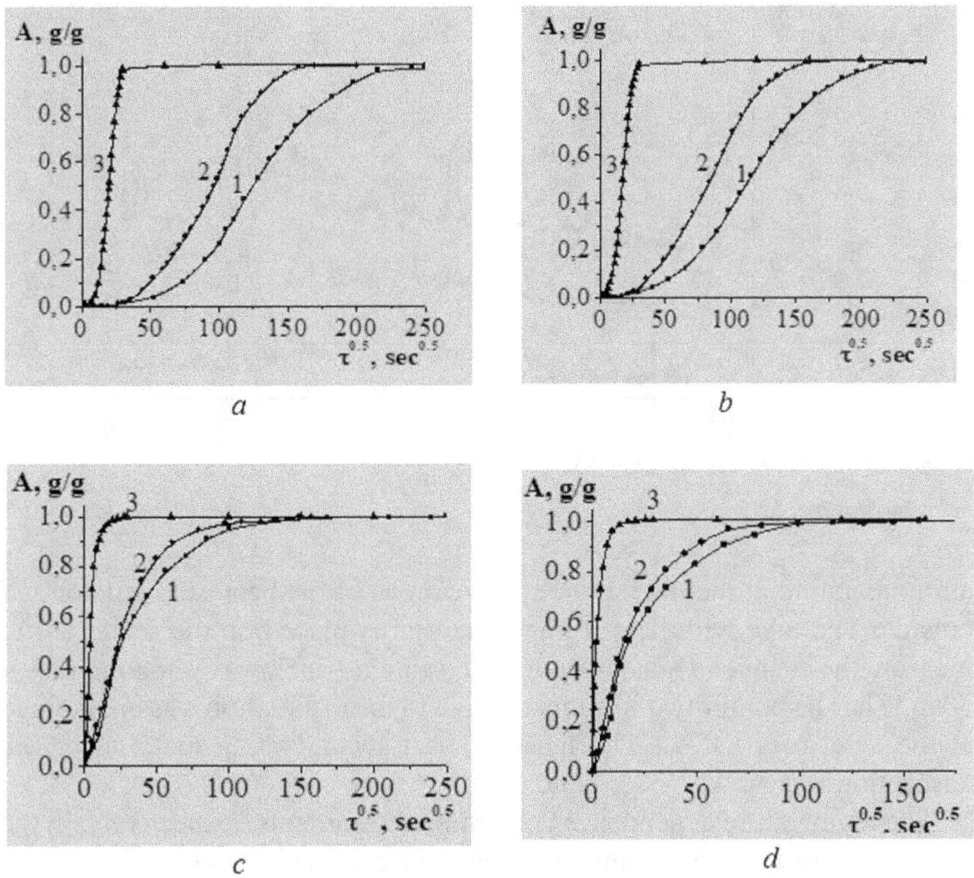

separated, in which the temperature, pressure and transport of sorbate are strictly regulated by the operating conditions of the heat accumulator.

To determine the efficiency of the synthesized composite sorbents, an experimental setup was created, the scheme of which is shown in Fig. 9.

The setup consists of a hollow heat-insulated pipe (1), into which the heat-accumulating material (2) is loaded, through which air is supplied by means of a compressor (3). Behind the compressor is a heater (4) to regulate the air temperature. Thermocouples are installed in the installation to measure the

Figure 7. WAXS profiles of the anhydrous samples: 1 – silica gel KSM; 2 – composite 'silica gel – sodium sulphate', 3 – sodium sulphate Sukhyy K. M., Belyanovskaya E. A. et al (2014)

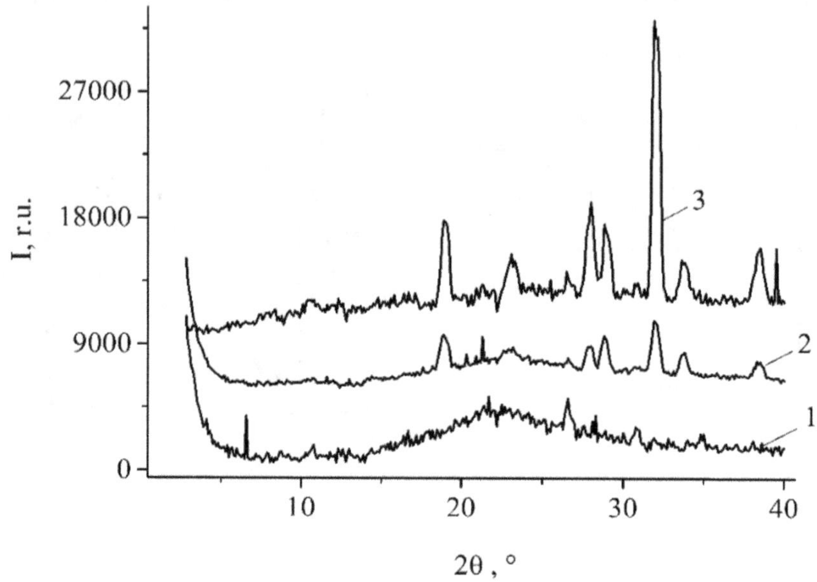

air temperature at the inlet, outlet, in the areas of the heat accumulator and outside. The tube with heat storage material is placed on the scales (5) to measure the change in mass during the research. Air flow was maintained at 50 m3 / h. The humidity of the air supplied to the installation was maintained at the level of 70–80%. The mass of heat-accumulating material before adsorption was 3.5 kg.

The temperature dependences of composite sorbents, respectively, in the process of water sorption and air temperature at the inlet and outlet of the system from time to time are given in Fig.10. It is established that the process of sorption at air temperature below 30°C, supplied to the system practically does not occur, which indicates that it is necessary to initially transfer a certain amount of heat to the system.

Analyzing the sorption process (Fig. 10), we can see that after a short heating of the accumulating material (15 min) is the sorption of water from the air, which is fed to the system with the release of heat. The temperature gradient of sorption moves through the material, reaching the saturation level after two hours. This is well illustrated by the temperature dependences at the inlet and outlet of the system. At the same time, the air temperature at the outlet of the

Figure 8. WAXS profiles of the hydrated samples: 1 – composite 'silica gel – sodium sulphate', 2 – sodium sulphate Sukhyy K. M., Belyanovskaya E. A. et al (2014)

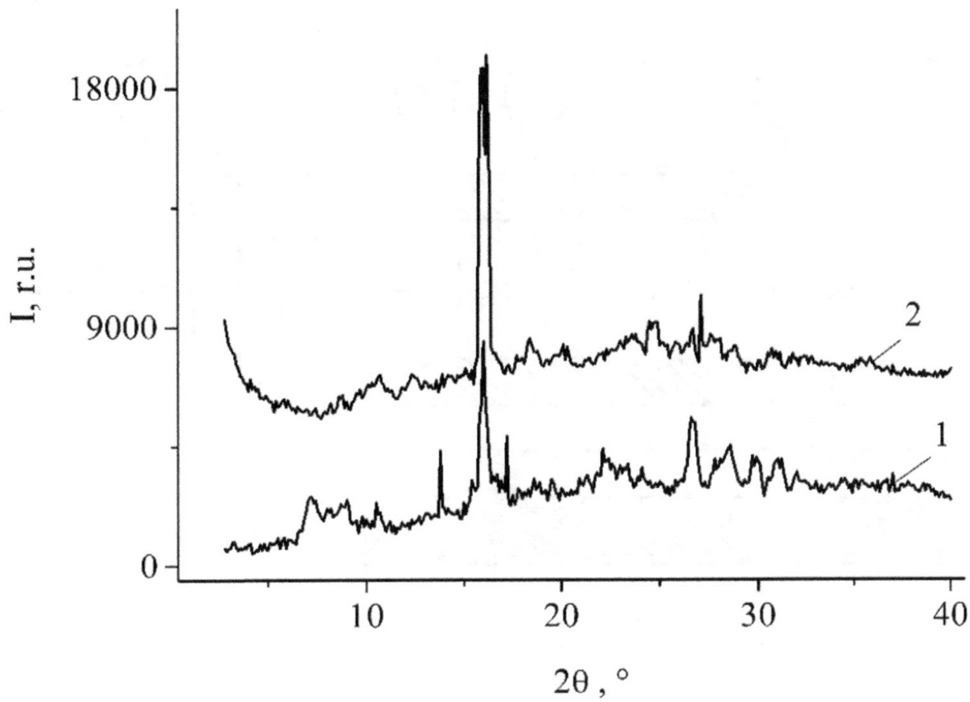

system is maintained in the range of 90-115°C for two hours. This allows to calculate the parameters of heat storage capacity of synthesized composite sorbents.In the process of sorption of moist air, the mass of the material increased by 80%, i.e. the moisture content in the composite sorbent is 44%, which almost coincides with the data of previous studies of measuring the moisture content of composite sorbents after sorption for laboratory samples.

The heat storage capacity of a composite sorbent of water can theoretically be calculated by the formula Sukhyy K.M. et al. (2014):

$$Q = \Delta h \cdot \rho \cdot C \cdot \left(\frac{1000}{Mw}\right) \cdot \left(\frac{1}{3600}\right) = 293, \quad (4)$$

where $\Delta h \cong 60$ is heat adsorption of water vapour, kJ/ mole; $\rho = 720$ is packed density of the composite sorbent, kg/m^3; C is maximum moisture-holding capacity; Mw = 18 is the molar mass of water, g/mole.

Figure 9. Experimental setup for heat storage material investigation Sukhyy K. M., Belyanovskaya E. A. et al (2014)

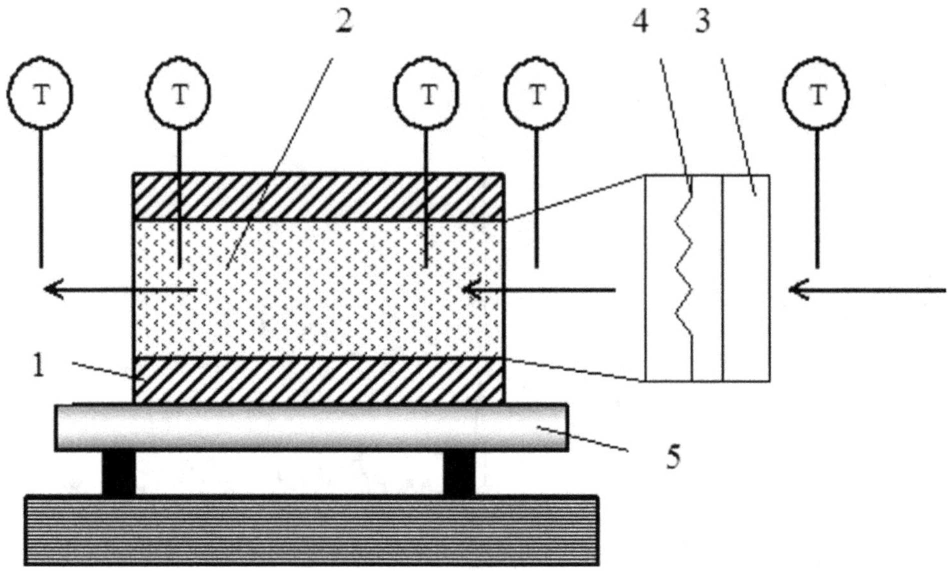

Figure 10. Temperature dependences of the composite sorbent 'silica gel –sodium sulphate' during water sorption in inlet (3) and outlet of system (4) and air in inlet (1) and outlet of system (2) on time Sukhyy K. M., Belyanovskaya E. A. et al (2014)

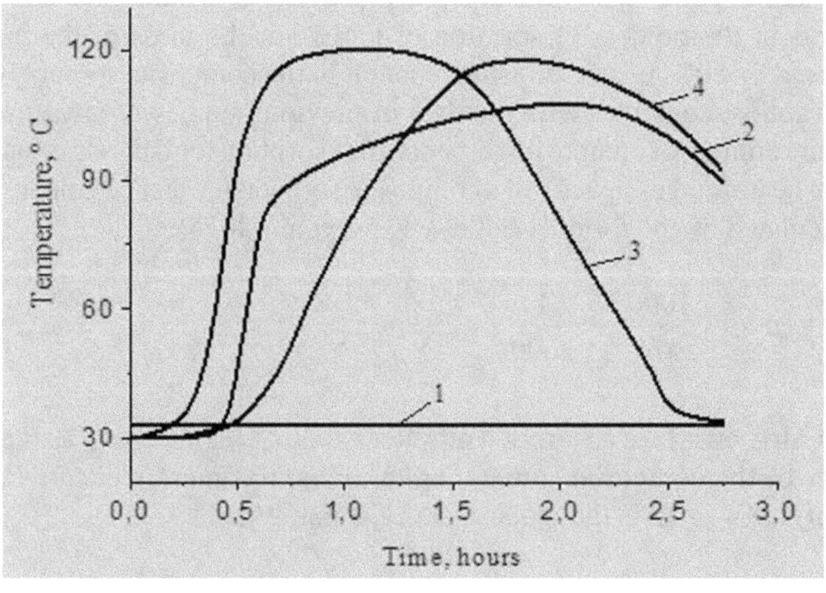

Theoretical value of heat storage density is estimated by Eq.(4) as 293 kW-h/m^3.

The amount of heat released during sorption by a composite sorbent of water can be calculated by the formula:

$$Q = Cp \bullet \rho \bullet \Delta T \bullet V \quad (5)$$

where Cp = 1,012 is medium isobaric heat capacity of dry air, kJ/(kg·K); ρ = 0.825 is medium air density, kg/m^3;

Substituting the data obtained for Eq.(5), we can calculate the amount of heat released during sorption by the composite adsorbent of moist air 1.25 kW-h.

Thus, it is possible to calculate the heat storage capacity of the synthesized composite sorbents, which is 0.36 (kWh/kg), or taking into account the density of the composite sorbent, which is 720 kg/m^3 - 258 [(kWh) / m^3].

The energy storage density for composite sorbents, calculated theoretically and found during the experiment, practically coincide, which testifies to the validity of the assumptions made regarding the mechanisms of heat storage capacity.

In the reverse desorption process (Fig. 11) at the initial stage the material is heated (1.5 h) with slow removal of water, followed by rapid removal of sorbed water, and the desorption process ends after 5.5 hours.

It should be noted that the presented studies were reproduced over ten cycles, and no negative changes in the appearance of composite sorbents were observed.

CONCLUSION

New composite adsorbents of water "silica gel – crystalline hydrate" were synthesized and their adsorption properties were studied. It is shown that such composite sorbents are characterized by high water adsorption at the level at least of 0.70-0.85 g / g. It is established that the adsorption properties of such composite sorbents differ significantly from those for massive salts. Adsorption properties of the synthesized composite sorbents "silica gel crystal hydrate" are shown to be a non-linear combination of the properties of silica gel and massive salt, and the rate of water sorption process is three times higher. The hydration process passes from the kinetic regime characteristic of the massive salt to the diffusion regime, in which the limiting stage is the

Figure 11. Temperature dependences of the composite sorbent 'silica gel –sodium sulphate' during water desorption in inlet (2) and outlet of system (3) and air in inlet (1) and outlet of system (4) on time Sukhyy K. M., Belyanovskaya E. A. et al (2014)

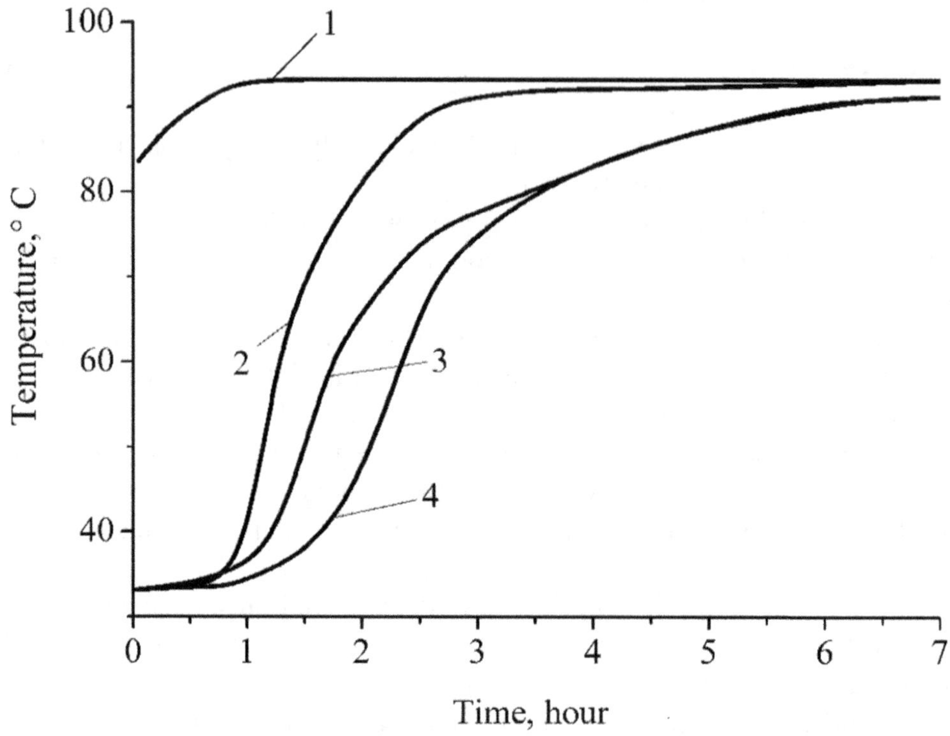

transport of water through the pore system of the silicon-oxygen matrix. By the method of wide-angle X-ray scattering, it was found that the difference between the properties of the massive salt of the crystal hydrate from that in the composite sorbent is the nanostructured salt in the pores of the silicon-oxygen matrix.

Thermodynamic analysis of the cycle of sorption heat storage module, which uses synthesized composite sorbents. The method of wide-angle X-ray scattering revealed that the properties of the massive salt of the crystal hydrate from that of the composite sorbent lies in the nanostructured salt in the pores of the silicon-oxygen matrix are in a strong juxtaposition. The heat storage density is at least of 0.36 kWh/kg, which indicates the high prospects of the synthesized composite sorbents.

REFERENCES

Cabeza, L. F., Solé, A., & Barreneche, C. (2017). Review on sorption materials and technologies for heat pumps and thermal energy storage. *Renewable Energy*, *110*, 3–39. doi:10.1016/j.renene.2016.09.059

Gordeeva, L., Grekova, A., Krieger, T., & Aristov, Yu. (2013). Composites "binary salts in porous matrix" for adsorption heat transformation. *Applied Thermal Engineering*, *50*(2), 1633–1638. doi:10.1016/j.applthermaleng.2011.07.040

Grekova, A. D., Gordeeva, L. G., & Aristov, Yu. I. (2017). Composite "LiCl/vermiculite" as advanced water sorbent for thermal energy storage. *Applied Thermal Engineering*, *124*, 1401–1408. doi:10.1016/j.applthermaleng.2017.06.122

Hiremath, C. R., & Kadoli, R. (2013). Experimental studies on heat and mass transfer in a packed bed of burnt clay impregnated with $CaCl_2$ liquid desiccant and exploring the use of gas side resistance model. *Applied Thermal Engineering*, *50*(1), 1299–1310. doi:10.1016/j.applthermaleng.2012.08.002

Jong, A. J., Trausela, F., Fincka, C., van Vlieta, L., & Cuypers, R. (2014). Thermochemical heat storage – system design issues. *Energy Procedia*, *48*, 309–319. doi:10.1016/j.egypro.2014.02.036

Scapino, L., Zondag, H. A., Van Bael, J., Diriken, J., & Rindt, C. C. M. (2017). Sorption heat storage for long-term low-temperature applications: A review on the advancements at material and prototype scale. *Applied Energy*, *190*, 920–948. doi:10.1016/j.apenergy.2016.12.148

Serhiienko, Ya. O., Sukhyy, K. M., Belyanovskaya, E. A., Kolomiyets, E. V., Gubynskyi, M. V., Tkalya, O. I., Sukha, I. V., & Zaichuk, O. V. (2019). Technology of obtaining new materials for adsorptive heat energy transformation type «silica gel – crystalline gydrate». *J. Chem. and Chem. Technology.*, *2019*(27), 239–246.

Sukhyy, K. M., Belyanovskaya, E. A., Kozlov, Ya. N., Kolomiyets, E. V., & Sukhyy, M. P. (2014). Structure and Adsorption Properties of the Composites 'Silica Gel – Sodium Sulphate', obtained by Sol – Gel Method. *Applied Thermal Engineering*, *64*(1-2), 408–412. doi:10.1016/j.applthermaleng.2013.12.013

Sukhyy, K. M., Belyanovskaya, E. A., & Sukhyy, M. P. (2020). *Technology Development for Adsorptive Heat Energy Converters: Emerging Research and Opportunities. San Fransisco*. IGI-GLOBAL.

Tanashev, Yu. Yu., Krainov, A. V., & Aristov, Yu. I. (2013). Thermal conductivity of composite sorbents "salt in porous matrix" for heat storage and transformation. *Applied Thermal Engineering*, *61*(2), 401–407. doi:10.1016/j.applthermaleng.2013.08.022

Zamengo, M., & Kato, Yu. (2017). Comparison of magnesium hydroxide/expanded Graphite composites for thermal energy storage in cogeneration nuclear power plants. *Energy Procedia*, *131*, 119–126. doi:10.1016/j.egypro.2017.09.463

Zhong, Y., Zhu, Z., & Wang, S.-Q. (2005). Synthesis and rheological properties of polystyrene/layered silicate nanocomposite. *Polymer*, *104*(9), 785–796. doi:10.1016/j.polymer.2005.02.014

Chapter 7
Creation of Polymeric Nanocomposite

ABSTRACT

The present chapter is focused on problems and challenges for creation of polymer composites using clay minerals. The main difficulties are shown to be the low thermal stability of modifying mineral additives, such as decomposition; of the most commercially available modifying additives, quaternary ammonium salts decompose below the processing temperatures of most polymers and their low long-term thermal stability during processing and operation. Extrapolation of research results to other objects is complicated because of insufficient understanding of the mechanism of this nanotechnology for specific polymer matrices. The main directions of development of this nanotechnology are shown to be the use of polymers and oligomers as heat-resistant modifying additives of clay minerals.

BACKGROUND

Sometimes it is difficult to distinguish between products of sol-synthesis and intercalation, this can be done only by their prehistory. For the production of such nanocomposites are most often used natural silicates - so-called 'smectites' which surfaces are described in the review (Srebnik & Ovadia, 2003).These are very common minerals, which include hectorite and montmorillonite with a structure of the mica type: crystals consist of layers. alternating cations and negatively charged layers of silicates. However, unlike mica, the cations in

DOI: 10.4018/978-1-7998-9728-6.ch007

these minerals can be easily exchanged, including for transition metal ions. The exchange takes place mainly in layers containing solvated sodium cations. Intracrystalline cavities can swell when filled with organic solvents, and the degree of swelling depends on the nature of the cations inside the layers, as well as on the density of the negative charge on the silicate layers. In addition, most inorganic oxides have surface hydroxyl groups that create a strongly polar surface that are active in the binding of metal ions. Of course, the metal ions included in such structures can be easily reduced under the action of chemical reducing agents, as well as thermally or photochemically, and form nanoparticles in such materials. There are many methods of introducing polymers into them. The main properties of such materials are determined by characteristics of their structure, especially pore size. Models and properties of intercalated systems are considered in review (Schöllhorn, 1996). Until the 90s of the last millennium, at least 5,000 publications were devoted to the problems of intercalation of organic, organometallic and inorganic compounds, as well as the properties of the products formed. However, the main surge in research is in the last 10 years. Quite well-studied crystalline solid matrices with a system of percolation atomic pores or nanometer-sized molecular structures which can be filled with atomic or molecurar guest-structures such as clusters, inorganic coordination polymers, large molecules similar to fullerenes (Kononova et al., 2021), polymer units synthesized through in situ process. These processes are characterized by low diffusion activation energy, and the resulting materials are usually metastable and cannot be synthesized in other ways, e.g., thermal synthesis, in particular, through the processes of delamination or detachment of the polymer from the inorganic component. The current trend in the development and application of polymeric materials is to find rational ways to give traditional high-capacity polymers new properties.

The development of polymer nanocomposites based on polymers for structural purposes is an urgent problem. By definition, a composite material is a system of two or more phases that differ in chemical composition and structure. The range of polymer matrices and fillers for the creation of polymer composites is very wide Yang Y. & Nogami M. & Huang Z., Li Y. (2013). Substances used as fillers can be divided into basic classes. Solid fillers (chalk, talc, hydroxyl aluminum of the glass sphere, sacks, graphite, etc.) are introduced mainly to increase the hardness, strength, dimensional stability and heat resistance and to reducing the flammability of gas permeability of materials Landel R. & Nielsen L. (1993). When elastic rubber particles used as filler for thermoplastic polymers, rubber viscosity increases (Valášek, Müller, & Ružbarský, 2014).

Nanocomposites represent a new direction in the field of polymer composite materials. Nanocomposites include composite materials in which at least one of the linear particle sizes of the discrete phase (length, width, or height) does not exceed 100 nm (Kumar & Krishnamoorti, 2010). Currently, many different types of nanocomposites have been fabricated, the components of which can be various metals, ceramics, polymers, semiconductors, and other inorganic and organic substances.

In the early 1990s, Japanese researchers in first obtained nanostructured polymer composites based on the clay mineral montmorntonite and polyamide-6 and studied its mechanical and thermophysical properties Yano K. et al. (1993). It turned out that the modulus of elasticity of the obtained nanocomposite is 1.7 times, with strength 1.4 times higher than similar characteristics for the original polyamide-6. Such a significant improvement in mechanical properties is accompanied by a 1.5-fold decrease in the coefficient of thermal expansion, an increase in the heat resistance by Vick from 65 to 152°C, which is observed when the filler content is only 4.7%. It should be noted that the resulting material exhibit much lower density as compared to traditional composites, for which the content of inorganic component is usually 20-30%. Another group of authors determined that the introduction of clay minerals in the polymer leads to a significant reduction in the diffusion coefficient for different gases Cui et al. (2015).

The results of Japanese researchers were an incentive to create polymer silicate nanocomposites based on other polymer matrices. Over the last 15 years, various groups of researchers have obtained nanocomposites based on several dozen polymers and clay minerals, many of which have interesting properties Franco-Urquiza E. A. (2021). Currently, the most advanced compositions are already used in industry for the manufacture of special coatings, packaging films with barrier properties, car parts and electronic devices.

During the formation of polymer-silicate nanocomposites, polymer molecules penetrate into the gaps between the structural layers of the mineral (into the interlayer space of clay particles) and are "trapped" in a narrow gap with height of several nm, i.e., in fact in two-dimensional (2D) space. Such systems provide a good opportunity to study the behavior of polymer molecules in confined spaces, which is one of the important tasks of basic science. Using nanocomposite samples, it is convenient to study the dynamics and structure of a polymer chain located in a narrow gap, the peculiarities of crystallization or formation of other ordered phases, as well as relaxation processes of such macromolecules Vaia R.A. et al. (1997).

Currently, there is a rapid surge of interest of researchers and engineers in polymer nanocomposites, the use of which will undoubtedly bring materials science to a qualitatively new level of properties while reducing costs for production, operation and disposal.

Work on the creation of polymeric nanocomposites of clay minerals has been carried out since the 1970s (Theng, 2012), but no significant improvement in properties has been achieved. The beginning of nanotechnology for the creation of materials based on polymers of clay minerals can be considered a study by concern Toyota. These studies have shown a significant improvement in a wide range of properties of polyamigs during delamination of clay mineral. Since then, in the field of polymeric nanocomposites followers have studied almost all available polymer matrices such as polypropylene (Newman, 1987), polyamides and polyptopylene (Kusmono.et al., 2009), polyurethanes (Mallikarjuna & Sivaram, 2013), epoxy resins (Truc et al., 2019), polystyrene (Soheil, Abdolahifar, & Goudarzian, 2015), polyethylene (Qian, Zhang, & Yang, 2008) and polyethyleneoxide (Shu et al., 2019), mixtures of polymers (Penaloza, 2019) and other polymer matrix (Zhu et al., 2022).

The applied technology strategy depends on the level of distribution of the clay mineral in the selected polymer matrix. This is a prerequisite for modification of clays before use. Generally, when water-soluble polymers used, there is no need for preliminary modification of water-soluble molecules easily intercalated with minerals, which under certain conditions results in delamination of structures.

Most hydrophobic polymers don't ensure compatibility with hydrophilic clays. The most popular modifying additives in this case are aminoacids, quaternary ammonium salts, phosphoid compounds, which allow to hydrophobizate surface of clays (Perelomov et al., 2021).

Additionally, it is possible to increase compatibility by means of grafted copolymerization of polar functional groups, for example, as in the case of copolymerization of maleic anhydride to polypropylene (Wang, Hong, & Chung, 2005).

There are three main methods of obtaining polymer-clay mineral nanocomposites: in-situ polymerization, mixing in solution, mixing in melt. The first two methods are not widespread due to the complexity of the synthesis, the selection of the necessary solvents, as well as their cost. The in-situ polymerization method is used for such polymers for which it is not possible to obtain by mixing in the melt. The use of clay minerals as additives to polymers requires in-depth study of the structure of the latter, which allows us to understand the processes that occur during the combination of polymer and clay mineral.

STRUCTURE OF CLAY MINERALS

The main elements of the structure of clay minerals are silicon-oxygen tetrahedra SiO_4 and alumina-oxygen octahedra $Al(O, OH)_6$. (Penaloza, 2019). SiO_4 is a regular tetrahedron with oxygen atoms at its vertices and a silicon atom at the center (Fig. 1). $Al(O, OH)_6$, is a regular octahedron with an A1 atom at its center and oxygen atoms at its vertices. and OH groups. These structural units form tetrahedral and octahedral two-dimensional grids, respectively. Due to the proximity of the sizes of the faces of the tetrahedral and octahedral grids, they can connect with each other through young oxygen and form layers of two, three, four, and more packets that form the basis of the clay mineral. Depending on the quantitative ratio of tetrahedral and octahedral layers in the package, clay minerals are denoted as 1:1, 2: 1, 2: 2, etc.

Due to the different lengths of the edges of octahedra and tetrahedra caused by isomorphic substitution of elements with different sizes of atomic radii, internal stresses arise in the silicate layers, which limit the sizes of the layers in the xy-plane. In the z direction, the structural layers are parallel to each other, forming crystallites (tactoids). Water molecules and metal cations may be present between the packets. Obtaining minerals capable of swelling (smectites) is of particular interest (Laird, 2006).

One of the representatives of smectites is montmorillonite, it's the characteristic feature being the ability to disperse into separate nanolayers under special treatment. Other advantages of montmorillonite are the high capacity of cation exchange, which allows you to easily change its properties, its wide distribution in nature and relatively low cost.

Montmorillonite belongs to the family of clay minerals 2: 1. His layers involve two external silicon-oxygen grids and an intermediate aluminium-oxygen mesh (Fig. 1). The thickness of the elementary package is 0.96 nm. The composition of the mineral can be represented by a chemical formula. $M_x(Al_{4-x}, Mg_x)Si_8O_{20}(OH)_4$ where M referes to a cation of alkali or alkaline earth metals, x is a degree isomorphic substitution, x = 0.5-1.3.

In montmorillonite, the Al^{3+} in the octaedric mesh can be isomorphically substituted with ions Mg^{2+} and sometimes with ions Fe^{2+} and Li^+. Substitution of cations Al^{3+} with ions of lower valency results in negative charge of

Figure 1. Structure 2 1 of clay minerals Penaloza, D. P. Jr. (2019)

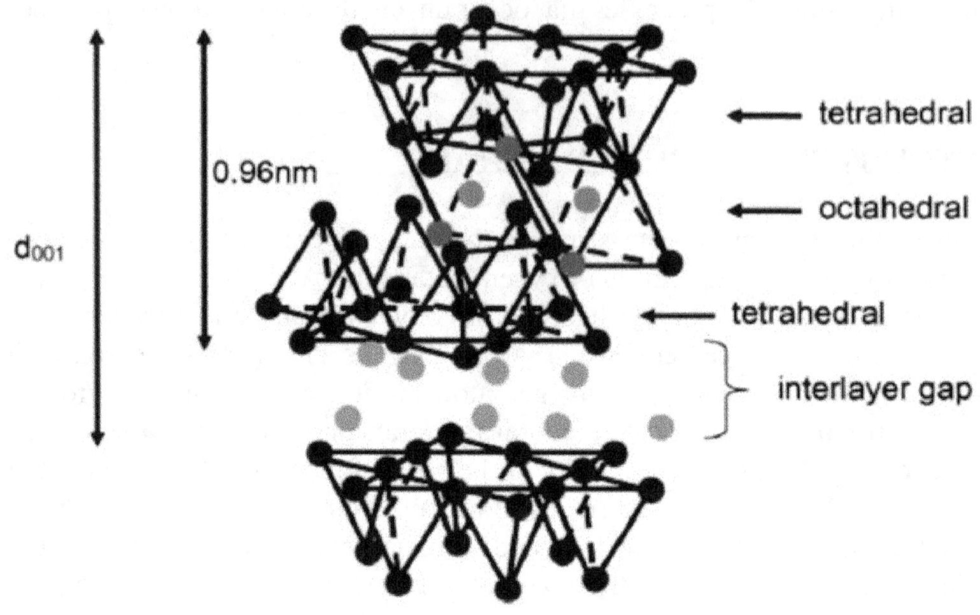

elementary layers which equals to 0.26 – 0.67 of the charge of electrone per one formula unit. In the gap between the motmorillonite packages, cations of metals (Na^+, Li^+, Ca^{2+}) which equilibrate negative charge of the layers are situated. Depending on the type of exchange cations there are Na+- or Li+-montmorillonite etc. Cathions of metals which found in natural minerals can be replaced by othere ions during reactions of ion exchange which allows to modify the clay with rhyme substances. All clay minerals have a certain variability of cation exchange. This is an important characteristic of the mineral and corresponds to the number of exchange cations which can be replaced by other types of cations. Montmorillonite has the highest cathion exchange capacity among clay minerals (up to 150 mg-equ/100 g)

In addition to exchange cations, molecules of bound water are always present in the interlayer. The content of moisture in the mineral depends on the valence of the interlayer cation and the relative humidity of the air. The value of the interval between packages for Na+-montmorillonite under relative humidity of 40 – 60% is 0.28 –0.30 nm, for Ca^{2+}-montmorillonnite it being 0.58 – 0.6 nm. Characteristic value for clay minerals is interplanar (interpackage) distance d_{001}. It depends on the thickness of the elementary plate and interpackage gap.

On the basis of numerous experiments for clay minerals range of substitutionary capability of cathions was composed (Kumari & Mohan, 2021).

$Al^{3+} > Ca^{2+} > NH_4^+ > Na^+ > Li^+$

As it turns out, NH_4^+ -ion is able to drive ions Na^+ and Li^+ from exchange positions. As a concequence, Li^+- and Na^+-montmorillonite often use various derivatives of ammonia which involve aminogroups and one or a few aliphatic chains of different length (Chou, & Lin, 2004). Aliphatic chains in amines often involve from 6 to 20 links and can be non-polar (hydrocarbon chans) or polar (alchohols, ethers or acids).

When molecules of modifying additive penetrated into interlayer space, hemosorption of aminogroups on negatively charged packages of clay occurs, and aliphatic "tails" are located in the interlayer space, which leads to splitting layers of clay minerals. Sometimes highly-molecular substances with polar groups such as amines which can interact with mineral surface are used. By filling the space between the packages of clay the molecules of the additive allow to increase the distance between structural layers of clay before its mixing with the polymer.

For preparation of the nanocomposites based on strongly polar polymers and monomers, e.g. polyethyleneoxide or polycaprolactam there is no need in preliminary processings. In these cases thermodynamically stable nanocomposites form without modification of clay minerals (Olad, 2011).

Experimental results and structural-mechanical models predict that the greatest effect of enhancing in filled polymers is stated when the hardness of filler exceeds matrix in many times. By two independent methods (Messersmith & Giannelis, 1995). it was stated that the module of clay minerals in basal area xy is about of 178 GPa, this being substantially higher than module of polymeric matrice (0.1 – 1 GPa). So, the important question for predicting the properties of the filled polymer is the characteristics of the fillers of polymer composites.

Influence of Filler Characteristics on Properties of Polymeric Composite

First and foremost, the capability of stregthening of filler is affected by three primary characteristics (Pal, 2016), such as particle size, structure, and surface activity. Improvement of physical propreties is directly related to the particle size of the filler. A significant increase in the surface area in contact with the polymer phase is likely to increase the strength of the composite. Reducing the particle size also leads to an increase the interaction with polymer matrix. Furthermore, distribution by size also affects the strength characteristics. Macroparticles with wide distriburon by size are beter distributed in polymeric matrices as compared with fillers with narrower distribution by size. Another important prerequisite to increase the strength is absence of big particles or agglomerates in the polymer.

Additionally, form of filler particle is an important parameter which strongly affects properties of polymer composites. Inorganic and mineral groups reveal significant differences in geometry of particles depending on crystal form of mineral. Minimal anisometria is observed for minerals which form isometric particles, i.e., with approximately the same size of particles in three directions. The plates can be considered as the most anisometric particles because one of the sizes is much more as compared with the other two ones. Compositions which contain fillers with identical surface area and chemical nature of particles but with different shape are known to exhibit simultatious increasing the strength and anisometry (Pal, 2016).

The size of the filler particle as discussed earlier can be considered as physical contribution into hardening. Reaction of filler and functional groups of polymer leads to hybridization which strongly enhances the strength of the composite. In the case of absence with strong interaction polymer molecules adsorb on surface of fillers which is accompanied with induced mobility of polymeric macromolecules near surface of the filler. The absence of active centers on the filler particles for adsorption of polymer molecules adversely effects on mechanical properties of composite.

Adhesion between polymers and fillers can also result from binding agent located on the interphase so-called primers. This mechanism of enhancing the interaction in the system polymer-filler is well-studied for most of the mineral fillers (Pal, 2016). Organosilane binders have been used successfully to enhance the physical and mechanical properties of a large number of fillers, including clay, mica, silica, and talc.

CHANGING THE STRUCTURE OF POLYMERS UNDER LOAD

It is known that the mechanisms of deformation of any materials depend on the speed and temperature acting on them. The type of fracture (brittle or plastic) at a given temperature is determined by the ratio between the rate of deformation and the relaxation time of individual structural elements of the system. If the speed of influence is such that the transformation of the internal structure is absent, the material collapses brittlely. For each sample at the drawing temperature there is a critical deformation rate, which depends on the structure of the sample and the temperature. Deformation, carried out at a rate less than critical, causes the restructuring of different levels of supramolecular structure.

The process of deformation in flexible-chain partially-crystalline fields (with less than critical stretchening) is done in three stages (Monasse, Queyroy, & Lhost, 2008). stretching to form a "neck", transformation of spherulitic structure into fibrillar and deformation of the fibrillar structure formed.

The first stage of deformation is reversible. When the load is removed, the specimen completely regains its shape. In the elastic region (up to the level of deformations corresponding to the limits of proportionality) the amorphous component is mainly deformed, the deformation resistance of which is much less than in the crystalline part. Further changes occur in the crystal region as well: phase transitions of the martensite type, sliding and rotations of lamellae, as well as sliding and inclinations in molecular chains in lamellae are observed (Rothon, 2017). The first stage of deformation ends when the upper yield strength is reached (Fig. 2).

In recent years, researchers (Rothon, 2017) have noted the importance of strain compression, ie the mutual influence of deformation of crystalline and amorphous components. After the initial stage of deformation, when the passage chains in the amorphous sections are fully stretched, the resistance of the amorphous component deformation increases significantly and there is a "deformation locking". Due to the high degree of connection of the "closed" amorphous component with the nearest crystallites, further deformation of the amorphous layer occurs only together with the deformation rearrangements in the crystalline component.

In the second stage of deformation in the polymer sample there is a complete restructuring of the original structure. Restructuring of the opherolite structure into microfibrillar often occurs with the formation of a visible "neck". In the

Figure 2. Typical stress-strain curve of partially crystallized polymer

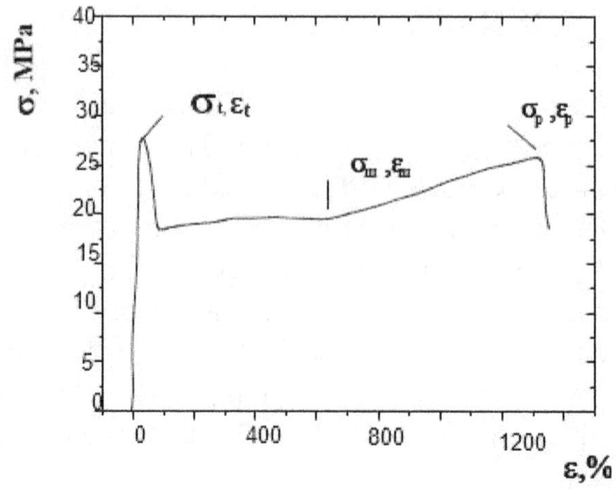

nominal stress-elongation diagram, the moment of formation of the “neck” is accompanied by a sharp drop in load, after which the stress remains almost constant fully fibrillated (Fig. 2)

The stress at which the “neck” of the sample propagates. called the lower yield strength or stress in the “neck” of the sample. At the end of this step, the material is a collection of microfibrils located approximately parallel to the deformation axis (Rothon, 2017).. the size is 100-200 A, and the z-axes are oriented in the direction of stretching.

In the spaces between the fibrils are interfibrillar molecules or fragments of macromolecules, some of which are intermediate chains which bind with the fibrils. Several models have been proposed to describe the transition from spherolite morphology to fibrillar. The most popular were three of them, melting-recrystallization, the Peterlin model, and the crystallographic model.

Proponents of the recrystallization model (Rothon, 2003) suggest that the work of plastic deformation in the “neck” region causes the microvolumes of the polymer to heat up to a temperature close to the melting point. Due to movability of the chains unbending of molecular ‘folds’ occurs, the sample passes into melt, crystallizes again in the form of microfibrils oriented in a given mechanical field direction.

The authors of the following models believe that the microfibrillar structure is a block of the old spherulite structure, only the crystals are newly oriented. According to Xu and An (2022), in the deformable polymer, as a result of local stress concentrations on any inhomogeneities of the structure, pores appear in which the nuclei of “micronecks” oriented along the tensile axis are formed. In such areas, fragmentation of lamellae into separate blocks is observed, while the connection between adjacent blocks is preserved due to separate partially straightened chains. In the process of extraction, the crystallites are rotated so that the direction of the folds coincides with the direction of stretching. The set of folded blocks connected by through chains forms a separate fibril.

In the crystallographic approach (Peterlin, 1975), it is assumed that polymer crystals are deformed according to classical crystallographic laws - by shifts along certain crystallographic elements. In this case, all textural rearrangements (reduction of lamella thickness, disintegration of lamellae into individual blocks, change in the structure of crystallites, etc.) are associated exclusively with crystallographic sliding processes in lamellae and occur without melting. The authors of the model suggest that although the uniaxial stretching and shear of polymers is almost always the formation of micropores, it is not essential for the formation of fibrillar structure, as accepted in the approach of Peterlin.

By the opinion of supporters of the crystallographic approach, transformation of spherullite morphology of the polymer in the fibrillar occurs by multifold repetition of the sliding processes of macromolecular chains.

Deformation of fibrillar structures is a special, yet insufficiently understood, area of plastic behavior of polymers. Fully fibrillated films and fibers can be streched several more times. At this stage of deformation there is a constant increase in stress - strain hardening up to the tensile stress σp (Fig. 2). It is considered that the plastic deformation of fibrillated samples is carried out mainly due to the slippage of microfibrils relative to each other (Peterlin, 1975).

Interfibrillar molecules in the shear motion of microfibrils can either be removed from the fibrils or ruptured. Additionally, a change in the shape of the crystallites (the phenomenon of distortion of crystallites (Haudin, 1982), which is observed in most partially crystalline polymers, as well as the deformation of amorphous layers, can make some contribution to the overall deformation of the fibrillated sample. Orientation elongation causes conformational transitions in amorphous regions (Peterlin, 1975). Increasing the length of amorphous layers with a stretching leads to an increase in the scatter of their longitudinal dimensions, while longitudinal dimensions of crystallites also slightly increase due to crystallization of amorphous component of flat ends of crystals.

Plastic deformation of the fibrillar structure is stopped due to the rupture of the sample when reaching a certain degree of elongation, the value of which depends on the type of polymer and tensile conditions. There are several approaches to explaining the reason for the rupture of the oriented sample (Peterlin, 1975). Ward and Sweeney (2004). assume that due to the increase in the scatter along the length of the amorphous layers in those of them that have the smallest dimensions, local overstrains occur, which lead to discontinuities of macromolecules. Biron (2016) attributes the main reason for the termination of the orientation movement to the suppression of segmental motion in amorphous regions when prestressing loads are reached (hereinafter referred to as "mechanical glass transition"). It has been suggested that the proportion of interfibrillar molecules that prevent microfiber slippage increases as the degree of extraction increases (Peterlin, 1975). Peterlin suggested that on final stages of deformation stages are the convergence of microfibrils, which causes a sharp increase in sliding friction forces and, as a consequence, a significant increase in internal stress. Many followers have found the researchers' approach Peterlin, according to which the destruction of oriented samples occurs by a thermofluctuation mechanism. The stress applied to the sample during deformation reduces the activation energy of the process deformation.

Currently, researchers have achieved great success in modeling the plastic properties of partially crystalline polymers (Wang, 1998). Three main approaches to modeling stress-elongation diagrams of partially crystalline polymers are proposed: a microcomposite model, a model based on the idea of stretching the macromolecular network of gears, and a continuous structural model.

OBTAINING POLYMER-SILICATE NANOCOMPOSITES

Depending on the degree of dispersion of the filler, composites based on clay minerals and polymers are usually divided into three main types (Chen et al., 1998): heterophasic, intercalated and exfoliated (or delaminated). The last two types are nanostructured composite materials, and the first conventional microcomposite.

At low thermodynamic compatibility of the filler and the matrix, the polymer molecules cannot penetrate into the gaps between the layers of the clay mineral. The material obtained in this case is a mechanical mixture of polymer and mineral with a dispersed phase size of up to 1 mm.

During the formation of the intercalated nanocomposite, the polymer molecules penetrate into the space between the layers of the mineral and expand them. The distance between the packages is increased, but the order in the rotation of the layers in the xy plane.

In the formation of the exfoliated nanocomposite, the polymer molecules not only penetrate into the gap between the layers of silicate, but also push them apart in such a way that the orderliness of the elementary packages and their binding are disordered.

In addition to the three main types, in practice composites are often found, the structure of which is intermediate between intercalated and exfoliated. In such samples, part of the elementary clay layers (mainly the crystallite surface layers) exfoliate and lose contact with the crystallite, and the order in the arrangement of other packages is to some extent preserved. Achieving one or another degree of intercalation of polymer macromolecules in the interlayer space of a clay mineral is primarily due to the method of obtaining polymeric silicate composites.

The literature describes three main methods of obtaining nanocomposites based on polymers and clay minerals (Chen et al, 1998): mixing in solution: mixing in the melt, polymerization filling. When using the first method of mixing is carried out in a medium of a common solvent, which dissolves the polymer well and is able to cause significant swelling of the clay mineral. After adsorption of the polymer on the surface of the mineral layers, the solvent is removed. Melt mixing is the simplest and fastest method of preparation of polymer-silicate nanocomposites, based on traditional polymer processing technology. In this case, the composite material is obtained by heating a mixture of polymer and clay mineral above the glass transition temperature (for vitreous) or melting point (for partially crystalline) polymers under a line of constant pressure or shear forces. The third method consists of two stages: fixing the desired catalyst (or initiator) on the inner surface of the mineral and then intercalating the corresponding monomer in the interlayer space of the silicate, followed by whispered polymerization. In some cases, the synthesis is carried out without the use of a catalyst or initiator, in such systems, the polymerization is carried out by heating or UV radiation. In addition to these methods, it is possible to obtain nanocomposites by synthesizing the silicate particles and silicic acid in polymer solution (Rahman & Padavettan, 2012).

To predict the structures of polymer composites, which are formed by coupling clay minerals with a polymer matrix it is important to consider in thermodynamic regularities of interactions of ingredients of composites.

THERMODYNAMIC ANALYSIS OF THE SYSTEM CLAY MINERAL-MODIFYING ADDITIVE-POLYMER

To analyze the factors influencing the process of formation of nanocomposites, it is necessary to theoretically describe the behavior of the components of the system of mineral-modifying additive-polymer when combined. Researchers Vaia and Giannelis started the first attempt (Vaia et al., 1997). They developed an averaged field, a description of which was published in Zhang et al. (2017) and considered the free energy of a system consisting of a clay mineral and a polymer melt. Changing of free energy due to intercalating the polymer and expanding the layers of mineral involves two parts:

$$AF = F(h)\text{-}F(h_0) = \Delta E - T\,\Delta S$$

where h_0 and h refers to the initial and final distances between layers of polymer minerals; ΔE is changing of the internal energy of system due to new intermolecular interactions, ΔS is change of the enthropy due to changing configuration of the system units, T refers to an absolute temperature.

Having been based the positions of statistic thermodynamics, authors obtained analytical equations for ΔE and ΔS according to which change of entropy when composite formed results from two reasons. For the first one, involvement of the molecules which earlier were in the melt in the narrow gap between mineral layers. The second one is changing the configuration of modifying additive. Change of the entropy of modifying additive ΔS_{ad} is positive when macromolecules intercalated into interlayer space because of molecules of modifying additive obtain more of configuration freedom as h increased. In contrast, changing the entropy of the polymer ΔS_{pol} will be negative because molecules of polymer penetrate from melt into narrow gap and are affected by limiting walls.

The evaluation carried out by the authors of the model showed that the total change in the entropy of the clay mineral-modifying additive-polymer system is negative. Based on this, the formation of the nanocomposite becomes unfavorable due to the entropy factor. The value of ΔS_{ad} largely depends on the surface density of the grafted chains and their length, and ΔS_{pol} - on the initial interlayer distance. The optimal selection of these parameters reduces the entropy barrier, which prevents the formation of nanocomposites.

The internal energy of this system consists of paired interactions of three components - polymer-mineral, polymer-modifying additive and mineral

$$E = E_{pol\text{-}min} + E_{pol\text{-}ad} + E_{min\text{-}ad}$$

Knowing the characteristics of the matrix polymer, it is possible to select the type of mineral and modifying additives so that the clay mineral modifying additive-polymer system is as compatible as possible. The interaction of the polymer with the modifying additive, without taking into account the structure of the layer formed by the adsorption of surfactants (surfactants) Vanderzande (1998), will be most favorable if the structure of surfactant molecules is identical to the structure of polymer molecules.

To form a thermodynamically stable nanocomposite, it is necessary that the energy interactions in the system compensate for the decrease in entropy ($\Delta F < 0$). If the energy interactions in the system are too weak and do not allow overcoming the entropy barrier ($\Delta F > 0$), phase separation of components will occur and a normal microcomposite will be formed.

The authors Balazs, Singh, Zhulina, and Lyatskaya (1999) simulated the interaction in the system of mineral modifying additive-melt polymer based on the theory of self-consistent field. The interaction between the components of the system in this approach was taken into account using the Flori-Huggins interaction parameter x. The type of interaction potential will be determined by the distribution density of the polymer and modifying additives, as well as the values of the paired interaction parameters.

The dependence of the free energy of the system on such characteristics as the chain length of the modifying additive and the polymer, the surface density of the modifier additive molecules, and the paired interaction parameters were analyzed. The results obtained by this group of researchers are in good agreement and complement the provisions of the averaged field model.

It was found that in the absence of a modifying additive, the polymer and the filler are thermodynamically incompatible, and in their contact, phase separation occurs, except in the case of strong gravity between the polymer and the mineral. Unequal amount of modifying additive and its absence complicates the penetration of polymers into the interlayer space.

According to calculations, increasing the length of polymer molecules contributes to the phase separation of components, because the entropy of long molecules when intercalated into a narrow gap decreases to a greater extent than the entropy of shorter molecules. Increasing the chain length of the modifying additive, on the contrary, promotes the formation of thermodynamically stable nanocomposite. In this case, there is better interpenetration of polymer molecules and modifying additives, which increases the configurational freedom of both components.

The results obtained in both approaches give a better understanding of the factors that control the process of the nanocomposite and can serve as a theoretical basis for experiments to obtain polymer-silicate compositions.

MODELING THE PROPERTIES OF FILLED POLYMERS

The bulk of research in the field of polymer-silicate nanocomposites has been for the last 15 years. During this time, nanocomposites based on polymers with different degrees of polarity and chain strength were obtained. Many of the obtained materials have interesting properties. The mechanical characteristics of polymer-silicate nanocomposites were studied under different strain modes. In all cases, there is a significant increase in the mechanical properties of nanocomposites in comparison with similar characteristics of the original polymers.

For example, the strength limit and Young's modulus in nanocomposites based on epoxy resin and montmorillonite (Balazs et al., 1999) increase more than ten times compared to the valuesfor the original polymer at a microfiller concentration of 15% wt.

Since silicate layers are impermeable to molecules of liquids and gases, the diffusion coefficient for different beds in polymer-silicate nanocomposites decreases several times in comparison with the diffusion coefficient of the initial polymers (Giannelis, Krishnamoorti, & Manias E. (1999).

The thermophysical properties of nanocomposites are also interesting. Data on the decrease in the coefficient of thermal expansion, increase in shape stability and heat resistance in nanocomposites in comparison with the original polymers are given (Çankaya, 2021).

In addition, polymer-silicate nanocomposites are characterized by increased resistance to combustion (Kojima et al., 1994). The introduction of silicate particles into the polymer matrix causes a significant reduction in such characteristics as the rate of heat release, the amount of smoke formed, the yield of CO_2 (Yano et al., 1993).

Since the average diameter of silicate plates (200-300 nm) is slightly smaller than the wavelength of visible light, the optical transparency of polymers after the introduction of silicate particles is preserved (Unnikrishnan et al., 2012). Data on the ability to biodegrade nanocomposites based on polymers such as poly-c-caprolactam and aliphatic polyether are given (Alexandre & Dubois, 2000). For nanocomposites based on polyelectrolytes, the ionic conductivity is increased tenfold. Composites containing clay minerals also have high resistance to punctures and cuts, high vibration-absorbing properties Theng (2012).

It is usually impossible to achieve a combination of all the described properties in one composite material. In addition, the introduction of filler particles into the polymer may be accompanied by the appearance of undesirable properties. For example, the presence of solid inclusions in the polymer matrix often leads to an increase in viscosity, which complicates the formation of products. Many composite materials are more brittle than the original polymers. In this regard, the task in the creation of composite materials is carried out taking into account the area of their possible application.

In the literature, the supramolecular structure of polymer molecules in the interlayer spaces of clay minerals is reported to be significantly different from the nalmolecular structure of the polymer in volume, changes in the mechanism of crystallization and vitrification of such molecules (Çankaya, 2021). The available experimental data, but their few analyzes, do not allow us to draw unambiguous conclusions about the peculiarities of the structure and properties of polymer molecules that are in conditions of limited volume.

Numerous experimental data obtained in determining the physico-mechanical properties of nanocomposites describe the properties of individual materials obtained in one way or another. Systematic studies of the dependence of the properties of nanocomposites on the degree of filling, the structure of the filler in the composite, the properties of the polymer matrix and other parameters are virtually absent, which complicates the creation of polymer nanocomposites with specified physical and mechanical properties. There are several factors that significantly affect the physicochemical properties of filled polymers (Olad, 2011): the size of the filler particles, the shape of the filler particles; the orientation of the filler particles and their relative position in the matrix degree of filling: the interaction between the filler and the polymer matrix; the nature of the polymer matrix.

In traditionally used fillers, the particle size varies from a few units to hundreds of micrometers. As the particle size decreases, their total surface area increases at a given volume fraction and, consequently, the contact area with the polymer matrix increases. Under the condition of favorable for this system of interaction on the face of the distribution of the matrix filler, this fact leads to improve a number of characteristics, including mechanical. Thus, the hardness increases significantly (several times), the yield strength for thermoplastics and strength for elastomers increases (Alexandre & Dubois, 2000). On the other hand, in publications devoted to the study of mechanical properties of nanocomponents, their increased fragility is noted. The effect of aluminum hydroxide particle size on the plastic properties of composites based on high-density polyethylene was investigated. Composites filled with medium-sized particles (8 μm) have the greatest deformation during rupture (several tens of μm) and too small (1 μm) particles. will definitely be the smallest (Tseng, Wu, Lee, & Chang, 2001).

Both isodiametric particles (glass spheres, rubber crumb) and anisodiametric particles (fibers, nanotubes, plate filler) are used as fillers. As noted above, to characterize the degree of anisodiametry of the filler particles use the concept of the factor of the shape of the FF, which is determined by the ratio of the largest particle size to the smallest (Shambayati et al., 1990). This value is included in the equation for mechanical characteristics, diffusion coefficients and thermal expansion. Theoretical (Bigg, 1987) and experimental (Yu et al., 2018) studies indicate that filler particles with a high value of the form factor have a large enhancing effect when introduced into the polymer matrix compared to isodiametric inclusions and particles with lower FF. The form factor for montmorillonite particles depends on its structure (tactoids or individual plates) and can reach several hundred.

For anisodiametric inclusions, the orientation of the masses in the matrix is of great importance. Thus, if there is a complete (or predominant orientation of anisodiametric particles along a certain direction), the modulus of elasticity in this direction is several times higher than the modulus of elasticity of a composite with chaotically located inclusions (Alexandre & Dubois, 2000).

In addition to the orientation factor, the mechanism for the introduction of the filler significantly depends on their mutual location. With an orderly arrangement of the particle is at a strictly defined distance from each other, and their centers form a spatial lattice. The calculation of the microstrained state with chaotic and regular arrangement of inclusions was performed in Halpin, Finlayson, and Ashton (1992). In real systems, the particles are in most cases irregular. In addition, when using ultrafine particles, their agglomeration in the matrix can occur due to the huge surface area and, consequently, the high surface energy of nanoscale inclusions (Halpin, Finlayson, and Ashton, 1992). Adhesion to agglomerates changes the mechanism of destruction of composites and leads to a change in the whole complex of their properties.

The level of interaction between the matrix and the filler has a significant effect on the mechanical and other properties of composites. During the deformation of composites with a high level of interfacial interaction, part of the load is transferred from the matrix to the solid particles, which causes an increase in the mechanical properties. At a low level of interfacial interaction, the separation of filler particles from the matrix occurs at the initial stages of deformation, which can lead to a decrease in mechanical characteristics.

Contradictory data on the effect of adhesion on the hardness of filled polymers are given in the literature (Verbeek, 2003). The authors Gordon and Fakley (2003) believe that adhesion does not affect the modulus of elasticity of the composite. This point of view is based on the fact that the modulus of elasticity is measured at the initial stage of the deformation curve (E ® 0) in the region of elastic deformations to the occurrence of cracks and delaminations. Therefore, in this region of elongations, the particles remain rigidly bound to the matrix and the value of the adhesive bond strength has no significant effect. On the other hand, there are publications on the sharp difference between the modules of composites with high and poor adhesion in the polymer-filler system (Halpin et al., 1992). Obviously, the discrepancy is due to the fact that infinitesimal deformations are difficult to measure in practice. As shown by measurements of the volumes of composites during deformation (Sato & Furukawa, 1963), in this area of elongation can already occur irreversible changes in the structure and composites, pores and tears can be formed. The modulus determined in this way is not a true modulus of elasticity, and if disconnections from the matrix occur at elongations less than the proportionality limit, this leads to lower values of the measured modulus.

Despite the large number of publications on micro- and nanocomposites, the influence of the above factors on various properties is not fully understood. In Yilmazer U. & Farris R.J. (1983), the effect of the average molecular weight of the matrix polymer on the deformation behavior of composites was studied. and mechanical characteristics (modulus, strength, strain hardening coefficient, etc.), which are directly included in the structural-mechanical models. Detailed study of the mechanical characteristics of the polymeric binder on the properties of composites will predict their deformation-mechanical properties.

The analysis of experimental results with the help of obvious theoretical models helps to explain the properties of composites, evaluate the impact of various factors which control the properties of materials, and also allows predicting the characteristics of the composite formed for given system matrix – filler. It is necessary to note that nowadays the models were developed for filled polymers with microfillers, the size of which being higher as compared with delaminated clay minerals. Additionally, there is no models which allow to describe the entire load curve. So, for the description of the module of elastity, strength, yield strength and tensile strength in their polymers, different equations are used.

Most of models for the modulus of elasticity of polymer composites have been developed for filled elastomers (Yilmazer & Farris, 1983). Theoretical analysis of composites on the basis of thermoplastic polymers is complicated with the crystalline phase, which affects the deformation elongation of both initial thermoplastics and composites on their basis. In literature references, there are only a few models for an adequate analysis of the modulus of elastity of filled thermoplastics. Equations of Halpin - Tsai and Mori-Tanaka are the most used (Gloaguen & Lefebvre, 2001).

In 1956, E. Kerner considered the system of spherical particles with the M_{fil} module, which are located in the middle of the polymer matrix with the M_o module and the Poisson coefficient μ_o with the strong adhesive bond between the phases. After analyzing the stress and strain fields, which occurs in a such two-phase system during stressing. Kerner obtained the equation for the expression of the elastity modulus, which was concretized by Halpin, Tsai, and Kardos, (1976)., Lewis and Buchdahl (1975):

$$\frac{M_E}{M_o} = \frac{1 + AB\varphi}{1 - B\psi\varphi}$$

Where M_E is module of the composite, M_o is the module of the polymeric matrix, A is the factor which reflects the shape of the filler particles and their orientation in the composite, B referes to the factor which reflects the ratio of modules of filler and matrix.

Even though it is not isotropic, the values of the modulus and other mechanical characteristics, which vary from different direct lines, vary. The modulus in the direction which coincides with the whole orientation of anisotropic inclusions M_p, shows the largest value, and the modulus determined on direction which is the perpendicular axis of particles orientation M_{perp}. being of lowest value. In Ahmed S. & Jones F. (1990), it was stated that the constant A in the first case is considered as A = 2FF, where FF is the factor of the shape of the particles, and in the other case, A does not depend on the shape of particles and equals to 2. Factor B reflects the ratio of elastity modulus of filler and matrix.

$$B=\frac{\frac{M_{fil}}{M_o}-1}{\frac{M_{fil}}{M_o}+A}$$

where $M_{fil}$is the modulus of the filler, Mo is the modulus of the polymeric matrix.

At very high values of hardness, the M_{fil}/M_o ratio is close to one. Parameter ψ depends on parts by volume of filler at their maximum dense packing Ahmed S. & Jones F. (1990)

$$\psi=1+\left(\frac{1-\phi_{fil}}{\phi_{fil}^2}\right)\varphi$$

where ф is the part by volume of the filler.

In Halpin J.C. & Finlayson K.M. & Ashton J. E. (1992) it was suggested to calculate the elastity modulus as solid inclusions orientated chaotically

$$M_{chaot}=0.49M_p+0.51M_{perp}$$

Authors Lewis and Nielsen (1970) obtained another equation for $M_{chaot}=\frac{3}{8}M_p$.

Mori-Tanaka, proposed in 1973, describes the modulus of elasticity of an isotropic polymer which involves ellipsoid-shaped hard inclusions (Brisard et al., 2018). As equation Halpin - Tcai, this model takes into accounts Young's modulus and Poisson's coefficient of the matrix and filler and shape factor inclusions. With the same parameters of the matrix system, the value of the modulus of composites, which is calculated according to Mori-Tanaka, are close to the value of the modulus, which are obtained from the equation of Halpin-Tsai.

Theoretical models for the analysis of the elasticity of composites Halpin - Tsai and Mori - Tanaka were used in this work, because they can rather effectively describe the mechanical properties of composites with inclusions of various natures and shapes, including nanocomposites based on clay minerals. The "Minimum Cross section Model", which describes the strength of composites based on thermoplastic polymers, was proposed by T. L. Smith in 1989 Schöllhorn R. (1996). This theory is well suited to the experimental values of strength of the filled thermoplastics, they were widely used in the works of previous researchers Bazhenov S.et al. (1994). According to this approach, during the deformation of polymers filled with hard particles, the matrix is detached from the inclusions and pores are formed, their size being approximitely equal to the size the inclusion . As a result of which the "working" crossection of the sample decreases by a value, as the total area of the projections of the particles in the crossection the strength of the composite changes according to the law of "two thirds":

$$\sigma_A = \frac{\sigma_o}{1-\beta\varphi^{2/3}}$$

σ_c and σ_0 are the strength of the composite and the matrix, β is the factor depending on the shape of the inclusions ($\beta = S_{perp}/V_{part}^{2/3}$), for spheric particles β = 1.21.

Apparently, there are only a few approaches to the theoretical description of the boundaries of the fluidity of the filled compositions. The authors of Gordon T. & Fakley M. (2003) admitted that in case of ideal adhesion to the polymer to the surface, during deformation, the bonding of the phases is saved and a part of the applied load in the composite is transferred to the parts of the surface. It was showed Gordon T. & Fakley M. (2003) that the dependence of the fluidity limit of the composite vs. a concentration of inclusions is an increasing function. In the case, as the adhesion between the polymer and the filler is fully absent, when the limit of fluidity reached, all the particles appear to be detached from the matrix and the dependence decreases gradually by the 'model of the minimal crossection'. For composites with intermediate values of adhesion, the deposit was ranked between two functions Onitiri M. & Ubi P. (2021).

The authors of Shalygina T.A. et al. (2021) admit that for a solid bond between a filler and a polymer, the fluidity boundary of a composite is equal to the fluidity limit of a pure polymer, and in the case of imperfect adhesion, the values of the yield strength are between the yield strengths of the matrix values, as the "minimal crossection model" is assumed Lewis T.B. & Nielsen L.E. (1970).

Apparently, there is only one theoretical approach for the lower fluidity limit (tension of the "neck" extension) proposed by the authors Lewis T.B. & Nielsen L.E. (1970). They assumed that as solid bond between the matrix and the polymer occurs, the expansion of the "neck" in the composite remains equal to the lowest fluidity limit the pure polymer. If pores are formed in the composite before the moment of reaching the lowest limit of fluidity, the tension of the expansion of the "neck" changes proportionally to the volume concentration of the filler:

$$\sigma_n = (\sigma_n)_o(1\text{-}1.25\alpha\varphi)$$

where α is the part of detached particales; σ_n is tension of the expansion of the "neck".

The fact that the lower yield limit of the composite does not depend on $\varphi^{2/3}$, as the upper yield limit and strength, but is determined by φ results from the averaging of the working cross section of the sample during the development of the "neck". The destruction of the sample and the initiation of plastic flow always occurs in a weaker section, and when the "neck" spreads, it passes both through weaker sections (pore centers) and through areas with a predominant content of the polymer matrix. Therefore, the stress of propagation of the "neck" in the composite will decrease more slowly in contrast to the tensile strength.

In composites with imperfect adhesion between phases, for a theoretical estimation of the behavior of composites when teared, An analytical equation obtained by of L.E. Nielsen is often used. According with it, neck is gradually decreased with an increase in the concentration of the filler Lewis T.B. & Nielsen L.E. (1970). The deformation when teared in filled elastomers is well-described by this function. However, in the composites based on thermoplastic matrices another situation is often observed. The breaking elongation slowly decreases up to some critical degree of filling φ_{crit}. When this value is reached, the samples are destroyed with a wider range of elongation values at break (there is a transition from plastic to brittle fracture). At filler concentrations higher than φ_{crit}, the rupture occurs when the yield limit is reached. The authors Lewis T.B. & Nielsen L.E. (1970) developed a model for this case. In this model, the dependence of the elongation load in the region of strain-hardening was approximated by a linear function which slope is equal to the value of the current modulus E_{cur}. It was also assumed that the strength of the composite decreases according to the "minimum cross-section model". The authors Lewis T.B. & Nielsen L.E. (1970) obtained the following expression for deformation at rupture of filled thermoplastics:

$$\varepsilon_c = \varepsilon_o - \frac{\beta\sigma_o}{E_{cur}} \cdot \varphi^{2/3} + \frac{\sigma_o}{E_{cur}} \cdot \varphi$$

where ε_c and ε_o are breaking elongation of the composite and the initial polymer, σ_o refers to the strength of matrices; E_{cur} is a current module at the site of strain-hardening, β is the factor from the 'model of minimal crossection'.

The expression for ε_c is valid under the condition $\varphi < \varphi_{crit}$. In this case, the 'neck' formed is stable and will extend to the entire sample. When the strength of the composite decreases to a value equal to the lower yield strength, the destruction of the samples can occur at any stage of the formation of the "neck". At a filler concentration greater than φ_{crit}, the 'neck' is not formed and the samples are brittle. The value of φ_{crit} was obtained from the condition of equality of strength and lower yield strength of the composite. The adhesive strength of the matrix and the filler has a noticeable effect on the value of φ_{crit}, the author of the model considers two cases with ideal and zero adhesion between the matrix and the filler. In the first case, the lower yield strength is considered equal to the lower yield strength of pure polymer, and the expression for calculating the critical concentration of the filled takes the form

$$\varphi_{crit} = \frac{1}{\beta} \cdot \left(1 - \frac{1}{k_{el}}\right)^{3/2}$$

where $k_{el} = \sigma_o/(\sigma_n)_o$

Since at zero adhesion the lower yield strength of the composite decreases in proportion to φ, the strength will reach a value equal to the lower yield strength at higher filling degrees than in the first case. As a result, the tendency to brittleness in composites with poor adhesion is observed at higher concentrations of filler, determined by the expression:

$$\varphi_{crit} = \frac{\beta^3 k_{el}^3}{27} \cdot \left(1 - 2\cos(\frac{\alpha}{3} + \frac{\pi}{3})\right)^3$$

Where $\alpha = \arccos(1 - \frac{27(k_{el} - 1)}{2\beta^3 k_{el}^3})$, $k_{el} = \sigma_o/(\sigma_n)_o$

CONCLUSION

Thus, various aspects of the development of solid state chemistry, scientifically based selection of new types of host and guest component gratings, search for ways to optimize intracrystalline host-guest interactions, and study of the influence of matrices on electronic properties of complex systems are widely studied. The orientational influence of interlayer formations on the degree of crystallinity and stereoregularity of the formed polymer is one of the fundamental problems of physicochemistry of polymers and composite materials.

The main difficulties facing nanotechnology in the creation of polymer composite using clay minerals is the low thermal stability of modifying mineral additives and insufficient understanding of the mechanism of this nanotechnology in cases of specific polymer matrices, which complicates extrapolation of research results to other objects.

Despite significant progress in the development of polymer nanocomposites using clay minerals, understanding the advantages and disadvantages of this nanotechnology, industrial implementation of this technology on a large scale is possible only with a deep understanding of mechanisms to improve the properties of polymers. Such nanocomposites have great potential as a replacement for traditional fiber-filled polymer composites and the development of new structural plastics.

The difficulty of producing heat-resistant modified clay minerals is as follows. The most commercially available modifying additives of clay minerals Quaternary ammonium salts decompose at temperatures below the processing temperatures of most polymers. Additionally, even if the thermal stability of such modifying additives is sufficient for processing, their long-term thermal stability during processing and operation is low. Heat-resistant modifying additives such as phosphonium compounds are too expensive for industrial use.

Therefore, the main expectations in the field of realization of this nanotechnology are connected with the use of polymers and oligomers as heat-resistant modifying additives of clay minerals. Based on a review of a set of theoretical and experimental data on the creation and study of polymer-clay mineral nanocomposites, accumulated at the moment in the scientific literature, the purpose and objectives of this study were formulated. view of our work on the creation of new nanocomposite materials based on thermoplastic polymers and clay mineral - montmorillonite, as well as a comparative study of the specifics of their structure and physical and mechanical properties.

REFERENCES

Affdl, J. C. H., & Kardos, J. L. (1976). The Halpin-Tsai equations: A review. *Polymer Engineering and Science*, *16*(5), 344–352. doi:10.1002/pen.760160512

Ahmed, S., & Jones, F. (1990). A Review of Particulate Reinforcement Theories for Polymer Composites. *Journal of Materials Science*, *25*(12), 4933–4942. doi:10.1007/BF00580110

Alexandre, M., & Dubois, P. (2000). Polymer-layered silicate nanocomposites: Preparation, properties and uses of a new class of materials. *Materials Science and Engineering R Reports*, *28*(1-2), 1–63. doi:10.1016/S0927-796X(00)00012-7

Balazs, A. C., Singh, C., Zhulina, E., & Lyatskaya, Yu. (1999). Modeling the Phase Behavior of Polymer/Clay Nanocomposites. *Accounts of Chemical Research*, *32*(8), 651–657. doi:10.1021/ar970336m

Bazhenov, S., Li, J. X., Hiltner, A., & Baer, E. (1994). Ductility of filled polymers. *Journal of Applied Polymer Science*, *52*(2), 243–254. doi:10.1002/app.1994.070520211

Bigg, D. M. (1987). Mechanical properties of particulate filled polymers. *Polymer Composites*, *8*(2), 115–122. doi:10.1002/pc.750080208

Biron, M. (2016). 2 - Thermoplastic Specific Properties. In *Material Selection for Thermoplastic Parts. Practical and Advanced Information* (pp. 39–75). William Andrew Applied Science Publishers. doi:10.1016/B978-0-7020-6284-1.00002-7

Brisard, S., Guilleminot, J., Sab, K., & Tran, V. P. (2018). Mori–Tanaka estimates of the effective elastic properties of stress-gradient composites. *International Journal of Solids and Structures*, *146*, 55–68. doi:10.1016/j.ijsolstr.2018.03.020

Buchdahl, R. (1975). Mechanical properties of polymers and composites – Vols. I and II, Lawrence E. Nielsen, Marcel Dekker, Inc. *Journal of Polymer Science. Part B: Polymer Letters*, *13*, 120–121.

Çankaya, N. (2021). Synthesis, characterization and thermal properties of oxo methacrylate-containing polymer/clay nanocomposites. *Politeknik Dergisi, 24*, 1017–1022.

Chen, B., Evans, J., Greenwell, C., Boulet, P., Coveney, P., Bowden, A., & Whiting, A. (2008). A critical appraisal of polymer-clay nanocomposites. *Chemical Society Reviews, 37*(3), 568–594. doi:10.1039/B702653F PMID:18224264

Chou, Ch.-Ch., & Lin, J.-J. (2004). One-Step Exfoliation of Montmorillonite via Phase Inversion of Amphiphilic Copolymer Emulsion. *Macromolecules, 38*. doi:10.1021/ma047761x

Cui, Y., Kumar, S., Kona, B., & Houcke, D. (2015). Gas Barrier Properties of Polymer/Clay Nanocomposites. *RSC Advances, 5*(78), 63669–63690. Advance online publication. doi:10.1039/C5RA10333A

Franco-Urquiza, E. A. (2021). Clay-Based Polymer Nanocomposites: Essential Work of Fracture. *Polymers, 13*(15), 2399. doi:10.3390/polym13152399 PMID:34372002

Giannelis, E., Krishnamoorti, R., & Manias, E. (1999). Polymer-Silicate Nanocomposites: Model Systems for Confined Polymers and Polymer Brushes. *Advances in Polymer Science, 138*, 107–147. doi:10.1007/3-540-69711-X_3

Gloaguen, J.-M., & Lefebvre, J.-M. (2001). Plastic deformation behaviour of thermoplastic/clay nanocomposites. *Polymer, 42*(13), 5841–5847. doi:10.1016/S0032-3861(00)00901-0

Gordon, T., & Fakley, M. (2003). The influence of elastic modulus on adhesion to thermoplastics and thermoset materials. *International Journal of Adhesion and Adhesives, 23*(2), 95–100. doi:10.1016/S0143-7496(02)00064-7

Halpin, J. C., Finlayson, K. M., & Ashton, J. E. (1992) Primer on Composite Materials Analysis. Taylor and Francis, 354.

Haudin, J.-M. (1982). Plastic Deformation of Amorphous and Semicrystalline Materials. *Les Editions de Physique*. 291-311.

Kojima, Y., Usuki, A., Kawasumi, M., Okada, A., Kurauchi, T., Kamigaito, O., & Kaji, K. (1994). Fine structure of nylon–6–clay hybrid. *Journal of Polymer Science. Part B, Polymer Physics, 32*(4), 625–630. doi:10.1002/polb.1994.090320404

Kononova, I., Kononov, P., Moshnikov, V., & Ignat'ev, S. (2021). Fractal-Percolation Structure Architectonics in Sol-Gel Synthesis. *International Journal of Molecular Sciences*, *22*(19), 10521. doi:10.3390/ijms221910521 PMID:34638862

Kumar, S. K., & Krishnamoorti, R. (2010). Nanocomposites: Structure, Phase Behavior, and Properties. *Annual Review of Chemical and Biomolecular Engineering*, *1*(1), 37–58. doi:10.1146/annurev-chembioeng-073009-100856 PMID:22432572

Kumari, N., & Mohan, C. (2021). Basics of Clay Minerals and Their Characteristic Properties. In (Ed.), Clay and Clay Minerals. IntechOpen. doi:10.5772/intechopen.97672

Kusmono, K., Ishak, M., Chow, W. S., Takeichi, T., & Rochmadi, R. (2009). Effect of Clay Modification on the Morphological, Mechanical, and Thermal Properties of Polyamide 6/Polypropylene/Montmorillonite Nanocomposites. *Polymer Composites*, *31*, 1156–1167. doi:10.1002/pc.20902

Laird, D. (2006). Influence of layer charge on swelling of smectites. *Applied Clay Science*, *34*(1-4), 74–87. doi:10.1016/j.clay.2006.01.009

Landel, R., & Nielsen, L. (1993). Mechanical Properties of Polymers and Composites. CRC Press, 580 doi:10.1201/b16929

Lewis, T. B., & Nielsen, L. E. (1970). Dynamic mechanical properties of particulate-filled composites. *Journal of Applied Polymer Science*, *14*(6), 1449–1471. doi:10.1002/app.1970.070140604

Li, Yu., Wang, S., Wang, Q., & Xing, M. (2018). A comparison study on mechanical properties of polymer composites reinforced by carbon nanotubes and graphene sheet. *Composites. Part B, Engineering*, *133*, 5–41. doi:10.1016/j.compositesb.2017.09.024

Mallikarjuna, S., & Sivaram, S. (2013). Influence of Organic Modification and Polyurethane Structure on Clay Dispersion. In: Thermoset Nanocomposites. doi:10.1002/9783527659647.ch3

Messersmith, P. B., & Giannelis, E. P. (1995). Synthesis and barrier properties of poly(ε-caprolactone)-layered silicate nanocomposites. *Journal of Polymer Science Part A*, *33*(7), 1047–1057. doi:10.1002/pola.1995.080330707

Monasse, B., Queyroy, S., & Lhost, O. (2008). Molecular Dynamics prediction of elastic and plastic deformation of semi-crystalline polyethylene. *International Journal of Material Forming*, *1*(S1), 1111–1114. doi:10.100712289-008-0174-1

Newman, A. C. D. (1987). *Chemistry of Clays and Clay Minerals*. Longman Scientific & Technical.

Olad, A. (2011). Polymer/Clay Nanocomposites. In *Advances in Diverse Industrial Applications of Nanocomposites*. doi:. doi:10.5772/14464

Onitiri, M., & Ubi, P. (2021). Failure modes in particle filled plastic matrix composites. *Journal of Engineering Science and Technology*, *5*, 79–96.

Pal, K. (2016). Effect of different nanofillers on mechanical and dynamic behavior of PMMA based nanocomposites. *Composites Communications*, *1*, 25–28. doi:10.1016/j.coco.2016.08.001

Penaloza, D. P. Jr. (2019). Modified clay for the synthesis of clay-based nanocomposites. *Journal of Silicate Based and Composite Materials*, *71*(1), 5–11. doi:10.14382/epitoanyag-jsbcm.2019.2

Perelomov, L., Mandzhieva, S., Minkina, T., Atroshchenko, Y., Perelomova, I., Bauer, T., Pinsky, D., & Barakhov, A. (2021). The Synthesis of Organoclays Based on Clay Minerals with Different Structural Expansion Capacities. *Minerals (Basel)*, *11*(7), 707. doi:10.3390/min11070707

Peterlin, A. (1972). Morphology and Properties of Crystalline Polymers with Fiber Structure. *Textile Research Journal*, *42*(1), 20–30. doi:10.1177/004051757204200105

Peterlin, A. (1975). Plastic deformation of polymers with fibrous structure. *Colloid & Polymer Science*, *253*(10), 809–823. doi:10.1007/BF01452401

Qian, Z., Zhang, S., & Yang, M. (2008). Effect of Clay Modification on Photooxidation of Polyethylene/Clay Nanocomposites. *Polymers & Polymer Composites*, *16*(8), 535–546. doi:10.1177/096739110801600807

Rahman, I. A., & Padavettan, V. (2012). Synthesis of Silica Nanoparticles by Sol-Gel: Size-Dependent Properties, Surface Modification, and Applications in Silica-Polymer Nanocomposites—A Review. *Journal of Nanomaterials*, *2012*, 132424. doi:10.1155/2012/132424

Rothon, R. N. (2003). Particulate-filled polymer composites. Shrewsbury, UK: Rapra Technology, 320

Rothon, R. N. (2017). Fillers for Polymer Applications. Springer Verlag., 211 doi:10.1007/978-3-319-28117-9

Sato, Y., & Furukawa, J. (1963). A Molecular Theory of Filler Reinforcement Based upon the Conception of Internal Deformation (A Rough Approximation of the Internal Deformation). *Rubber Chemistry and Technology*, *36*(4), 1081–1106. doi:10.5254/1.3539632

Schöllhorn, R. (1996). Intercalation Systems as Nanostructured Functional Materials. *Chemistry of Materials*, *8*(8), 1747–1757. doi:10.1021/cm9601421

Shalygina, T. A., Rudenko, M. S., Nemtsev, I. V., Parfenov, V. A., Voronina, S. I., & Simonov, I. D. (2021). Influence of the Filler Particles' Surface Morphology on the Polyurethane Matrix's Structure Formation in the Composite. *Polymers*, *13*(22), 3864. doi:10.3390/polym13223864 PMID:34833164

Shambayati, S., Schreiber, S. L., Blake, J. F., Wierschke, S. G., & Jorgensen, W. L. (1990). Structure and basicity of silyl ethers: A crystallographic and ab initio inquiry into the nature of silicon-oxygen interactions. *Journal of the American Chemical Society*, *1990*(112), 697–703. doi:10.1021/ja00158a031

Shu, G., Zhao, J., Zheng, X., Xu, M., Liu, Q., & Zeng, M. (2019). Modification of Montmorillonite with Polyethylene Oxide and Its Use as Support for Pd0 Nanoparticle Catalysts. *MDPI Polymers*, *11*(5), 755. doi:10.3390/polym11050755 PMID:31035647

Soheil, S., Abdolahifar, H., & Goudarzian, N. (2015). Preparation and Modification of Polystyrene/Clay Nanocomposite Using Ethylene Octene Copolymer and SBS Block Copolymer. Conference: Conference of Polymer Processing Society At: Graz, Austria

Srebnik, S., & Ovadia, L. (2003). Theoretical Investigation of Imprinted Crosslinked Silicates. *Journal of Sol-Gel Science and Technology*, *26*(1/3), 107–113. doi:10.1023/A:1020701721529

Theng, B. K. G. (2012). *Formation and Properties of Clay-Polymer Complexes* (2nd ed.). Elsevier.

Truc, T., Thuy, T., Oanh, V., Hang, T., Nguyen, A., & ... . (2019). 8-hydroxyquinoline-modified clay incorporated in an epoxy coating for the corrosion protection of carbon steel. *Tribology - Materials, Surfaces & Interfaces. Maney Publishing.*, *14*, 26–33.

Tseng, C., Wu, J., Lee, H., & Chang, F. (2001). Preparation and crystallization behavior of syndiotactic polystyrene-clay nanocomposites. *Polymer*, *42*(25), 10063–10070. doi:10.1016/S0032-3861(01)00568-7

Unnikrishnan, L., Mohanty, S., Nayak, S., & Singh, N. (2012). Synthesis and characterization of polysulfone/clay nanocomposite membranes for fuel cell application. *Journal of Applied Polymer Science*, *124*(S1), E309–E318. Advance online publication. doi:10.1002/app.34355

Vaia, R. A., Sauer, B. B., Tse, O. K., & Giannelis, E. P. (1997). Relaxations of confined chains in polymer nanocomposites: Glass transition properties of poly(ethylene oxide) intercalated in montmorillonite. *Journal of Polymer Science. Part B, Polymer Physics*, *35*(1), 59–67. doi:10.1002/(SICI)1099-0488(19970115)35:1<59::AID-POLB4>3.0.CO;2-Q

Valášek, P., Müller, M., & Ružbarský, J. (2014). Using Recycled Rubber Particles as Filler of Polymers. *Applied Mechanics and Materials*, *616*, 260–267. doi:10.4028/www.scientific.net/AMM.616.260

Vanderzande, C. (1998). *Lattice Models of Polymers (Cambridge Lecture Notes in Physics)*. Cambridge University Press. doi:10.1017/CBO9780511563935

Verbeek, J. (2003). The Influence of Interfacial Adhesion, Particle Size and Size Distribution on the Predicted Mechanical Properties of Particulate Thermoplastic Composites. *Materials Letters*, *57*(13-14), 1919–1924. doi:10.1016/S0167-577X(02)01105-9

Wang, M.-J. (1998). Effect of Polymer-Filler and Filler-Filler Interactions on Dynamic Properties of Filled Vulcanizates. *Rubber Chemistry and Technology*, *71*(3), 520–589. doi:10.5254/1.3538492

Wang, Z. M., Hong, H., & Chung, T. C. (2005). Synthesis of Maleic Anhydride Grafted Polypropylene with High Molecular Weight Using Borane/O_2 Radical Initiator and Commercial PP Polymers. *Macromolecules*, *38*(22), 8966–8970. doi:10.1021/ma0516182

Ward, I. M., & Sweeney, J. (2004). *An Introduction to the Mechanical Properties of Solid Polymers* (2nd ed.). Wiley.

Xu, M., & An, B. (2022). Plastic deformation of polymer matrix composites reinforced with beaded fibers. *International Journal of Solids and Structures*, *256*, 111985. doi:10.1016/j.ijsolstr.2022.111985

Yang, Y., Nogami, M., Huang, Z., & Li, Y. (2013). Engineering Nanostructures of Inorganic Materials for Optical and Chemical Applications. *Journal of Nanomaterials*, *2013*, 157869. doi:10.1155/2013/157869

Yano, K., Usuki, A., Okada, A., Kurauchi, T., & Kamigaito, O. (1993). Synthesis and properties of polyimide–clay hybrid. *J. Polym. Sci. Part A.*, *31*(10), 2493–2498. doi:10.1002/pola.1993.080311009

Yilmazer, U., & Farris, R. J. (1983). Mechanical behavior and dilatation of particulate-filled thermosets in the rubbery state. *Journal of Applied Polymer Science*, *28*(11), 3369–3386. doi:10.1002/app.1983.070281106

Zhang, G., Wu, T., Lin, W., Tan, Y., Chen, R., Huang, Z., Yin, X., & Qu, J. (2017). Preparation of polymer/clay nanocomposites via melt intercalation under continuous elongation flow. *Composites Science and Technology*, *145*, 157–164. doi:10.1016/j.compscitech.2017.04.005

Zhu, Y., Iroh, J., Rajagopolan, R., Aykanat, A., & Vaia, R. (2022). Optimizing the Synthesis and Thermal Properties of Conducting Polymer–Montmorillonite Clay Nanocomposites. *Energies*, *15*(4), 1291. doi:10.3390/en15041291

Chapter 8
Development of Elastomeric Compositions of Adhesive Application

ABSTRACT

The present chapter is focused on the development of elastomeric compositions for adhesive application. The modification of montmorillonite is shown to result in changing the chemical composition of the surface of montmorillonite, its reactivity, and the structure due to increasing the interplanar distance. Montmorillonite modified with polymeric quarternary ammonium salt and quarternary ammonium salt was studied in model elastomeric compositions. The strength of adhesive bond of elastomeric compositions with montmorillonite modified with quarternary ammonium salt or polymeric quarternary ammonium salts exceeds compositions with convenient additives at least by 25 – 37 after various types of ageing. Quarternary ammonium salts are revealed to be vulcanization-active compounds for sulfur vulcanization of elastomeric compositions.

BACKGROUND

In order to determine the conditions of processing and predict the effect of modified with nitrogen-containing compounds montmorillonite such as quaternary ammonium salt (QAS) or polymer quaternary ammonium salt on the formation of the properties of elastomeric compositions, physicochemical

DOI: 10.4018/978-1-7998-9728-6.ch008

characteristics of montmorillonite modified with nitrogen containing compounds were studied.

Before studying the physicochemical characteristics of montmorillonite modified with nitrogen-containing compounds, the samples were subjected to thorough washing. The samples were extracted with appropriate solvents in a Soxhlet apparatus to remove ungrafted nitrogen-containing compounds. Samples of montmorillonite modified with amino esters, extracted in isopropyl alcohol at a temperature of 80°C and dried to constant weight. Samples of montmorillonite modified with PQAS were extracted in water at a temperature of 80 °C and dried to constant weight.

The content of modifiers adsorbed on montmorillonite was determined by the difference in weight loss during calcination between modified and base montmorillonite: Mm (QAS) - 9.4 wt. %, Mm (QAS) - 10.0 wt. %, Mm (PQAS) - 12.5 wt. %.

Analysis of the results of X-ray photoelectron spectroscopy of the surface layer of montmorillonite showed that the sample Mm (PQAS) has the smallest number of nitrogen groups in contrast to other forms modified with Quaternary ammonium salts (Table 1). This probably indicates a more massive penetration of nitrogen-containing groups into the interlayer space of the structure of montmorillonite as a result of ion exchange sorption. Indeed, the study of montmorillonite modified with Quaternary ammonium salts using wide-angle X-ray analysis (Fig. 1 – 3) showed the presence of a shift of basal reflexes in the area of small angles compared to base montmorillonite and corresponding modifying additives.

The shift of basal reflexes after modification indicates an increase in the interlayer space in the structure of montmorillonite. The calculation according to the formula a (Schmidt et al., 2002) of Wolf-Bragg (Table 2) showed that as a result of modification of montmorillonite with the proposed Quaternary ammonium salts there is an increase in its interlayer space by 56 – 62%. Montmorillonite modified with polymeric Quaternary ammonium salt has the maximum size of the interlayer space.

Determination of the specific surface area of base and modified montmorillonite by thermal desorption of nitrogen showed that the modification of Quaternary ammonium salts increases the size of the specific surface area in 2.4 - 6.9 times, namely: base montmorillonite 1.73 m^2/ g, modified QAS - montmorillonite 11.27 m^2/g, modified QAS-U montmorillonite - 10.19 m^2/g, modified with PQAS montmorillonite - 11.93 m^2 /g h.

IR-spectral studies have shown that after thorough washing, the spectra of modified montmorillonite have peaks characteristic of modifying additives.

Table 1. Chemical composition (%) of the surface layer of the base and modified with Quaternary ammonium salts of montmorillonite

Montmorillonite	Chemical composition of surface, %			
	O	Al	Si	N
Basic	61.5	10.8	27.7	-
Modified with QAS	60.3	10.9	26.1	1.7
Modified with QAS-U	60.4	10.6	26.4	1.4
Modified with PQAS	62.4	10.2	26.4	0.8

This indicates the presence of a chemisorption bond between the molecules of modifying additives and the surface of montmorillonite. IR spectra of Quaternary ammonium salts have different absorption bands (Fig. 4), but the most characteristic are: 3300-3500 cm^{-1} - υ oscillations of OH groups; 3229 cm^{-1} - υ oscillations of HH groups; 2800-3000 cm^{-1} - υ oscillations of CH groups; 1582-1610 cm^{-1} - δ oscillations of C_6H_6 groups; 1350-1480, 830 cm^{-1} - δ oscillations of CH groups; 1180-1300 cm^{-1} - υ fluctuations of CO groups (groups C – O – C, O = C – O).

Treatment of montmorillonite with Quaternary ammonium salts leads to changes in its spectrum (Fig. 5-7). On the crest of the wide band Me-O (from max. 1036 cm^{-1}) a number of weak bands appear, which belong to the modifying additives: 817,880; 1170; 1234 cm^{-1}. The last lanes are hidden by this lane.

The presence of other bands in the region 2800-3000 cm^{-1} and in the region 1800-1300 cm^{-1}, which correspond to ν and δ fluctuations of CH, N-H, C = O, C-N indicates the presence of organic matter in montmorillonite. Thus, the IR spectral analysis of basic and modified montmorillonite indicates the chemical inoculation of modifying additives on the surface of montmorillonite.

Table 2. Results of calculation of interplanar distances in the structure of the base and modified Quaternary ammonium salts of montmorillonite

Montmorillonite	Scatter angle, 2θ, °	Space between layers, d, nm
Base	8.6	1.03
Modified with QAS	5.4	1.63
Modified with QAS-U	5.5	1.61
Modified with PQAS	5.3	1.67

Figure 1. WAXS-profiles of base (1), modified QAS (2) montmorillonite and individual QAS (3)

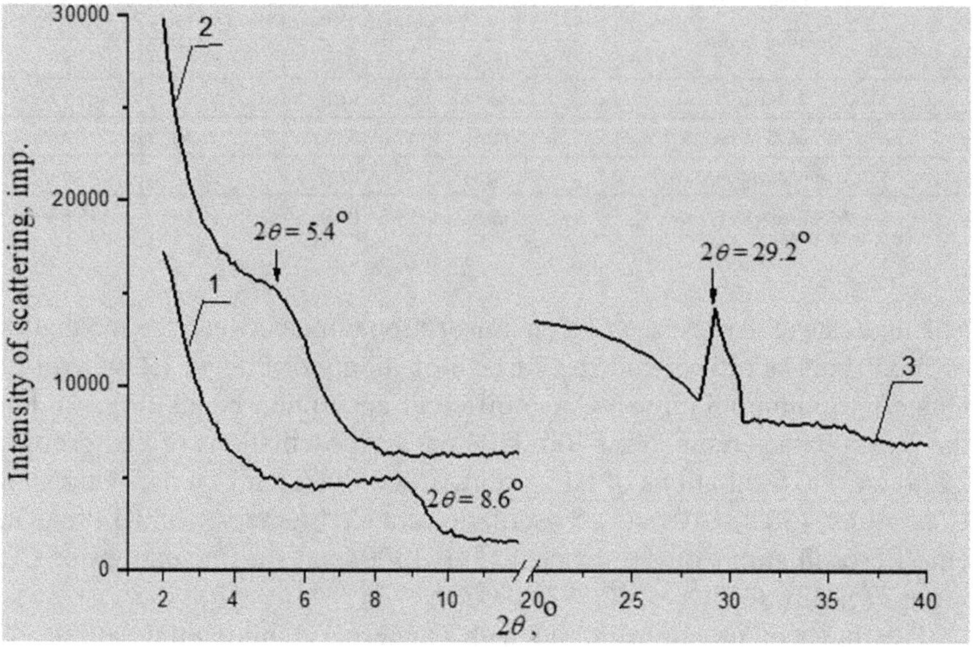

DTA data (Table 3) show that Quaternary ammonium salts grafted on montmorillonite are stably bound to it up to 110°C - the mixing temperature in rubber mixing in stage II with the introduction of modified montmorillonite. A significant number of these modifying additives begin to desorb from the surface of montmorillonite at 120-150°C, i.e. modified montmorillonite is thermally stable at the mixing temperature in the rubber mixer and is able to actively desorb during heating and vulcanization of elastomeric compositions.

Thus, the modification of montmorillonite allows to change the chemical composition of the surface of montmorillonite and, probably, its reactivity, and, at the same time, allows to change the structure of aluminosilicate, increasing the interplanar distance Ovcharov V.I. et al (2005). Increasing the interplanar distance in the structure of montmorillonite with modification of QAS may be a prerequisite for the destruction of its structure during mixing with elastomer and as a result lead to more efficient dispersion of montmorillonite in elastomeric matrix, increase contact surface with elastomer and language in rubber-reinforced systems.

It was found that the modifying additives are stably bound to the mineral base of montmorillonite at processing temperatures of rubber mixtures

Figure 2. WAXS profiles of base (1), modified QAS-U (2) montmorillonite and individual QAS-U (3)

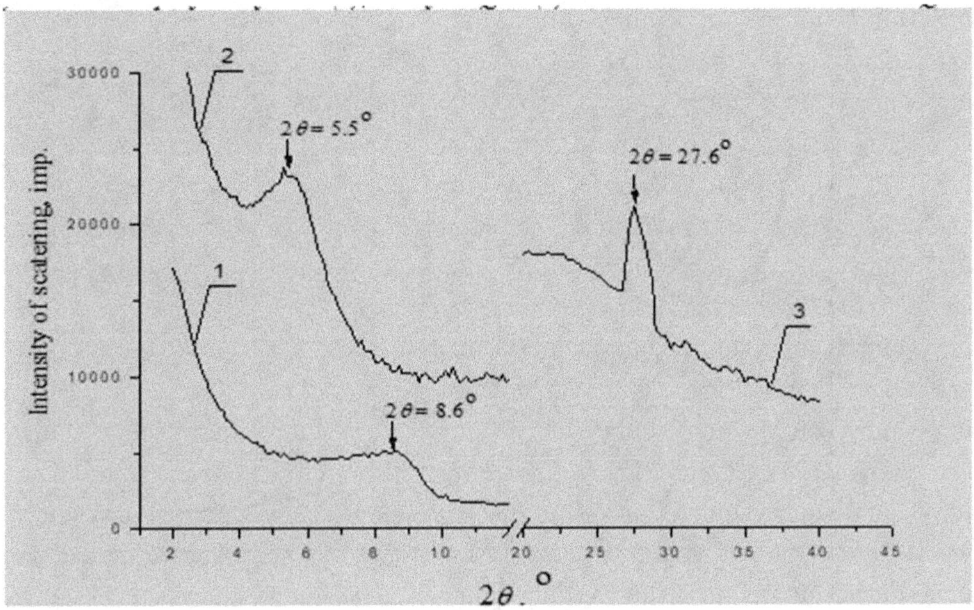

and are capable of desorption at vulcanization temperatures of elastomeric compositions. Influence of organomodified montmorillonite on the formation of the general complex of properties of elastomeric compositions

Since the 1960s, work has been carried out on the use of nitrogen-containing compounds as adhesive additives of elastomeric compositions in rubber-metal cord and rubber-textile cord systems. But one of the main disadvantages of such additives is their toxicity (ammonia release), reduced resistance to vulcanization (Greene, 2021), so a promising area of work is the search for new nitrogen-containing compounds capable of increasing the adhesive strength in systems "rubber-reinforcing material".

Due to the ability of Quaternary ammonium salts to anticorrosive action against metals (Avdeev et al., 2021), as well as their vulcanization activity in sulfur vulcanization systems (de Araújo et al., 2022), there was interest in using these compounds as adhesive additives in systems: rubber - metal cord and rubber - textile cord.

With the introduction of polymeric ammonium salt there are significant difficulties in its uniform distribution in the rubber matrix. Therefore, montmorillonite was used as a carrier of Quaternary ammonium salts, which, as described in the literature (Gallo, Rizzo, & Guerra, 2019), is capable of

Figure 3. WAXS-profiles of base (1), modified PQAS (2) montmorillonite and individual PQAS (3)

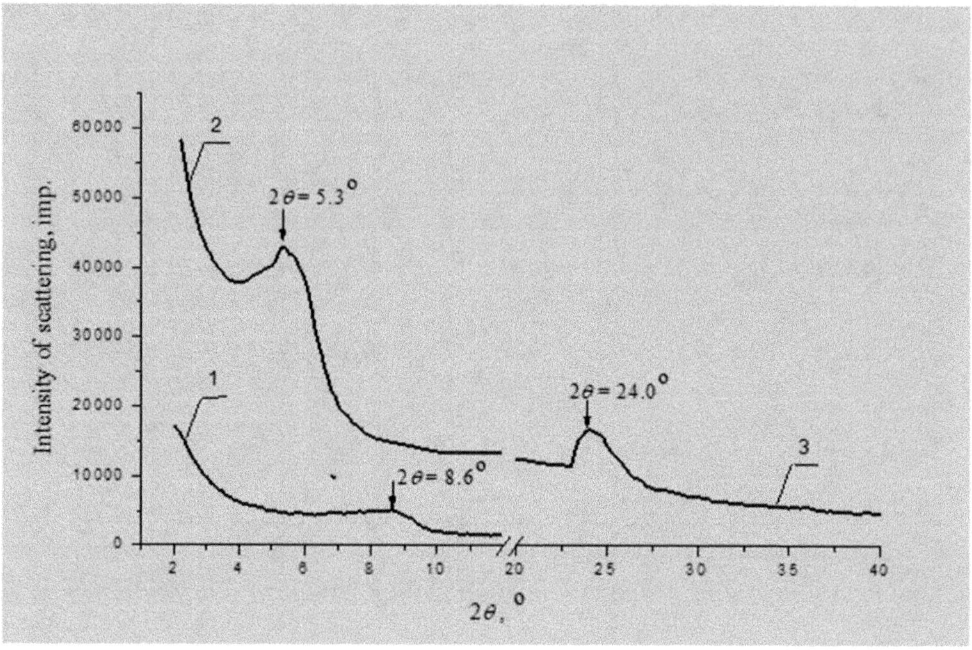

Figure 4. IR spectra of Quaternary ammonium salts

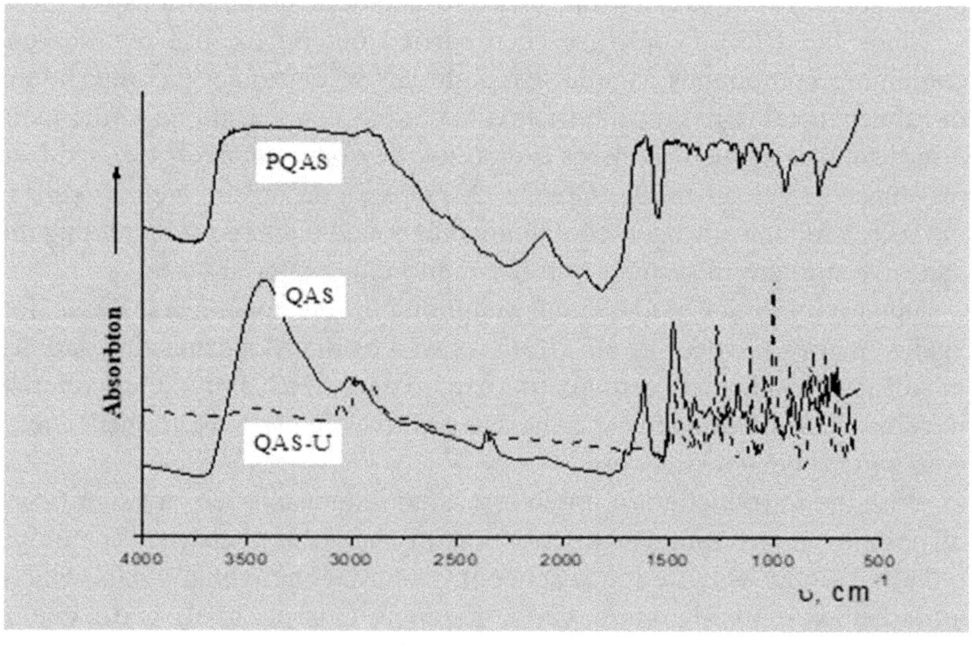

intercalation and exfoliation as a result of its modification and processing in polymer compositions.

Figure 5. IR spectra of modified QAS (1), basic montmorillonite (2) and modifying additive QAS (3)

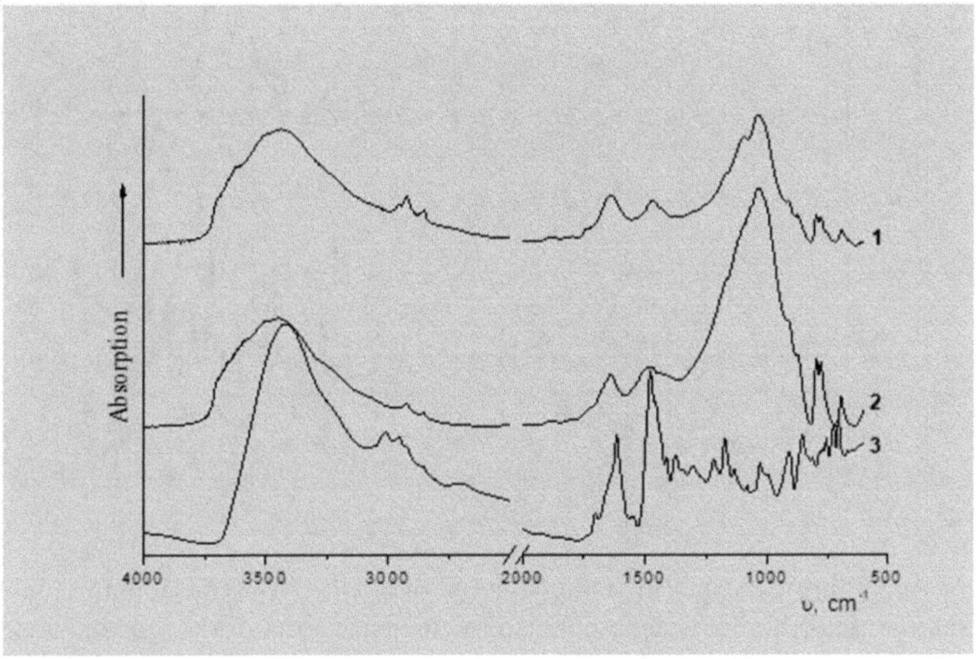

STUDY OF MONTMORILLONITE MODIFIED WITH QUATERNARY AMMONIUM SALTS IN STANDARD ELASTOMERIC COMPOSITIONS

In order to study the effect of modified Quaternary ammonium salts of montmorillonite on the formation of the overall complex properties of elastomeric compositions based on polydiene rubbers, its tests were performed as part of model elastomeric compositions.

The study of montmorillonite modified with polymeric ammonium salt PQAS was carried out in model elastomeric compositions based on SKMS-30 ARC in comparison with individual components of modified montmorillonite (montmorillonite base and PQAS) and their mechanical mixture 6 and 4.

Figure 6. IR spectra of modified QAS-U (1), basic montmorillonite (2) and modifying additive QAS-U (3)

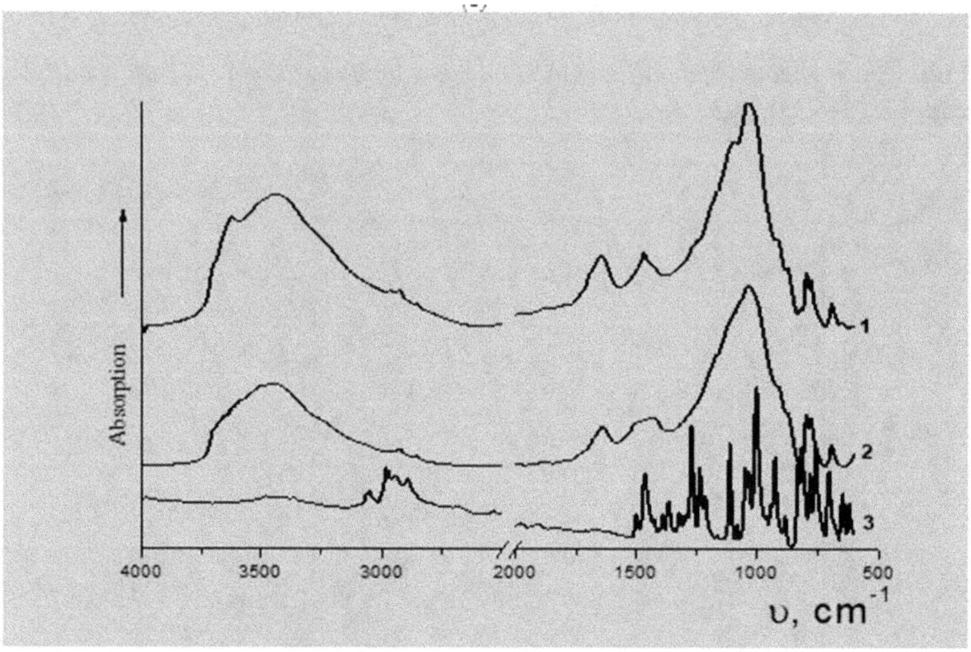

According to the data in Table. 4, the use of modified polymeric ammonium salt montmorillonite is accompanied by intensification of the vulcanization process: reducing the time to achieve the optimum vulcanization to 30% and increasing the vulcanization rate while maintaining a sufficient induction period at 155°C. The study of physical and mechanical characteristics of rubber showed that the optimal set of properties has a composition containing 3.0 parts by wt. modified with polymeric ammonium salt of montmorillonite. This rubber has a conditional stress at 300% elongation 30-32% higher than rubber with basic montmorillonite and rubber with a mechanical mixture (Mm + PQAS), and the tensile strength of this rubber is 42% higher than base montmorillonite and 32% compared to the mechanical mixture while maintaining the elastic characteristics of the rubber; rubber with modified polymeric ammonium salt montmorillonite has an advantage over other compositions in the complex of physical and mechanical characteristics under normal conditions and after aging (Table 4).

Thus to form the optimal set of properties of standard elastomeric compositions based on SKMS-30 ARC allows 3.0 part by wt. montmorillonite

Figure 7. IR spectra of modified PQAS (1), basic montmorillonite (2) and modified PQAS additive (3)

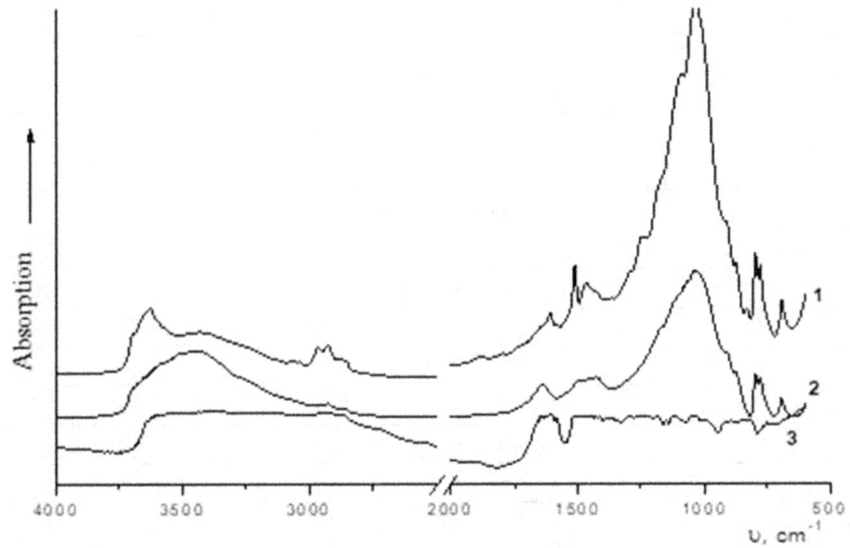

modified with polymeric ammonium salt. At this concentration, the rubbers have satisfactory vulcanizing properties and maximum elastic strength characteristics.

The study of the concentration dependence of the strength of the adhesive bond: breaker rubber - metal cord (4 L 30 NT) at close values of the conditional

Table 3. Results of differential thermal analysis of base and modified Quaternary ammonium salts of montmorillonite

Temperature, °C	Mm	Mm(QAS)	Mm(QAS-U)	Mm(PQAS)
	Mass loss, wt. %			
100	0.1	0.8	0.8	0.2
110	0.1	0.9	0.9	0.7
120	0.2	3.2	5.2	3.5
150	0.3	4.4	6.8	5.9
175	0.5	8.9	7.7	9.1
200	0.8	9.2	8,8	10.8

Table 4. Composition (parts by weight) and properties of standard elastomeric compositions based on SKMS-30 ARC (zinc oxide-5.0; stearic acid-1.5; sulfur-2.0; altax-3.0) with different research additives

Additive, characteristic	Experiment					
	1	2	3	4	5	6
Composition						
Mm	3.0	-	2.6	-	-	-
PQAS	-	0,4	0,4	-	-	-
Mm(PQAS	-	-	-	1.0	3.0	5.0
Technological properties						
ŋ, c. u. by Mooney	54	53	53	53	53	54
Vulcanization characteristics at 155°C						
t_s, min.	15.7	13.6	14.1	13.1	12.7	12.5
t_{C90}, min.	38.0	35.2	36.6	31.5	26.0	22.8
V_c, dH·m/min.	4.48	4.66	4.51	4.97	5.15	5.39
Physico-mechanical parameters						
f_{300}, MPa	1,5	1,6	1,5	1,6	2,0	1,9
f_p, MPa aging120 °C×12 h.	2.6 1.6	2.9 1.7	2.8 1.7	3.1 1.4	3.7 1.9	3.5 1.7
ε, % aging 120 °C×12 h	410 200	430 210	420 200	440 210	435 205	425 200
B, kN/m	8	9	9	9	10	10

Investigation of the effect of montmorillonite modified with Quaternary ammonium salts on the formation of adhesive bonds of rubber-metal cord systems

stress at 300% elongation (Fig. 8) showed that the optimal concentration of modified Quaternary ammonium salt (QAS) montmorillonite is 3.0 part by wt. At this concentration, the bond strength of the rubber with the metal cord is on a par with the rubber containing the ISS adhesion modifier, which is currently used in the manufacture of tires.

It was found that rubbers with a mechanical mixture of montmorillonite with Quaternary ammonium salts at concentrations of 1.0-7.0 pts. wt. do not reach the strength of the bond with the metal cord, characteristic of rubber with modified Quaternary ammonium salts montmorillonite. Individually introduced montmorillonite in concentrations of 1.0-7.0 pts. wt. leads to a decrease in the strength of the bond, the adhesive bond. Individually introduced Quaternary ammonium salt in concentrations of 1.0-7.0 pts. wt. also does not provide a high level of bond strength in the adhesive bond: breaker rubber - metal cord.

Figure 8. Concentration dependence of strength (bond strength - H-method) of the adhesive bond, connection: breaker rubber - metal cord (4 L 30 NT) (f_{300} ~ 14,5 ÷ 15,8 MPa)

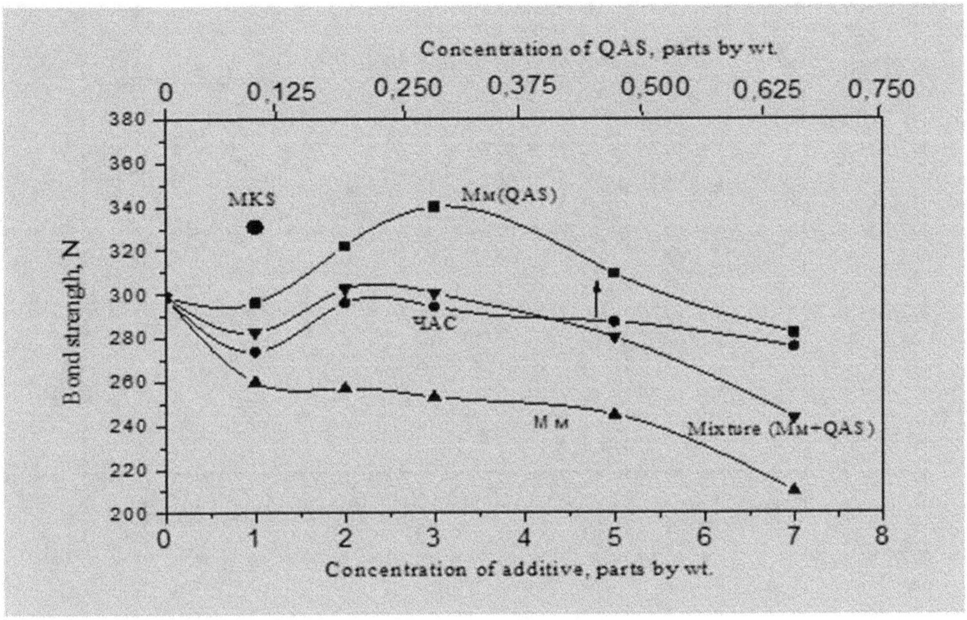

The study of the concentration dependence of the strength of the adhesive bond: breaker rubber containing modified QAS montmorillonite - metal cord at conditional stress at 300% elongation 14.5 - 16.0 MPa (Fig. 8) showed that the optimal concentration is 3, 0 wt.h.

At the optimal concentration, the strength of the adhesive bond of rubber containing modified QAS-U montmorillonite is 8% higher than that of rubber with ISS modifier. It was found that rubber with a mechanical mixture (montmorillonite + QAS-U), individual montmorillonite and QAS-U do not reach the adhesive strength of rubber with modified QAS-U montmorillonite.

The study of the concentration dependence of the strength of the adhesive connection: breaker rubber - metal cord (Fig. 9, 10) at close values of the stress at 300% elongation (14.5-15.6 MPa) showed that the optimal concentration of modified PQAS montmorillonite is 3.0 wt. .h. Rubber containing 3.0 wt.h. modified PQAS montmorillonite has a bond strength with metal cord 20% higher than rubber containing the ISS modifier (Fig. 10).

Figure 9. Concentration dependence of strength (bond strength, H-method - adhesive), connection: breaker rubber - metal cord (4 L 30 NT) (f_{300} ~ 14.5 ÷ 16.0 MPa)

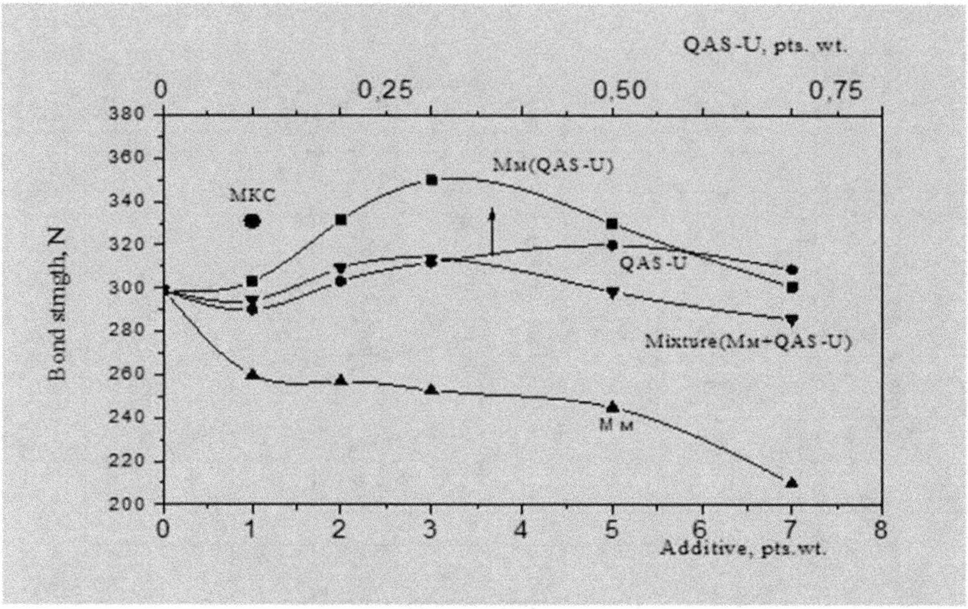

Comparative analysis of the strength of the adhesive bond: breaker rubber - metal cord (Fig. 11) showed that the maximum adhesion strength is characterized by rubber containing montmorillonite modified with PQAS.

The study of the concentration dependence of the strength of the adhesive bond after different types of aging (Fig. 12) showed that it is 3.0 pts. wt. modified PQAS montmorillonite allows to provide the maximum indicators of durability of adhesive connection. The bond strength of the rubber, which contains 3.0 pts. wt. modified PQAS montmorillonite, with a metal cord brand 4 L 30 NT 25-37% higher than rubber with a modifier of the ISS after various types of aging. Comparative analysis of the coefficients of efficiency of adhesive additives in the system: breaker rubber - metal cord showed that to increase the efficiency of the additives should be arranged in a row (Fig. 13): Mm (QAS)> Mm (QAS-U)> Mm (PQAS).

This sequence is preserved even after various types of aging (Fig. 14).

It should be noted that the efficiency factor Mm (PQAS) is 3 - 3.5 times higher than the industrial modifier ISS. Analysis of technological, vulcanization and physical and mechanical properties of breaker elastomeric compositions with modified Quaternary ammonium salts montmorillonite showed that the

Figure 10. Concentration dependence of strength (bond strength - H-method) of the adhesive connection: breaker rubber - metal cord (4 L 30 NT) (f300 ~ 14.5 ÷ 15.6 MPa)

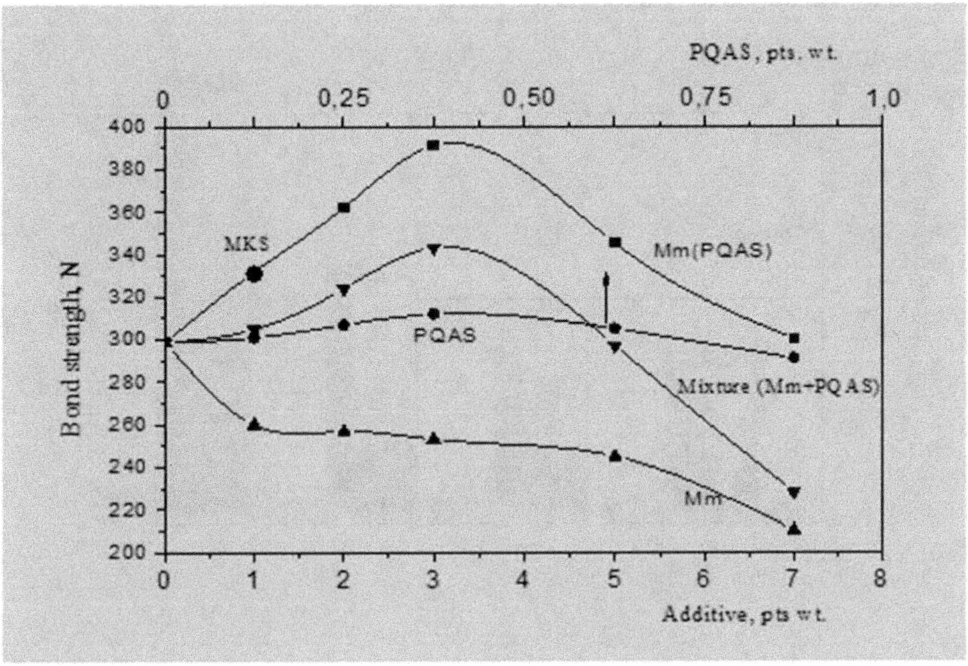

use of modified montmorillonite (Tables 5 - 7) is accompanied by an increase of 1.2 to 1.7 times the cohesive strength of rubber compounds. In particular, the maximum cohesive strength is inherent in rubber mixtures containing 3.0 to 5.0 pts. wt. modified with Quaternary ammonium salts of montmorillonite. With increasing concentration of modified montmorillonite from 1.0 to 7.0 pts. wt. there is an intensification of the process of sulfur vulcanization, which is expressed by a reduction in the time to reach the optimum of vulcanization and the time of onset of vulcanization; at its concentrations up to 5.0 pts. wt. reduction of the induction period is insignificant.

The optimal set of physical and mechanical properties has a rubber content of 3.0 pts. wt. modified with Quaternary ammonium salts of montmorillonite, which predominates in these properties of rubber with individual components of the modified form, rubber with a mechanical mixture of components and rubber with an MKS modifier. These benefits are observed both under normal test conditions and after aging.

Figure 11. Diagram of the comparison of the strength (bond strength, bond - H-method) of the adhesive connection: breaker rubber - metal cord (4 L 30 NT) (f_{300} ~ 14.5 ÷ 15.8 MPa)

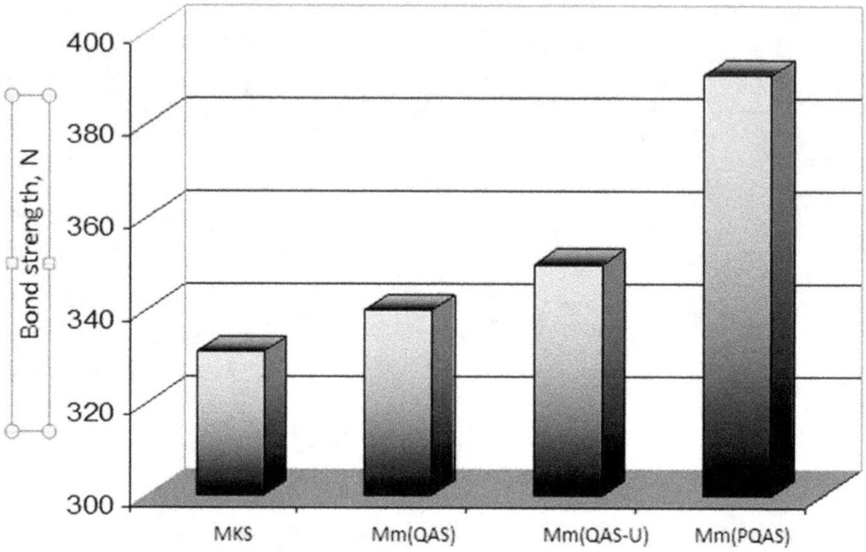

Thus, it was found that the most effective research additive is modified polymeric Quaternary ammonium salt montmorillonite, which, along with providing maximum strength of adhesive bond, breaker rubber - metal cord, allows to form an optimal set of technological, vulcanizing and physico-mechanical properties.

The study of the adhesion activity of montmorillonite modified with polymeric ammonium salt was carried out in elastomeric compositions based on NK for input control of the metal cord (Table 8). According to Table 8, the introduction of 3.0 and 5.0 pts. wt. modified with polymeric ammonium salt of montmorillonite slightly affects the plasticity and viscosity of rubber mixtures, but 60-67% increases the cohesion strength compared to the control composition and 26-29% compared to the MKS modifier. The rate of vulcanization of elastomeric compositions with modified polymeric ammonium salt montmorillonite is at the level of the ISS modifier, but it should be noted that the introduction of even 5.0 pts. wt. it reduces the time of the beginning of vulcanization less. The advantage of rubbers with modified polymeric ammonium salt montmorillonite is noted in terms of physical and

Figure 12. Concentration dependence of strength (bond strength - H-method) of the adhesive connection, connection: breaker rubber - metal cord (4 L 30 NT) after aging (f_{300} ~ 14.5 ÷ 15.6 MPa)

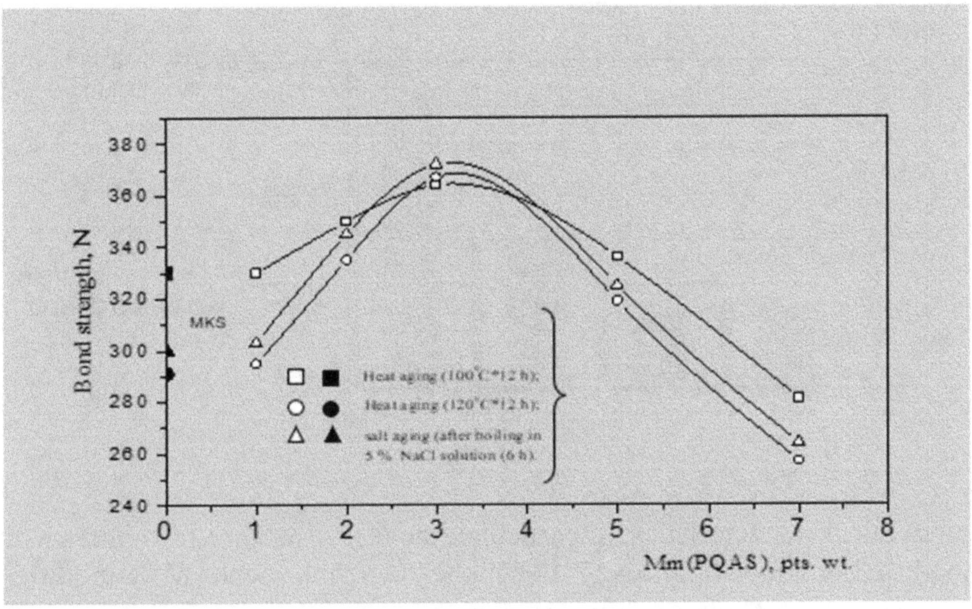

Figure 13. Diagram of comparison of the coefficients of efficiency of adhesive-active additives in the system: breaker rubber-metal cord (4 L 30 NT)

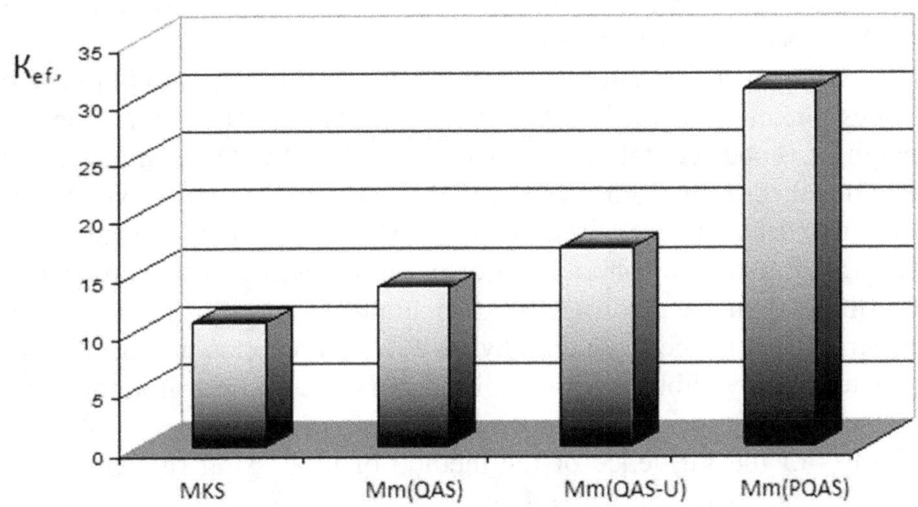

Figure 14. Diagram comparing the efficiency coefficients of adhesive-active additives in the system: breaker rubber-metal cord (4 L 30 NT) after aging

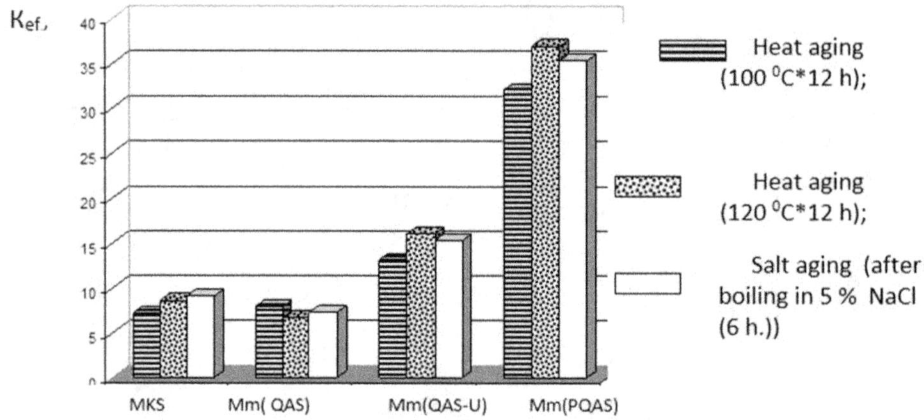

mechanical characteristics, in particular, the tear resistance of rubbers with modified montmorillonite is 17-19% higher than rubbers with MKS modifier.

The content of bound sulfur of rubber with modified polymeric ammonium salt montmorillonite is close to rubber with a modifier of the ISS, which may be the cause of increased adhesive bond in the rubber-metal cord system.

Conditional stresses of rubber with ISS modifier and modified polymeric ammonium salt montmorillonite are almost identical, so it is possible to compare their adhesive activity. The adhesion activity of modified montmorillonite in rubber-metal cord systems was studied using cords of grades 9 L 20/35 NT and 4 L 30 NT (Figs. 15 and 16).

According to Fig. 15 and 16 use 3.0 wt.h. modified PQAS montmorillonite provides maximum adhesive strength with both brands of cord. The bond strength of rubber containing 3.0 wt.h. modified PQAS montmorillonite, with a metal cord 9 L 20/35 NT 7-13% higher than the rubber with the ISS modifier under different test conditions. More significant advantages are defined at formation of communication of rubbers with a metal cord of 4 L 30 NT. In this system, the modified PQAS montmorillonite allows to increase the bond strength of rubber with cord by 14-21% compared to the ISS modifier. Characteristically, rubbers with modified PCHAS montmorillonite are quite stable to various types of aging, which also demonstrates its advantage.

Studies of the influence of the method of processing rubber mixtures on the formation of the general set of properties of breaker elastomeric

Table 5. Composition (parts by weight) and properties of breaker tires based on the combination of SKI-3 + NK (60 + 40) (diafen FP-2.0; stearic acid-2.0; styrene-indene resin-3.0 ; pine rosin-2.0; zinc oxide-8.0; carbon black N-339-60.0; oil PN-6 W-7.0; ground sulfur-1.7; polymer sulfur-5.7; sulfenamide Ts-1,0) santogard PVI-0,3) with different content of experimental additives

Additive, characteristic	Experiment								
	1	2	3	4	5	6	7	8	9
Composition									
Modifier MKS	-	1,0	-	-	-	-	-	-	-
Mm	-	-	2,7	-	2,7	-	-	-	-
QAS	-	-	-	0,3	0,3	-	-	-	-
Mm(QAS)	-	-	-	-	-	*1,0*	*3,0*	*5,0*	*7,0*
Technological properties									
Cohesive strength, MPa	0,69	0,87	0,68	0,84	0,75	1,03	1,12	1,10	0,96
Vulcanization characteristics at 155 °C									
t_s, min.	4,4	3,9	4,4	3,9	4,1	4,0	3,8	3,5	3,0
t_{C90}, min.	10,1	9,2	10,0	9,2	9,4	9,4	9,0	8,7	8,0
Physico-mechanical parameters									
f_{300}, MPa	13,0	14,6	14,2	14,4	14,3	14,6	15,3	15,9	16,4
f_p, MPa aging 120 °C×12 h	25,4 16,8	26,3 16,8	25,5 16,8	27,4 17,5	26,3 17,2	26,1 17,3	26,5 17,5	26,6 17,6	26,2 17,2
ε, % aging 120 °C×12 h.	495 300	500 295	495 280	495 280	490 275	490 290	485 280	470 270	480 275
B, kN/m	116	110	117	119	118	118	120	116	110
N (ε=150%), thousand c.	10,5	8,9	8,6	10,8	9,3	10,0	9,7	9,2	8,1

compositions based on SKI-3 and NK, as well as the formation of adhesive bonds in rubber-metal cord systems (Appendix B), showed that processing elastomeric compositions in rubber mixers with the best technological and physico-mechanical properties, as well as increased bond strength in the rubber-metal cord system under different test conditions compared to the compositions, the processing of which was carried out on laboratory rollers.

Together with CZL Rosava CJSC (Bila Tserkva) the possibility of replacing the cobalt-containing adhesion promoter (ISS modifier) with montmorillonite modified with Quaternary ammonium salts in elastomeric compositions based on diene rubbers (3) and SC P, because according to the test results, the experimental compositions on the set of properties were not inferior to serial rubbers. So, the replacement of 1.0 wt.h. ISS modifier by 3.0 wt.h. of montmorillonite modified with polymeric Quaternary ammonium salt (Mm (PQAS) in tires for tire breaker with sulfur vulcanization system allows to ensure a high level of physical and mechanical characteristics and increase the adhesive strength of the metal bond with rubber 4 L 30 by 25-37% compared with the cobalt-containing modifier.

Table 6. Composition (parts by weight) and properties of breaker tires based on the combination of SKI-3 + NK (60 + 40) (diafen FP-2,0; stearic acid-2,0; styrene-indene resin-3,0 ; pine rosin-2.0; zinc oxide-8.0; carbon black N-339-60.0; oil PN-6 W-7.0; ground sulfur-1.7; polymer sulfur-5.7; sulfenamide Ts-1,0) santogard PVI-0,3) with different content of experimental additives

Additive, characteristic	Experiment								
	1	2	3	4	5	6	7	8	9
Composition									
Modifier MKS	-	1,0	-	-	-	-	-	-	-
Mm	-	-	2,7	-	2,7	-	-	-	-
QAS	-	-	-	0,3	0,3	-	-	-	-
Mm(QAS)	-	-	-	-	-	*1,0*	*3,0*	*5,0*	*7,0*
Technological parameters									
Cohesive strength, MPa	0,69	0,87	0,68	0,89	0,77	1,05	1,2	1,15	0,92
Vulcanization parameters at 155 °C									
t_s, min.	4,4	3,9	4,4	3,7	3,9	3,8	3,5	3,1	2,7
t_{c90}, min.	10,1	9,2	10,0	8,9	9,2	9,2	8,7	8,1	7,4
Physical amd mechanical parameters									
f_{300}, MPa	13,0	14,6	14,2	14,5	14,4	14,5	15,8	16,3	17,0
f_p, MPa	25,4	26,3	25,5	27,5	26,0	26,2	27,3	26,8	26,5
aging 120 °C×12 h	16,8	16,8	16,8	17,6	17,3	16,9	17,6	17,3	16,7
ε, %	495	500	495	495	495	475	465	470	470
aging 120 °C×12 h	300	295	280	280	280	260	255	245	240
B, kN/m	116	110	117	119	118	120	125	111	104
N (ε=150%), thousand. c.	10,5	8,9	8,6	10,8	10,1	10,3	9,8	9,2	8,0

Comparing the chemical composition of the proposed salts, it should be noted that the salts QAS and PQAS have a significant number of identical elements, but QAS does not have a significant positive effect on adhesive strength, this suggests that hydroxyl groups in the structure of PQAS have a positive effect on adhesion bond in the system "rubber-brass metal cord".

Thus, montmorillonite modified with a polymeric Quaternary ammonium salt can be recommended as a promising adhesion promoter, which has an advantage over the cobalt-containing modifier both under normal conditions and after various types of aging. To ensure the optimal overall set of properties, it is recommended to use 3.0 part by wt. montmorillonite modified with polymeric Quaternary Ammonium Salt.

Investigation of the effect of montmorillonite modified with Quaternary ammonium salts on the formation of the adhesive bond of rubber-textile cord systems in the wedge belt element

In order to study the adhesion activity of montmorillonite modified with Quaternary ammonium salts, in the system "rubber-textile cord" were used industrial elastomeric compositions for stretching wedge belts based on a

Table 7. Composition (parts by weight) and properties of breaker tires based on the combination of SKI-3 + NK (60 + 40) (diafen FP-2,0; stearic acid-2,0; styrene-indene resin-3,0 ; pine rosin-2.0; zinc oxide-8.0; carbon black N-339-60.0; oil PN-6 W-7.0; ground sulfur-1.7; polymer sulfur-5.7; sulfenamide Ts-1,0) santogard PVI-0,3) with different content of experimental additives

Additive, characteristic	Experiment								
	1	2	3	4	5	6	7	8	9
Composition									
Modifier MKS	-	1,0	-	-	-	-	-	-	-
Mm	-	-	2,6	-	2,6	-	-	-	-
QAS	-	-	-	0,4	0,4	-	-	-	-
Mm(QAS)	-	-	-	-	-	*1,0*	*3,0*	*5,0*	*7,0*
Technological parameters									
Cohesive strength, MPa	0,69	0,87	0,70	0,80	0,72	0,95	1,05	1,10	1,00
Vulcanization parameters at 155 °C									
t_S, min.	4,4	3,9	4,4	4,0	4,2	4,0	4,0	3,9	3,2
t_{C90}, min.	10,1	9,2	10,0	9,4	9,7	9,4	9,2	9,0	7,4
Physical and mechanical parameters									
f_{300}, MPa	13,0	14,6	14,1	14,4	14,2	14,6	15,6	15,2	14,2
f_p, MPa aging 120 °C×12 h	25,4 16,8	26,3 16,8	25,8 17,0	27,2 17,4	26,5 17,2	25,9 18,2	27,0 19,1	26,6 18,7	26,2 17,4
ε, % aging 120 °C×12 h	495 300	500 295	495 280	495 285	485 280	510 300	505 295	505 285	490 270
B, kN/m	116	110	117	119	118	119	122	120	100
N (ε=150%), thousand c.	10,5	8,9	8,5	10,9	9,0	10,5	10,1	9,8	8,3

combination of rubbers SKMS-30 ARKM-15 and composition SKI-3 2.3) and textile anide cord brand A-10-2-3 T.

Equilibrium replacement of the adhesive-active system, which consists of white carbon black BS-100 (4.5 parts by weight) and modifier MP (2.0 parts by weight), modified with Quaternary ammonium salts montmorillonite leads to vulcanization of rubber compounds and, accordingly, reducing the complex properties of elastomeric compositions. The study of the concentration dependence of the influence on the formation of the general set of properties showed that the optimal set of properties have elastomeric compositions containing 3.0 wt.h. modified with Quaternary ammonium salts of montmorillonite.

The study of the adhesive activity of rubbers containing montmorillonite showed that the basic montmorillonite reduces the bond strength of the rubber with the textile cord during delamination (Fig. 17-19). Rubbers containing montmorillonite modified with Quaternary ammonium salts have a bond strength of 2.3 to 2.5 times higher than the textile cord in the wedge belt

Table 8. Composition (parts by weight) and properties of elastomeric compositions for input control of metal cord based on NC (phthalic anhydride - 0.5; zinc oxide - 5.0; stearic acid - 2.0; diafen FP - 2.0; pine rosin 3.0, special bitumen oil ASMG - 2.0, TV P-234 - 35.0, TV K-354 - 15.0, ground sulfur - 2.8, altax - 0.4, captax - 0.4) with different content of montmorillonite

Additive, characteristic	Experiment					
	1	2	3	4	5	6
Composition						
Modifier MKS	-	1,0	-	-	-	-
Mm	-	-	3,0	5,0	-	-
Mm(PQAS)	-	-	-	-	3,0	5,0
Technological parameters						
Ductility	0.55	0.54	0.55	0.54	0.56	0.57
Viscosity at 100°C, s.u. by Mooney	62	55	65	66	57	55
Cohesion strength, MPa	0.87	1.15	1.04	1.09	1.45	1.48
Vulcanization parameters at 155 °C						
ΔM, dN*m	23.4	27.1	24.1	24.2	26.5	26.8
t_s, min.	2.3	1.7	2.3	2.2	1.9	1.8
t_{c90}, min.	11.0	10.0	10.8	10.6	10.2	10.0
R_v, min.$^{-1}$	11.49	12.05	11.74	11.95	12.05	12.20
Physical and mechanical parameters						
f_{300}, MPa	8.6	10.2	9.0	9.0	9.4	9.8
f_p, MPa aging 120 °C×12 h.	25.4 17.2	27.3 17.2	25.8 17.6	25.8 17.3	27.6 19.1	27.3 18.6
ε, % aging 120 °C×12 h	580 365	560 350	555 340	545 340	570 370	560 350
B, kN/м	125	110	115	113	129	131
H, c. r.	65	67	65	67	63	64
S, %	30	28	30	30	30	30
Total Sulfur, %	1.78	1.96	1.80	1.80	1.88	1.89
Combined Sulfur, %	1.6	1.87	1.61	1.62	1.84	1.84
Free sulfur, %	0.18	0.09	0.19	0.18	0.04	0.05

element compared to rubber with base montmorillonite and 1.6 to 1.7 times higher than rubber without adhesive additives.

The bond strength with the textile cord of rubber with modified Quaternary ammonium salts of montmorillonite is on a par with rubber containing an adhesive-active system (BS-100 and MP modifier).

Figure 15. Diagram of comparable bond strength under different test conditions of metal cord brand 9 L 20/35 NT with rubbers containing test additives: 1 – without additive2 – 1,0 pts.wt. modifier MKS; 3 – 3,0 pts.wt. Mm; 4– 5,0 pts.wt. Mm; 5– 3,0 pts.wt. Mm(PQAS); 6– 5,0 pts.wt. Mm(PQAS)

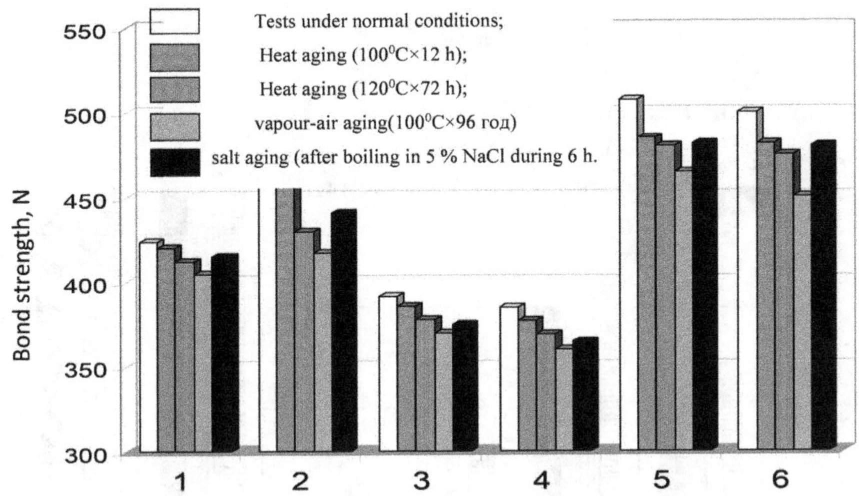

Comparative analysis of the bond strength of rubber with textile cord showed that the maximum value of strength has rubber containing modified polymeric ammonium salt montmorillonite (Fig. 20).

The study of technological, vulcanization and physico-mechanical properties of elastomeric compositions for the tensile layer of wedge belts (Tables 9-11) showed that modified with Quaternary ammonium salts montmorillonite provides improved technological properties in terms of viscosity, viscosity of rubber mixtures and base systems. BS-100 + modifier MP).

The use of montmorillonite modified with Quaternary ammonium salts is accompanied by intensification of sulfur vulcanization, increasing the crosslinking rate while maintaining the induction period, as well as increasing the relative degree of crosslinking, which in turn affects the strength characteristics of rubber. Elastomeric compositions containing modified QAS-U and PQAS montmorillonite have the optimal complex of vulcanization and physico-mechanical characteristics (under normal conditions and after aging).

Together with CZL CJSC "Inter-GTV" (Bila Tserkva) the possibility of replacing the modifying adhesion system white carbon black BS-100 + MP

Figure 16. Diagram comparing the strength of the connection under different test conditions of the metal cord brand 4 L 30 NT with rubbers containing experimental additives: 1 - without additives, 2 - 1.0 wt.h. MKS modifier; 3 - 3.0 pts.wt. Mm; 4– 5.0 pts.wt. Mm; 5– 3.0 pts.wt. Mm (PQAS); 6– 5.0 pts.wt. Mm (PQAS)

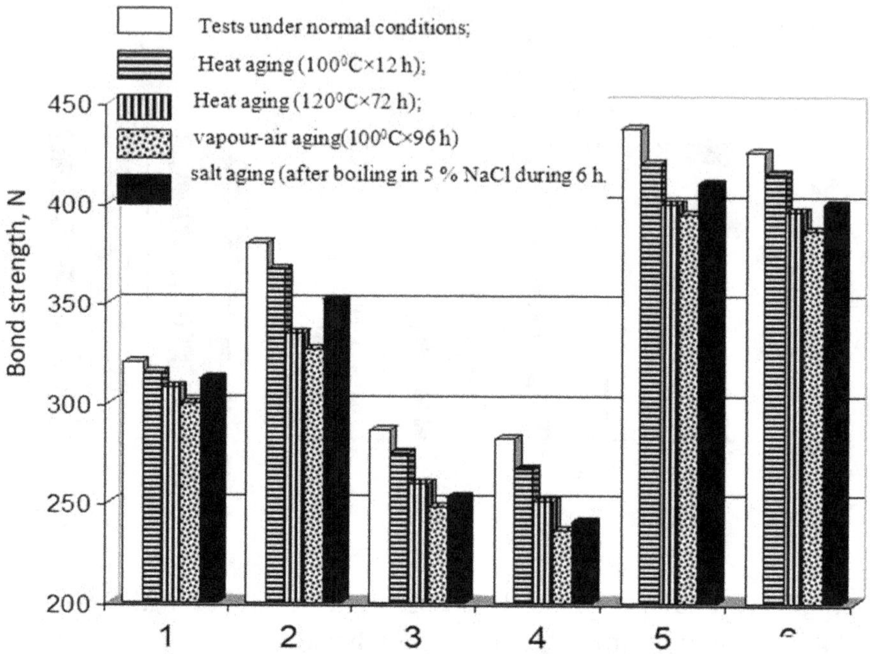

modifier with montmorillonite modified with Quaternary ammonium salts in elastomeric compositions based on diene rubbers 15-SCM and SKI-3) for the tensile layer of V-belts, because according to the test results, the test compositions are not inferior to serial rubbers in terms of the set of properties. Yes, the replacement of 4.5 parts by weight white carbon black BS-100 and 2.0 wt.h. MP modifier by 3.0 wt.h. of montmorillonite modified with Polymeric Quaternary ammonium salt (Mm (PQAS)) in rubbers for the sulfur vulcanization tensile layer allows to provide a high level of elastic-strength characteristics and adhesive strength of rubber-to-rubber bonds. -10-2-3 T in stratification.

Thus, montmorillonite modified with Quaternary ammonium salts can be used as a promising promoter of rubber adhesion to textile cord. The most positive results in the formation of the overall set of properties of elastomeric compositions for the tensile layer of wedge belts and their connection with the textile cord are achieved using 3.0 pts.wt. montmorillonite modified with polymeric Quaternary ammonium salt.

Figure 17. Diagram of comparison of bond strength for the wedge belt element: tensile layer rubber - textile cord (A-10-2-3 T)

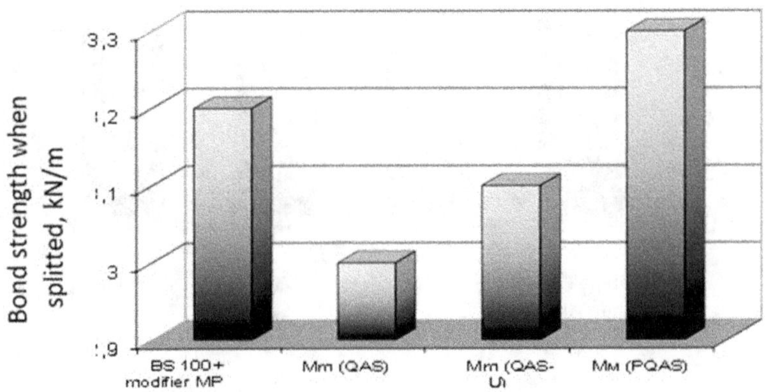

Figure 18. Diagram of comparison of the strength of the bond for the element of the V-belt: rubber tensile layer - textile cord (A-10-2-3 T)

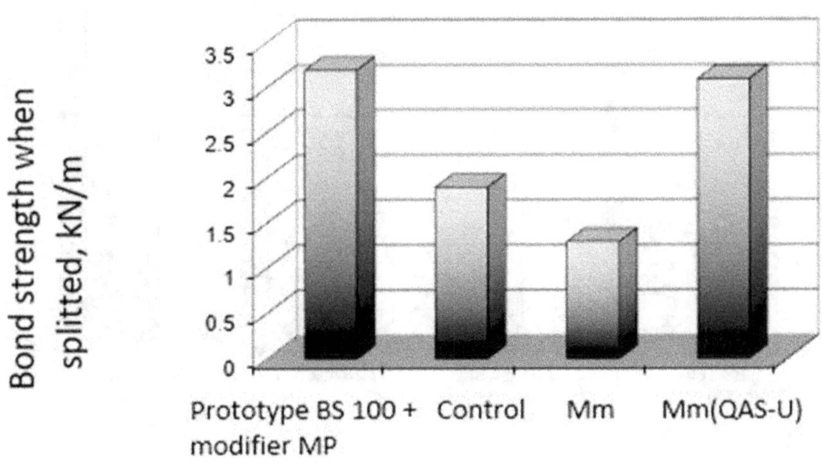

Figure 19. Diagram of comparison of the strength of the bond for the element of the V-belt: rubber tensile layer - textile cord (A-10-2-3 T)

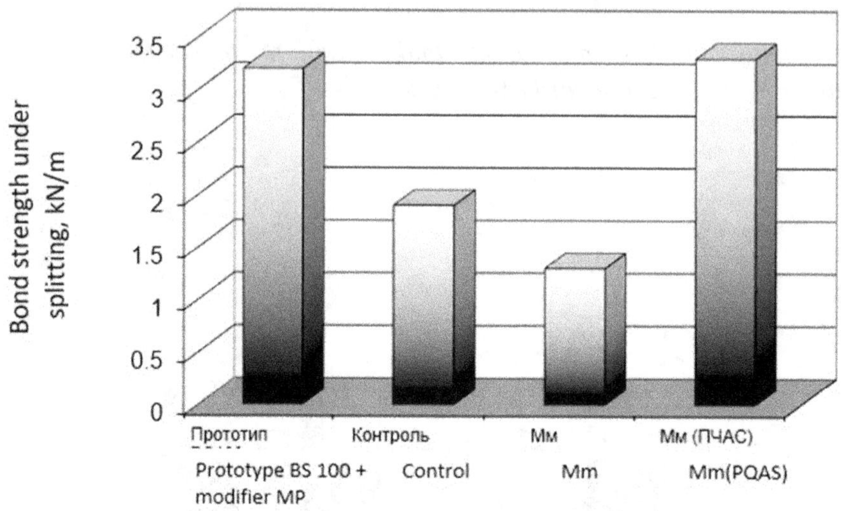

Figure 20. Diagram of comparison of bond strength for the wedge belt element: rubber of the tensile layer - textile cord (A-10-2-3 T)

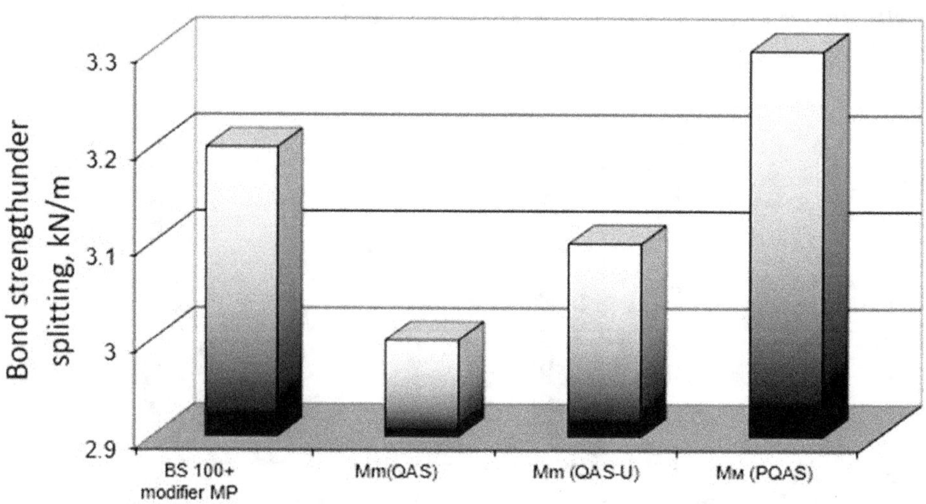

Table 9. Composition (parts by weight) and properties of rubbers based on the combination SKMS-30 ARKM-15 + SKI-3 (60 + 40) (white zinc-3,0; acetonanil H-1,5; diafen FP-1, 0; TV P-772-125,0; oil I-20-14,34; rosin-7,3; BS-100-4,5; phthalic anhydride-1,5; composite ABO-4-20,0 ; sulfenamide C-2,0; sulfur-2,0) for the tensile layer of the V-belt with various research additives

Additive,parameter	Experiment		
	1	2	3
Composition			
BS-100	4,5	-	-
Modifier MP	2,0	-	-
Mm	-	3,0	-
Mm(QAS)	-	-	3,0
Technological properties			
ŋ, c. u. by Moonet	72	72	68
Vulcanization characteristics at 155⁰C			
ΔM, dN*m	51,46	48,18	52,00
t_s, min.	12,5	16,9	12,0
R_V, min^{-1}	11,29	9,57	11,95
Phusical and mechanical properties			
f_p, MPa aging 120⁰C×12 h aging 120⁰C×24 h	14,2 13,8 13,4	11,1 10,5 9,0	13,8 13,2 12,8
ε, % aging 120⁰C×12 h aging 120⁰C×24 h	450 280 230	505 310 270	460 290 250
B, kN/m	40	33	38

Based on the studies, the optimal concentration of use in standard elastomeric compositions based on SKMS-30 ARC is 3.0 pts. wt. montmorillonite modified with polymeric Quaternary ammonium salt. At this concentration, the rubbers have satisfactory vulcanizing properties and maximum elastic strength characteristics.

Montmorillonite modified with polymeric Quaternary ammonium salt can be recommended as a promising adhesion promoter in rubber-metal cord systems, which has an advantage over the cobalt-containing modifier ISS, and rubber-textile systems, which has an advantage over the modifier B-adhesion modifier. To ensure the optimal overall set of properties of industrial elastomeric compositions, it is recommended to use 3.0 pts. wt. montmorillonite modified with polymeric Quaternary ammonium salt.

Table 10. Composition (parts by weight) and properties of rubbers based on the combination of SKMS-30 ARKM-15 + SKI-3 (60 + 40) (white zinc-3,0; acetonanil H-1,5; diafen FP-1, 0; TV P-772-125,0; oil I-20-14,34; rosin-7,3; BS-100-4,5; phthalic anhydride-1,5; composite ABO-4-20,0 ; sulfenamide C-2,0; sulfur-2,0) for the tensile layer of the V-belt with various research additives

Additive,parameter	**Experiment**		
	1	**2**	**3**
Composition			
BS-100	4,5	-	-
Modifier MP	2,0	-	-
Mm	-	3,0	-
Mm(QAS-u)	-	-	3,0
Technological properties			
ŋ, c. u. by Mooney	72	72	69
Vulcanization characteristics at 155^{0}C			
ΔM,dN*m	51,46	48,18	55,30
t_s, min.	12,5	16,9	11,3
R_v, min.$^{-1}$	11,29	9,57	13,24
Physical and mechanical characteristics			
f_p, MПa aging 120^{0}C×12 h aging 120^{0}C×24 h	14,2 13,8 13,4	11,1 10,5 9,0	14,7 14,0 13,6
ε, % aging 120^{0}C×12 h aging 120^{0}C×24h	450 280 230	505 310 270	450 275 245
B, kN/m	40	33	44

Investigation of montmorillonite and Quaternary ammonium salts as individual components of polydiene-based elastomeric compositions

When using modified montmorillonite in rubber compounds, the reactivity of all its components should be taken into account. Therefore, the study investigated montmorillonite in concentrations from 1.0 to 100.0 pts. wt. and Quaternary ammonium salts as active components of sulfur vulcanizing systems of elastomeric compositions based on various diene rubbers.

Table 11. Composition (parts by weight) and properties of rubbers based on the combination SKMS-30 ARKM-15 + SKI-3 (60 + 40) (white zinc-3,0; acetonanil H-1,5; diafen FP-1, 0; TV P-772-125,0; oil I-20-14,34; rosin-7,3; BS-100-4,5; phthalic anhydride-1,5; composite ABO-4-20,0 ; sulfenamide C-2,0; sulfur-2,0) for the tensile layer of the V-belt with various research additives

Additive, parameter	Experiment		
	1	2	3
	Composition		
BS-100	4,5	-	-
Modifier MP	2,0	-	-
Mm	-	3,0	-
Mm(PQAS)	-	-	3,0
	Technological properties		
η, c. u. by Mooney	72	72	71
	Vulcanization characteristics at 155°C		
ΔM, dN*m	51,46	48,18	55,00
t_s, min.	12,5	16,9	11,8
R_V, min.$^{-1}$	11,29	9,57	12,98
	Physical and mechanical parameters		
f_p, MPa aging 120°C×12 h aging 120°C×24 h	14,2 13,8 13,4	11,1 10,5 9,0	14,5 13,9 13,7
ε, % aging 120°C×12 h aging 120°C×24 h	450 280 230	505 310 270	450 280 240
B, kN/m	40	33	45

EVALUATION OF MONTMORILLONITE AS A COMPONENT OF ELASTOMERIC COMPOSITIONS BASED ON DIENE RUBBERS

It is known that the place of extraction (deposit), the depth of extraction of minerals has a significant impact on the properties of natural ingredients of elastomeric compositions. Therefore, it is necessary to conduct a set of studies to study the effect of unmodified montmorillonite on the overall set of properties of elastomeric compositions.

Evaluation of the strengthening effect of montmorillonite and traditional fillers in elastomeric compositions based on rubbers SKI - 3, SKMS - 30 ARC and SKN-26 M was performed on a complex indicator - gain (I3) taking into account the level of conditional stress at elongation and tear resistance. Thus, according to the level of rubber gain (Fig. 21) on the reinforcing effect of the tested fillers should be placed in a row: TV P-803> kaolin> montmorillonite> chalk> zeolite. The reinforcing effect of montmorillonite, depending on the rubber base of the composition increases in the following sequence: SKI-3 <SKN-26 M <SKMS-30 ARC. The rubber base of rubber also affects the dependence of the gain index on the content of montmorillonite (Fig. 22). Thus, in vulcanizates based on SKN-26 with an increase in the content of montmorillonite from 25 to 100 wt.h. the gain index also increases. In tires based on SKMS-30 ARC there is an extreme dependence for the gain index with a maximum at 50 pts. wt. montmorillonite. With increasing content of montmorillonite in the studied range in tires based on stereoregular cis-1,4-polyisoprene, its enhancing effect decreases.

Thus, the study of montmorillonite as a filler in elastomeric compositions based on carbochain diene rubbers shows that the maximum activity of montmorillonite is found in elastomeric compositions based on butadiene-methylstyrene rubber. The optimal concentration of montmorillonite should be considered a concentration of up to 50 pts. wt. per 100 pts. wt. rubber.

It should be noted that with increasing concentration of montmorillonite from 1 to 10 pts. wt. there are differences in the process of sulfur vulcanization of standard elastomeric compositions based on SKMS-30 ARC in comparison with traditional mineral ingredients of elastomeric compositions. With the introduction of 10 pts. wt. montmorillonite significantly intensifies the process of sulfur vulcanization (Fig. 23), but the expected qualitative improvement of the properties of elastomeric compositions does not occur. However, it should be noted the general trend of increasing the elastic strength characteristics of tires containing 3.0 and 5.0 pts. wt. montmorillonite (Table 13)

It should be noted that the effect of montmorillonite on the vulcanization process is most likely due to its elevated pH of the aqueous suspension compared to traditional mineral additives of elastomeric compositions.

Thus, it was found that montmorillonite has the maximum activity in elastomeric compositions based on butadiene-methylstyrene rubber. It actively influences the process of sulfur vulcanization, but almost does not change the physical and mechanical characteristics of rubber.

Figure 21. Augmentation ratio (I) of standard rubbers based on different rubber substances vs. the filler type

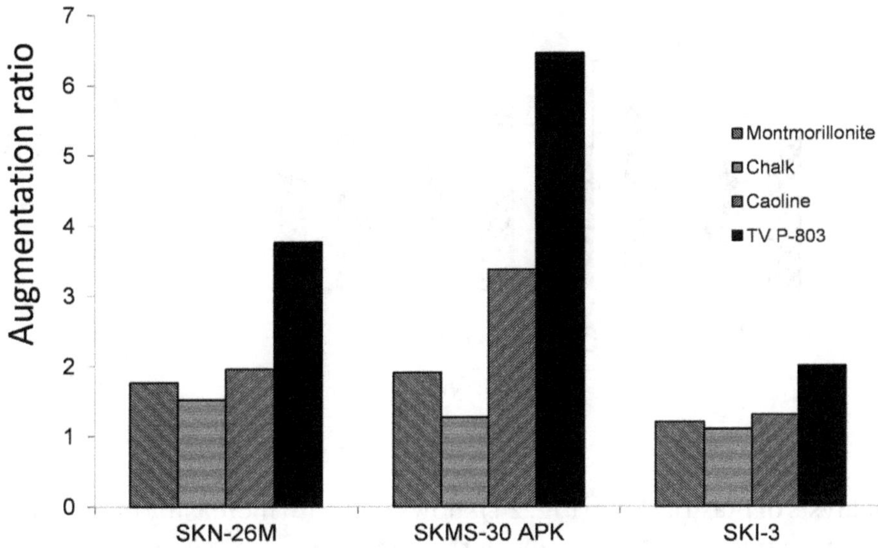

INVESTIGATION OF THE EFFECT OF MONTMORILLONITE MODIFIED WITH QUATERNARY AMMONIUM SALTS ON THE RHEOLOGICAL PROPERTIES OF ELASTOMERIC COMPOSITIONS

It is known that layered silicates are capable of destroying their structure (delamination) by mechanical action on them, especially for montmorillonite. The bundling ability of montmorillonite is used to obtain nanostructured composite materials, but it was found (Baki et al., 2022) that only after increasing the interlayer space in the structure of montmorillonite as a result of its modification aluminosilicate is able to nanoscale delamination. In this regard, it was of interest to study modified with Quaternary ammonium salts montmorillonite in the composition of elastomeric matrices in the plastic recorder "Brabender", where it is possible to easily implement changes in the mixing conditions of the elastomeric composition.

Study of base and modified polymeric Quaternary ammonium salt of montmorillonite in concentrations up to 10.0 wt.h. carried out in a matrix of natural rubber (NC) at temperatures of 80, 90, 100 0C and the speed of rotation of the rotors of the plastic cord 20, 30, 40 rpm

Figure 22. Augmentation ratio (I) of rubbers based on different rubber substances on the content of montmorillonite (C)

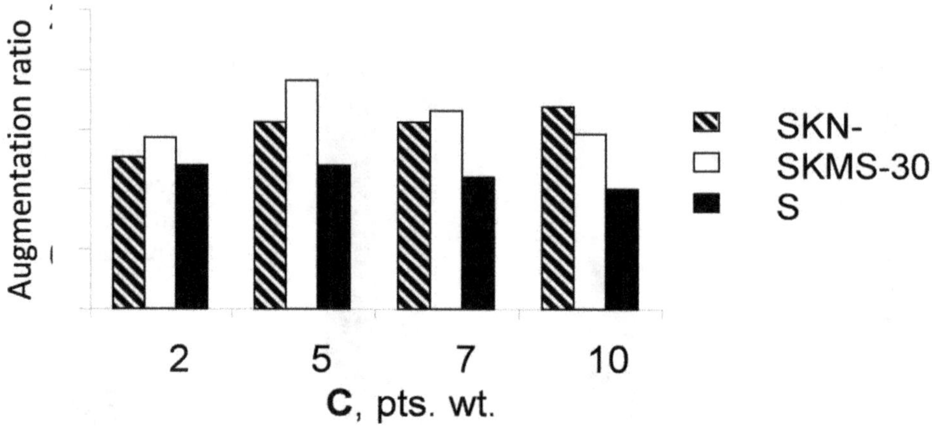

Analysis of plastograms (Fig. 24) shows that with increasing concentration of base and modified polymeric Quaternary ammonium salt of montmorillonite to 10.0 pts. wt. the value of the torque (M) increases to 60%. The torsional moments of the compositions filled with Mm (PQAS) are »40% higher than

Figure 23. Kinetic curves of sulfur vulcanization at 155°C model elastomeric compositions based on SKMS 30 ARC with an equilibrium concentration (10.0 parts by weight) of various mineral additives

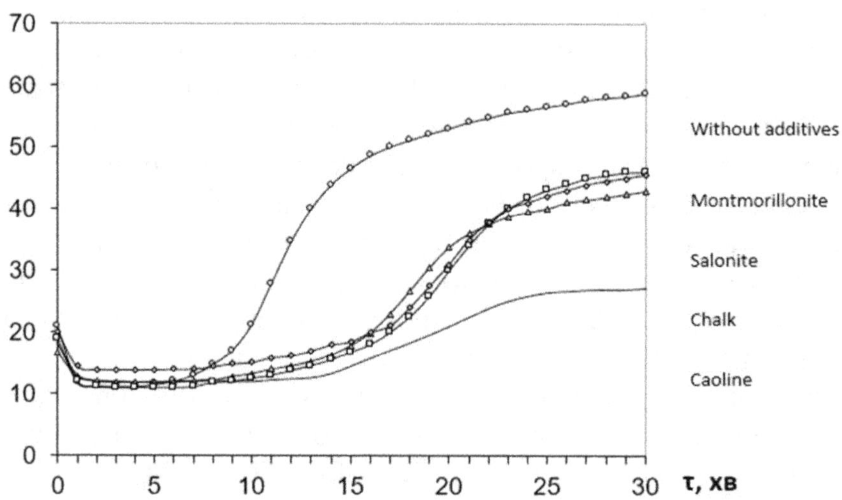

Table 12. Properties of standard elastomeric compositions based on SKMS-30 ARC with different concentrations of montmorillonite

Parameters	Bentonite, pts.wt.				
	-	1,0	3,0	5,0	10,0
Physical and mechanical parameters of rubber					
f_{300}, MPa	1,3	1,3	1,5	1,5	1,6
f_p, MPa in.cond. 120 °C ×12 h	2,1 1,4	2,5 1,5	2,6 1,6	2,8 1,7	2,5 1,4
B, kN/m	8	8	8	9	9
ε, % in. cond. 120 °C ×12 h.	455 215	420 190	410 200	410 200	400 180
H, c.u.	40	40	43	45	47
S, %	54	52	50	50	49

the torsional moments of the compositions filled with basic montmorillonite at equal concentrations. This may be due, firstly, to the destruction of the layered structure Mm (PQAS) and better dispersion in the rubber matrix, and secondly, the interaction between the organic and mineral components of the elastomeric composition with the formation of mineral rubber gel. It is established that with increasing rotational speed of rotors at identical temperatures the values of torsion moments of compositions with basic and modified PQAS montmorillonite at equilibrium filling decrease. Naturally, the values of torque decrease with increasing temperature of the material in the chamber of the plastic recorder. Accelerating the rotation of the rotors and lowering the temperature leads to an increase in the difference (ΔM) between the values of the torques of the compositions with modified PQAS and basic montmorillonite at the same concentration. This is especially confirmed with the introduction of test compounds from 3.0 to 10.0 pts. wt. The value (ΔM), determined by Eq.(1) as for compositions containing mineral compounds 5.0 pts.wt. (numerator) and 10.0 pts.wt. (denominator), presented in Table 13.

$$\Delta M = M_{B/PQAS} - M_{B/bas}, \text{ at } v = const; T = const; n = const, \quad (1)$$

where $M_{B/PQAS}$ is the torque of the composition containing B/PQAS, N*m;

$M_{B/bas}$ is the torque of the composition containing B/bas, N*m;
v refers to the content of the experimental additives in the compositions, pts. wt;

Table 13. The value of the difference in torque of the compositions with different content of base and montmorillonite with modified PQAS

Temperature of the composition, °C	Rotor speed, rev/min.		
	20	30	40
	ΔM, N*m		
80	8,9 / 12,6	12,4 / 13,7	17,3 / 18,3
90	5,5 / 11,7	8,3 / 12,4	8,9 / 16,8
100	2,7 / 4,2	4,4 / 8,0	5,6 / 10,9

Note: the numerator presents the values of ΔM compositions containing 5.0 pts.wt. mineral compounds; in the denominator - 10 pts.wt. mineral compounds.

T is the temperature of the composition, K;
n is the speed of rotation of the rotor of the plasticorder, rpm.

The maximum difference of torques have the experimental compositions at the minimum temperature (80°C) and the highest speed of rotation of the rotors (40 rpm), ie under conditions of maximum shear forces. This dependence is in favor of the destruction of the layered structure Mm (PQAS) with increasing shear forces, and in the polymer matrix is likely to form an inorganic nanophase. Thus, irreversible mechanisms in rubbers with layered minerals are associated with the sequential destruction of the layered structure of the modified mineral with increasing shear forces.

Analysis of plastograms (Fig. 25) of rubber SKN-18 with montmorillonite modified with different QAS showed that the introduction into the rubber of both basic and modified montmorillonite is accompanied by increased torque compared to the composition without additives. Compositions with Mm (PQAS) have the highest values of torque, especially at 70 °C. At this temperature, the torsional moment of the composition with Mm (PQAS) is 30% higher compared to the composition without additives, 18% as compared to the basic montmorillonite and 10% -compared with the modified QAS-U montmorillonite. This trend probably results from increasing the contact surface Mm (PQAS) with rubber in the destruction of the structure of modified

Figure 24. Influence of the content of basic (◊, Δ, ○) and modified PQAS (♦, ▲, ●) montmorillonite on the values of torques (M) of NC-based compositions at rotational speeds of rotors of plastic recorder: 20 rpm and 40 rpm (b) and different temperatures of the material in the chamber¨♦, ◊, - temperature of material 80 ^{0}C; ▲, Δ - temperature of material 90 ^{0}C; ●○ – temperature of material 100 ^{0}C.

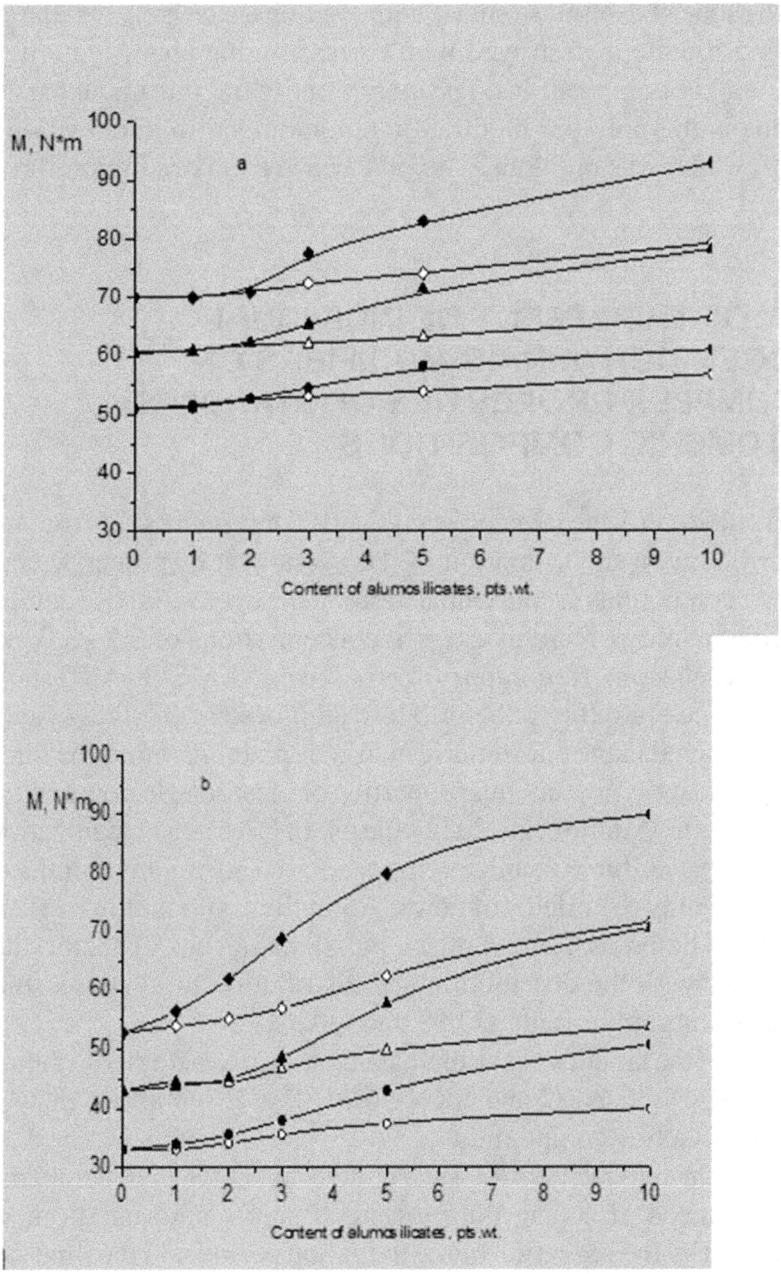

PQAS montmorillonite, as well as increasing the degree of interaction of organic and mineral components of the elastomeric matrix with the formation of mineral rubber gel.

Thus, the results of these tests indicate that with increasing shear forces in the process of mixing montmorillonite modified with Quaternary ammonium salts with rubber-based elastomeric compositions, it is likely that the structure of montmorillonite is destroyed with corresponding changes in rheological properties of the compositions. The most prone to fracture under harsh mixing conditions is the polymer-modified ammonium salt montmorillonite (mm (PQAS)), which was previously found to have a maximum interlayer space (1.67 nm)].

STUDY OF THE EFFECT OF INDIVIDUAL NITROGEN-CONTAINING MODIFIERS ON THE COMPLEX PROPERTIES OF STANDARD ELASTOMERIC COMPOSITIONS

In order to determine the effect of all components of the modified montmorillonite in the formation of the overall set of properties, nitrogen-containing compounds as individual ingredients of elastomeric compositions were studied. Study of amino esters at concentrations of 0.2 - 3.0 pts. wt. as part of model elastomeric compositions based on SKMS-30 ARC showed that when introduced into the rubber mixture aminoesters crystallize and are not evenly distributed in the elastomeric matrix, remaining individual inclusions, which significantly impairs the properties of elastomeric compositions.

It is known de (Araújo et al., 2022), the use of QAS as a vulcanization-active additives in the sulfur vulcanization system, in this regard, conducted a set of studies to study the effects of new QAS in the composition of elastomeric compositions based on diene rubbers. PQAS has an unsatisfactory degree of compatibility with the distribution and distribution in it, so we studied the Quaternary ammonium salts - QAS and QAS-U.

Analysis of the kinetics of sulfur vulcanization of elastomeric compositions based on SKMS-30 ARC and SKN-26 M (Fig. 26) showed that QAS are vulcanization-active compounds.

Equilibrium replacement of stearic acid by QAS is accompanied by a more pronounced effect on the process of sulfur vulcanization, which is demonstrated by the reduction of the induction period and the time to achieve

Figure 25. Plastigrams of SKN-18 with 5,0 pts.wt. montmorillonite

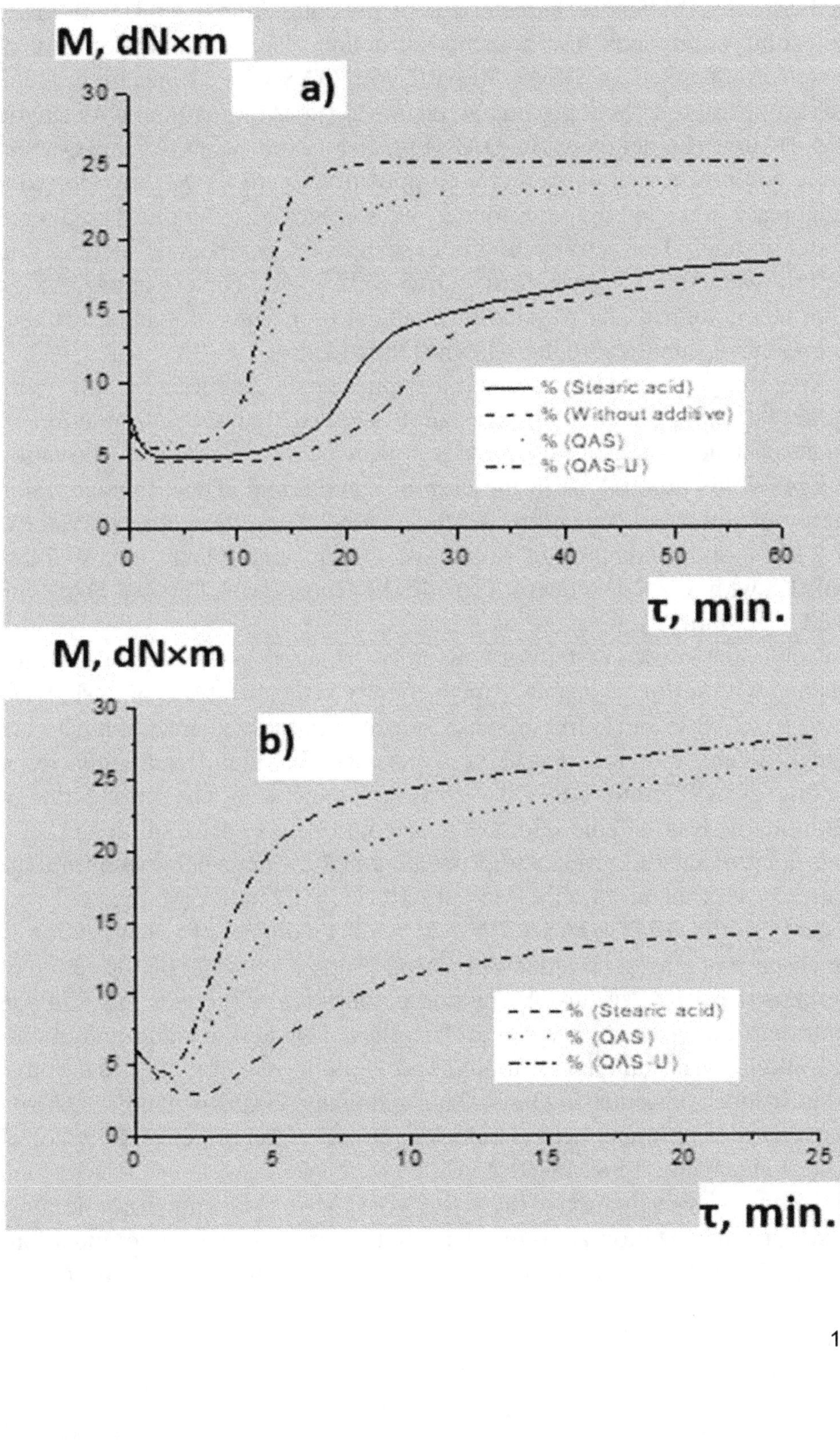

the optimum vulcanization. The maximum activity in compositions based on the studied rubbers is characterized by the compound QAS-U with lower molecular weight and less branched structure. The vulcanization rate of compositions based on SKMS-30 ARC with QAS-U is 6 times higher than the composition without organic activators, 5 times higher than with stearic acid, 1.4 times higher than with QAS (Fig. 26 a). The use of QAS and QAS-U in the composition of elastomeric compositions with SKN-26 M provides an increase in the maximum torque by an average of 2 - 2.5 times compared to stearic acid. The activity of the experimental compounds increases in the following series: stearic acid <QAS <QAS – U. The effect of QAS on other technological and physico-mechanical properties of standard rubber compounds is identical to the known stearic acid.

The study of the effectiveness of Quaternary ammonium salts with equimolar replacement of stearic acid, showed that the introduction of Quaternary ammonium salts there is a decrease in the effective activation energy of this process, as in the case of stearic acid. Thus, the activation energy of sulfur vulcanization SKMS-30 ARC without organic activator is 97.2 kJ / mol; introduction of 1.5 pts. wt. stearic acid reduces it to 89.7 kJ / mol; 1.3 wt.h. QAS-U reduces it to 74.6 kJ / mol, and 3.2 pts.wt. QAS - up to 63.3 kJ / mol.

Unlike the known system of activators “zinc oxide-stearic acid”, which together with sulfur is not able to form effective vulcanization bonds (Fig. 27, curve 6), QAS-B and sulfur are able to form a liquid vulcanization network at low vulcanization rate (Fig. 27, curve 5), which is significantly inferior to the thiazole accelerator DBTD - altax (Fig. 27, curve 3). The introduction of traditional activators - zinc oxide and stearic acid in the system “sulfur-QAS-U” does not provide a significant improvement in the kinetic parameters and the degree of vulcanization of SCMS-30 ARC (Fig. 27, curve 4). And only the system “sulfur-DBTD-QAS-U” provides a deep degree of vulcanization in the absence of almost no induction period (Fig. 27, curve 2) on the action of the known system “sulfur-DBTD-zinc oxide” (Fig. 27, curve 1) . Thus, in contrast to the claims about the high efficiency of Quaternary ammonium salts with alkyl, aryl and alkylaryl groups as accelerators of sulfur vulcanization of natural rubber, presented in Zhong, Zhu, and Wang (2005), we studied QAS-U based on urotropin and benzyl chloride. However, an effective activator of sulfur vulcanization SKMS-30 ARC.

Comparative evaluation of the activity of QAS-U with known accelerators of sulfur vulcanization at equimolar amounts in elastomeric compositions based on SKI-3 showed that at vulcanization temperatures of 165 and 175 °C

Figure 26. Kinetics of sulfur vulcanization at 153 0C standard elastomeric compositions based on SKMS-30ARC (a) and SKN-26M (b) with different surfactants

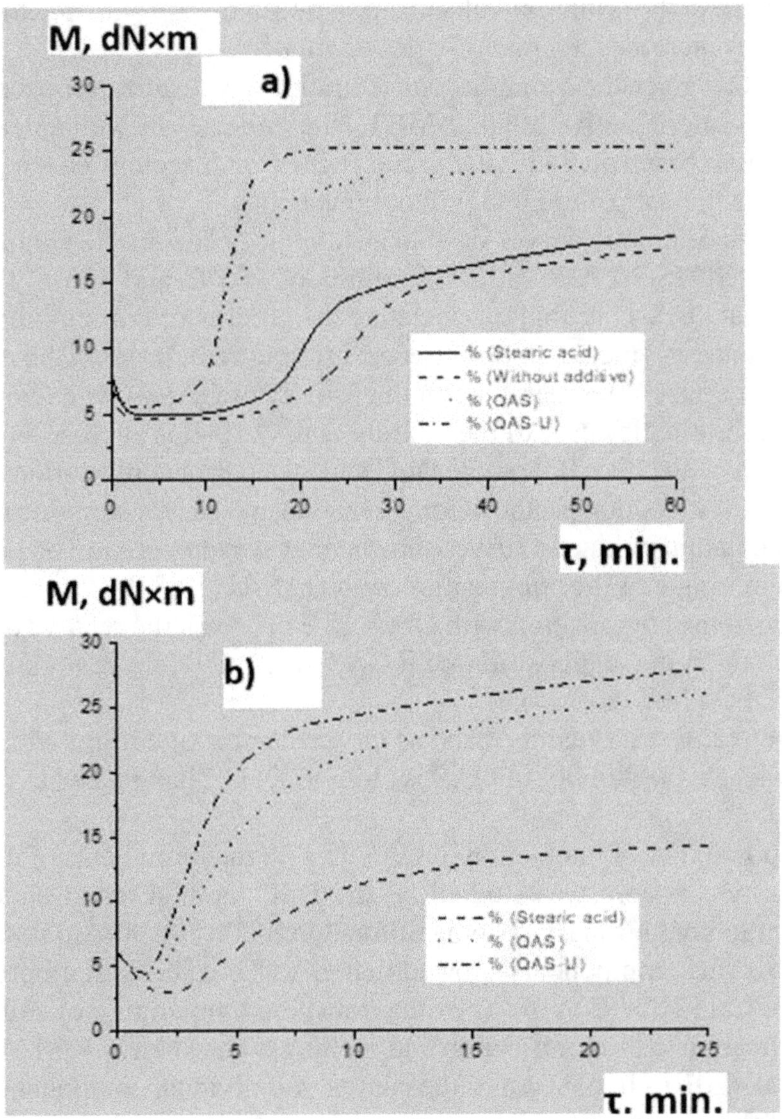

QAS-U is more active than 2-mercaptobenzthiazole (2-MBT), but inferior diphenylguanidine (DFG) (Fig. 28). The nature of the kinetic curve with QAS-U is identical to the curve with 2-MBT, ie the reversal process is not observed. As a result of calculating the kinetic curves, it was found that at a temperature of 165°C the highest value of the maximum torque has a composition with QAS, this value is 15% higher than the composition with 2-MBT. At a temperature of 175 °C, the introduction of QAS-U reduces the time to achieve the optimum vulcanization by 1.5 times compared to 2-MBT and 2 times increases the rate of vulcanization.

Therefore, given the equimolar concentration of vulcanization accelerators QAS-U is more active than 2-MBT, the process of vulcanization of compositions based on SKI-3 in the presence of QAS at elevated temperature (175 °C) is in contrast to DFG without reversion.

Analysis of kinetic curves of vulcanization of elastomeric compositions based on SKMS-30 ARC at temperatures of 165°C and 175°C (Fig. 29) showed that QAS-U is able to accelerate the process of crosslinking rubber in the absence of vulcanization activator - stearic acid. It should be noted the significant effect of temperature on the activity of QAS-B: at a temperature of 175°C the kinetic curve of the mixture with QAS-B is as close as possible to the curve with DFG. But unlike DFG and the control composition, QAS-U increases the induction period of vulcanization and has a lower vulcanization rate. Calculation of kinetic curves showed that at a temperature of 175°C the maximum torque of the composition with QAS-U (18.5 dN * m) is close to the value of the composition with DFG (18.7 dN * m), the relative degree of crosslinking in the composition with QAS (15, 1 dN * m) is slightly higher than in DFG (14.7 dN * m).

Therefore, QAS-U can be used as an accelerator of sulfur vulcanization of polydienes, the activity of QAS is close to DFG Zhang C., Cui Y., Lin S. & Guo J. (2022).

The possibility of interaction of QAS-U with the ingredients of the sulfur vulcanization system was studied by the DSC method. Analysis of DSC thermograms of the binary system "sulfur - QAS-U" (Fig. 30 a) in comparison with individual components showed that QAS-U in this system provides a decrease from 107 0C to 98 °C of the transition temperature of sulfur from α-modification to β- modification and melting point Sukhyy K.M., Privalko E.G. et al. (2010). In contrast to the curves of individual components on the curve of the binary system at a temperature of 139°C exothermic peak was observed, which indicates the reaction of sulfur interaction with QAS-U before the melting of the salt (175°C). The reaction of the interaction of sulfur and QAS-U is quite active, as evidenced by the intensity of the peak. In addition, it should be noted that the reaction of sulfur and altax - DBTD according to the literature Sukhyy K.M., Privalko E.G. et al. (2010) is recorded only by the endoeffects of eutectic melting.

Figure 27. Kinetics of sulfur vulcanization at a temperature of 155°C standard elastomeric compositions based on SKMS-30 ARC with different composition of sulfur vulcanization system

1 - sulfur (1.5 parts by weight), DBTD (3.0 parts by weight), zinc oxide (5.0 parts by weight);
2 - sulfur (1.5 parts by weight), DBTD (3.0 parts by weight), QAS-U (5.0 parts by weight);
3 - sulfur (1.5 parts by weight), DBTD (3.0 parts by weight);
4 - sulfur (1.5 parts by weight), zinc oxide (5.0 parts by weight), stearic acid (1.5 parts by weight), QAS-U (3.0 parts by weight);
5 - sulfur (1.5 parts by weight), QAS-U (3.0 parts by weight);
6 - sulfur (1.5 parts by weight), zinc oxide (5.0 parts by weight), stearic acid (1.5 parts by weight).

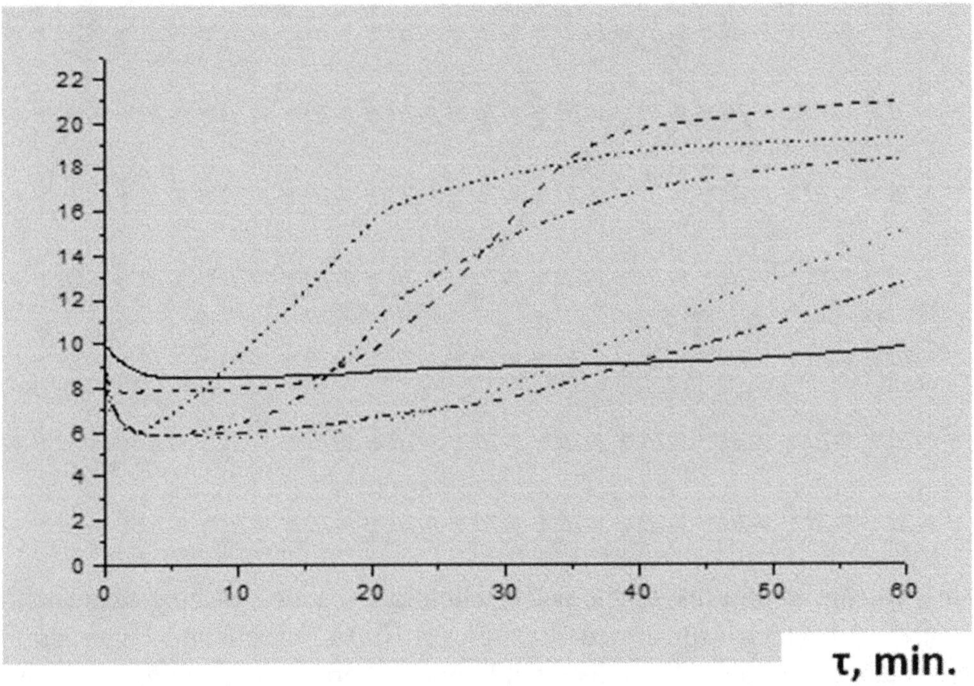

τ, min.

On the thermogram of the binary system "DBTD - QAS-U" (Fig. 30 b) the first observed endothermic peak at a temperature of 137°C, which indicates a decrease in melting temperatures of both components of this system. Then there is an intense exopic (168 °C), which characterizes the chemical interaction of DBTD and QAS.

Comparing the diagrams of the binary systems "sulfur - QAS-U" and "DBTD - QAS-U" it should be noted that QAS-U reacts faster with sulfur than with DBTD, but its interaction with the DBTD accelerator is more vigorous, as evidenced by the intensity peak. A study of the DSC diagrams of the binary systems "zinc oxide - stearic acid" and "zinc oxide - QAS-U"

Figure 28. Kinetic curves of sulfur vulcanization at 175°C standard elastomeric compositions based on SKI-3 using equimolar concentrations of different types of vulcanization accelerators: 1 - 2-MBT (0.42 parts by weight); 2 - DFG (0.52 parts by weight); 3 - QAS-U (0.67 parts by weight)

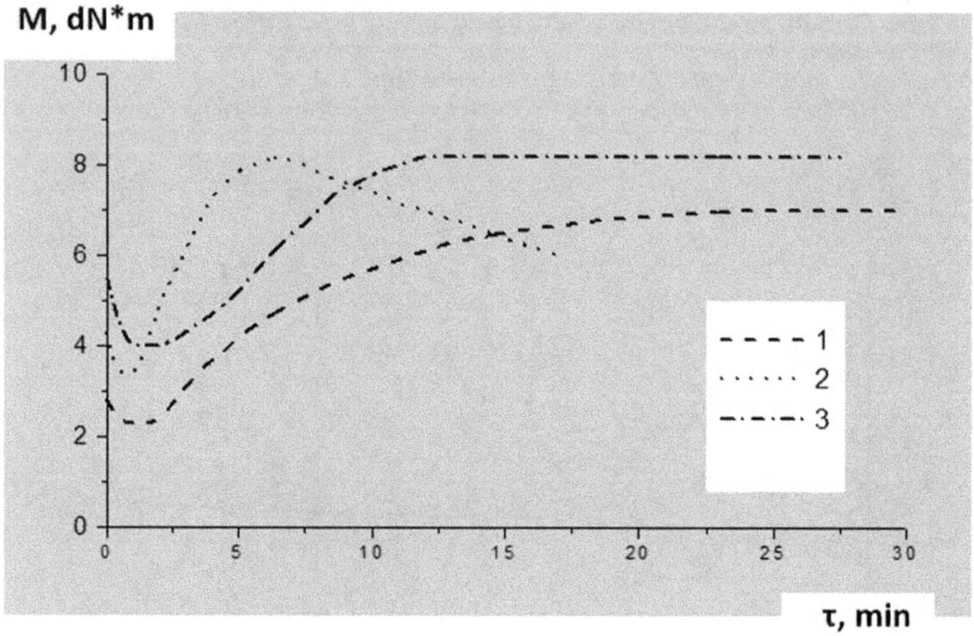

showed that in contrast to the active chemical reaction of zinc oxide with stearic acid, which is illustrated by exopics at 108 0C, the chemical interaction of zinc oxide with QAS -U does not happen; there is only a slight decrease in the melting temperature of QAS, probably due to the slight adsorption of salt on the surface of zinc oxide (Sukhyy et al., 2010).

Analysis of the thermogram of the reference mixture of components of the complete standard sulfur vulcanization system with DBTD (Fig. 31, curve 1), confirmed the vulcanization process according to the classical scheme (Sukhyy et al., 2010): endoeffect at 52°C corresponds to the melting of stearic acid; exoeffect 60°C - a consequence of the formation of zinc stearate, endoeffects at 101°C and 108°C caused by adsorption of system components on the surface of zinc oxide, the beginning of their melting and the formation of a real vulcanizing agent (RVA); the reason for the formation of the endoeffect at 130°C is the reaction of the interaction of the decomposition products of the system.

Figure 29. Kinetic curves of sulfur vulcanization at 175 0C standard elastomeric compositions based on SKMS-30 ARC with variation in the composition of the vulcanization system

Control. - sulfur (1.5 parts by weight), DBTD (3.0 parts by weight), stearic acid (1.5 parts by weight), zinc oxide (5.0 parts by weight);

1 - sulfur (1.5 parts by weight), stearic acid (1.5 parts by weight), zinc oxide (5.0 parts by weight);

2 - sulfur (1.5 parts by weight), QAS-U (1.5 parts by weight), zinc oxide (5.0 parts by weight);

3 - sulfur (1.5 parts by weight), DFG (1.5 parts by weight), zinc oxide (5.0 parts by weight).

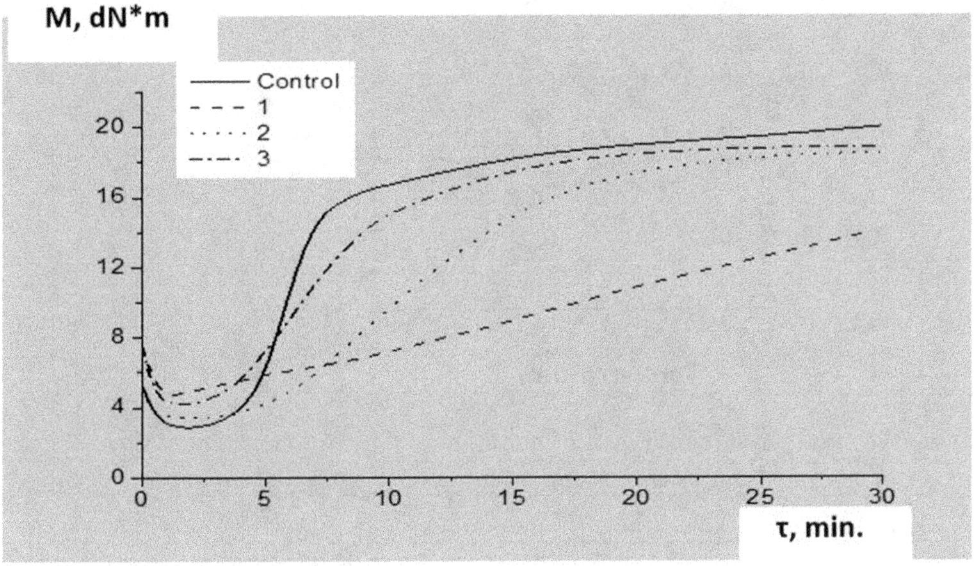

The thermogram of the mixture of components of the sulfur vulcanizing system, in which stearic acid is replaced by QAS-U (Fig. 31, curve 2), contains two eutectic endopics (96 and 104 °C) in nature similar to the endopics of the reference mixture thermogram (curve 1) but at lower temperatures, i.e., probably, the adsorption of the components of the mixture on the surface of zinc oxide and their direct interaction is more active. The chemical reaction of QAS-U with DBTD, as in the binary system, is expressed on the thermogram by intense exopics, but with a decrease in temperature from 168 to 149°C.

The thermogram of the mixture, in which QAS-U is used as a vulcanization accelerator (Fig. 31, curve 3), has a form close to the reference mixture. A feature of the thermogram is the significant intensity of the endopic at 110°C, which corresponds to the adsorption of system components on the surface of zinc oxide, the beginning of their melting and the formation of a real vulcanization agent (DAV).

Figure 30. Thermograms of individual components of sulfur vulcanization system and binary combinations of QAS-U with sulfur (a) and DBTD (b)

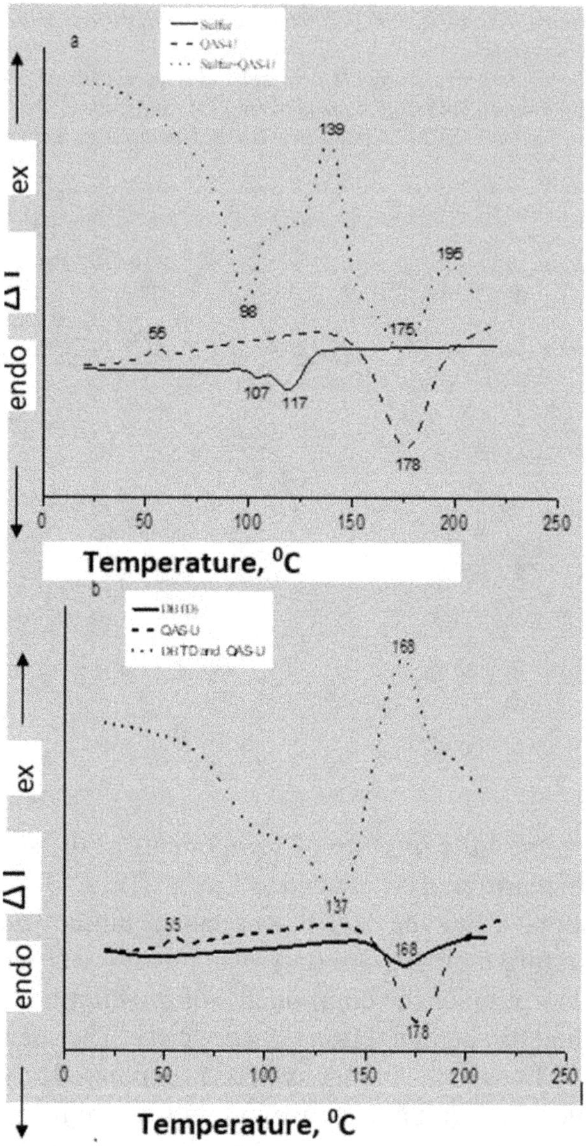

Analysis of the DSC test complex showed that with increasing temperature, QAS-U first reacts with sulfur; QAS-U is inert to zinc oxide. Given the presence in the composition of two accelerators (QAS-U and DBTD), their interaction is quite active at 168°C, but first there is a reaction of QAS-U with sulfur at 139°C.

Figure 31. Thermograms of mixtures of sulfur vulcanization system components
1 - sulfur (1.5 parts by weight), DBTD (3.0 parts by weight), zinc oxide (5.0 parts by weight), stearic acid (1.5 parts by weight);
2 - sulfur (1.5 parts by weight), DBTD (3.0 parts by weight), zinc oxide (5.0 parts by weight), QAS-U (1.5 parts by weight);
3 - sulfur (1.5 parts by weight), QAS-U (3.0 parts by weight), zinc oxide (5.0 parts by weight), stearic acid (1.5 parts by weight).

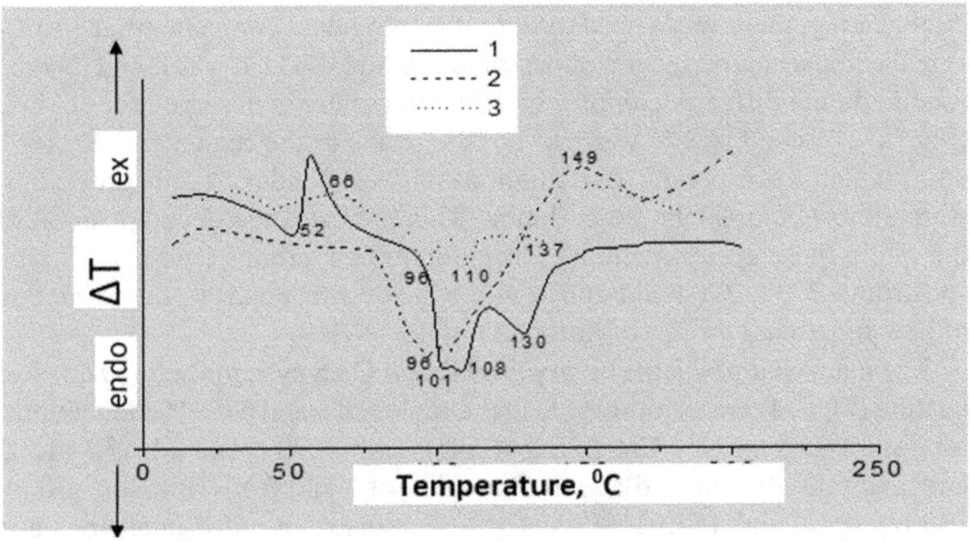

The chemistry of the sulfur vulcanization process in the presence of QAS was studied by IR spectroscopy of samples of individual components of the vulcanization system, their binary mechanical mixtures under normal conditions and after heating.

Comparative analysis of the spectra of individual substances with the spectrum of the binary system showed no interaction of QAS-U with sulfur under normal conditions. After heating this binary system at 140°C, changes in the spectrum of the heated sample were observed, which indicates the chemical interaction of sulfur with QAS-U (Fig. 32). The spectrum of the binary system before warming up has a wide band with a maximum of 1660 cm^{-1}, and after warming up the system on the spectrum with, narrow bands with maxima of 1670 and 1637 cm^{-1}, corresponding to δ oscillations of —NH and —NH_2 groups, and a band with a doublet 622 and 615 cm^{-1}, which corresponds to υ fluctuations of C-S groups, i.e. between sulfur and QAS-U there is an interaction and new reactive compounds are formed with groups —NH, —NH_2, C—S.

The IR spectrum of the binary composition of zinc oxide and QAS-U under normal conditions and after heating at 110°C has bands of valence and deformation oscillations of the salt itself, in contrast to the binary system of zinc oxide and stearic acid. This trend indicates the lack of chemical interaction of QAS-U with zinc oxide.

In the study of the IR spectra of the composition of DBTD and QAS-U after heating, there was a need to study the individual components at 170°C. When comparing the spectra of base and heated DBTD, it was found that the spectra do not differ. A completely different picture is observed for QAS-U (Fig. 33). When QAS-U warmed up, the peaks in the region of 600–1600 cm^{-1}, which correspond to δ oscillations of C—H bonds of cyclic groupings, were almost completely redistributed. This indicates the decomposition of the salt in these groups with the formation of new groups (and peaks in the spectrum): 3339, 3204 and 1667 cm^{-1}, which corresponds to the formation of new functional groups of nitrogen (—NH, —NH_2).

The main changes in the binary DBTD and QAS systems before and after heating (Fig. 34) are as follows: 1) in the region of ν and δ C—H connections, groups (groups —CH, —CH_2, CH_2=C—H) 2600-3070 cm- 1 (ν), 1300-1500 cm^{-1} and 600-900 cm $^{-1}$ (δ); 2) in the region of ν and δ N—H bonds - 3000-3300 cm^{-1} (ν) and 1540-1690 cm^{-1} (δ); 3) redistribution of intensities and location of peaks in the range of 900-1200 cm^{-1} indicates a change in ring substitutions in DBTD and QAS under the action of temperature during their decomposition, i.e., in the process of heating to 170°C decomposition of salt with the formation of new groups —NH and —NH_2 and new —CH connections. Changes also occur with DBTD, as its main bands are not registered on the spectrum, in particular, in the region of 400-500 cm^{-1}, where ν oscillations of S—S bonds of individual DBTDs are registered, significant changes have taken place. On the IR spectrum of the heated binary system there are no peaks 430 and 493 cm^{-1}, which indicates the rearrangement of S—S bonds. Therefore, at the vulcanization temperature there is a chemical interaction between QAS-U and DBTD with the formation of reaction products containing groups —NH, — NH_2, —CH, —CS, —SN.

The active chemical interaction of the two accelerators is the basis for studying the efficiency of QAS in binary systems with known classes of vulcanization accelerators. The study was conducted according to the scheme that allows to assess the presence of possible synergism in the action of combinations of organic accelerators according to The Science and Technology of Rubber Fourth Edition (2013). Based on the notions of synergism and semisynergism The Science and Technology of Rubber Fourth

Edition (2013), the following formula was used to quantify the synezrigma (S) of the accelerator combination:

Figure 32. IR spectra of binary sulfur + QAS-U systems under normal conditions (1) and after heating at 140°C (2)

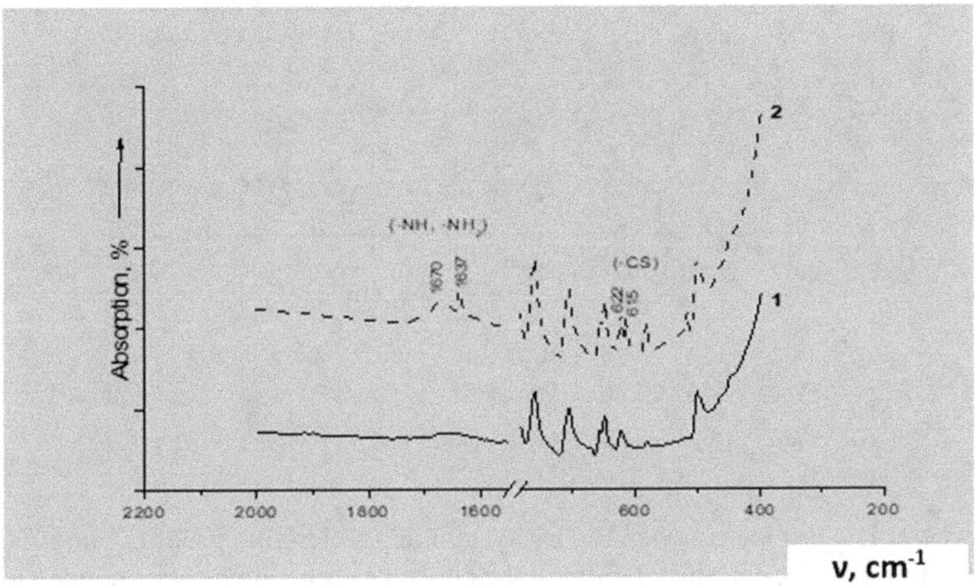

Figure 33. IR spectra QAS-U before and after heating at 170°C

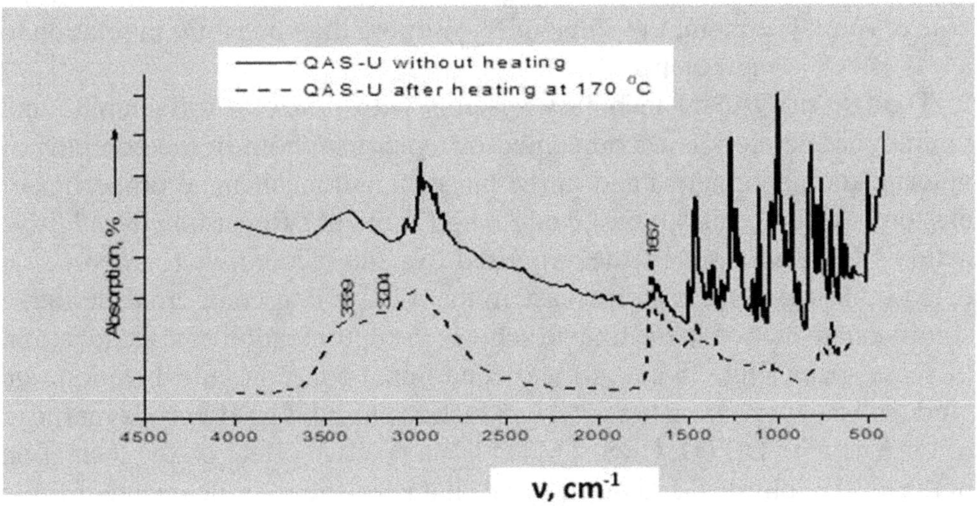

Figure 34. IR spectra of binary systems QAS-U + DBTD under normal conditions and after heating at 170 °C

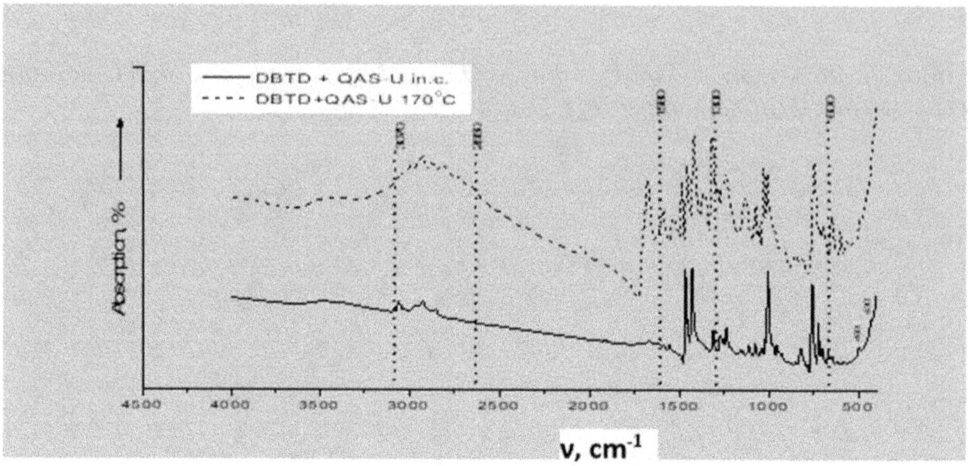

$$S = \frac{E_{AB} - (E_{2A} + E_{2B})/2}{(E_{2A} + E_{2B})/2}, \qquad (2)$$

where: E_{AB} is a factor when binary system of accelerators used; E_{2A} and E_{2B} are the same for each accelerator at the concentration which corresponded to their doubled molar of binary system. The existence of practical synergistic effect was estimated from the ratio: $S_{syn} = E_{AB}/E_{2A}$ or $S_{syn} = E_{AB}/E_{2B}$.

The value of practical synergistic effect is always more than one, and the effect achieved exceeds the effect of the most effective ingredient. In the case of semisynergism, the value of S_{syn} is more than one only in relation to a less effective ingredient.

The use of QAS-U in binary systems with thiazole, sulfenamide and thiuram accelerators is accompanied by synergism both in the complex of vulcanization parameters and in the physical and mechanical properties of elastomeric compositions based on SKI-3 (Table 14). Regarding the additive action of accelerators in the proposed systems according to rheometric data (Table 14) there is an increase in the relative degree of crosslinking, a significant reduction in the time to achieve the optimization of vulcanization, increase speed (Fig. 35 a, c, d) and conditional constant vulcanization rate while satisfactory parameters of the induction period. The effect of synergism of QAS-U with DBTD, CBS, TMTD has a positive effect on the formation of the level of physical and mechanical characteristics of vulcanizates (Table

The study of technological, vulcanization and physico-mechanical properties of elastomeric compositions for the tensile layer of wedge belts (Tables 9-11) showed that modified with Quaternary ammonium salts montmorillonite provides improved technological properties in terms of viscosity, viscosity of rubber mixtures and base systems. BS-100 + modifier MP).

The system of accelerators QAS-U - DFG in terms of vulcanization of SKI-3 corresponds to the effect of antagonism (Table 15, Fig. 35 b) and is positive only in some physical and mechanical parameters. Thus, montmorillonite in concentrations from 1.0 to 100.0 pts. wt. has the maximum activity in elastomeric compositions based on butadiene-methylstyrene rubber. In small concentrations from 1.0 to 10.0 pts. wt. it actively affects the process of sulfur vulcanization, but almost does not change the physical and mechanical characteristics of rubber.

The most reactive is the quaternary ammonium salt QAS-U, which can be used as an independent accelerator of sulfur vulcanization. Given the equimolar concentration of vulcanization accelerators, QAS-U is more active than 2-MBT and slightly inferior to DFG. QAS-U can be used as an accelerator of sulfur vulcanization in the absence of organic activator, the activity of QAS-U is close to DFG.

Significant intensity of chemical interaction of QAS-U with DBTD allows to use QAS-U as an effective co-accelerator of sulfur vulcanization. The established ability of QAS-U to form synergistic systems with thiazole, sulfenamide and thiuram accelerators, for most vulcanization characteristics and physical and mechanical properties of rubber QAS-U tends to detect practical synergism or semi-synergism.

The sequence of reactions of formation of the real vulcanization agent in the presence of QAS is defined. When sulfur interacts with QAS-U, compounds with reactive groups —NH, — NH_2, —CS, —SN are formed, which can affect the vulcanization of rubber and the formation of an adhesive bond in rubber-reinforced composites.

Table 14. Vulcanization and physico-mechanical characteristics of elastomeric compositions based on SKI-3 with different accelerators and their binary systems with QAS-U

Parameter	Accelerator or system of vulcanization accelerator								
	A	B				1/2 A + 1/2 B			
	QAS-U	DBTD	DFG	CBS	TMTD	½ DBTD	½DFG	½CBS	½ TMTD
Vulcanization parameters at 151 ^{0}C									
t_s, min.	4,5	9,0	3,0	15,0	3,0	$8,0^{2*}$	$7,0^{4}$	$11,0^{2}$	$4,4^{2}$
$\Delta M_{H, HF, HR} - M_L$, dN*m	6,3	12,1	9,3	14,1	16,3	$12,1^{1-2}$	$6,9^{4}$	$12,9^{2}$	$13,3^{2}$
$t_{c\,90}$, min.	26,5	23,5	13,0	21,7	6,5	$14,4^{1}$	$30,5^{4}$	$15,4^{1}$	$5,7^{1}$
V_c, dN*m/min.	0,3	0,8	0,9	2,1	4,7	$1,9^{1}$	$0,3^{4}$	$2,9^{1}$	$10,2^{1}$
k×10, $min.^{-1}$	0,37	1,45	1,09	4,46	4,76	$3,76^{1}$	$0,33^{4}$	$3,36^{2}$	$7,2^{1}$
t_{dis}, min.	4,5	19,5	6,0	20,0	4,5	$12,5^{2-3}$	$9,5^{4}$	$13,0^{2-3}$	$5,0^{2-3}$
Physical and mechanical characteristics									
f_{300}, MPa	0,2	0,6	0,4	0,7	1,0	$0,5^{2}$	$0,4^{2-3}$	$0,7^{2}$	$0,7^{2}$
f_{500}, MPa	0,2	1,3	0,6	1,4	1,9	$1,1^{2}$	$0,5^{2-3}$	$1,3^{2}$	$1,3^{2}$
f_p, MPa	0,9	12,1	7,8	14,1	18,9	$11,1^{2}$	$3,3^{4}$	$16,2^{1}$	$14,1^{2}$
ε, %	1020	900	1120	900	850	920^{2}	1140^{4}	950^{2}	940^{2}
B, kN/m	2	21	15	22	35	20^{2}	4^{4}	27^{1}	23^{2}
H, c.u.	8	29	20	30	35	28^{2}	17^{2}	29^{2}	32^{2}
S, %	48	62	58	58	64	60^{2}	58^{2}	54^{2}	66^{1}

Note: * - designation of the effects of the observed indicators: ^{1} - synergy (practical); ^{2} - semisynergism; ^{3} - additive action; ^{4} - antagonist of the binary system of accelerators.

Figure 35. Change in the characteristics (1-t_{c90}, 2- V_C, 3 -$M_{H, HF, HR} - M_L$) of sulfur vulcanization SKI-3 at 151 °C for different types of binary accelerator systems: a) QAS-U - DBTD; b) QAS-U - DFG; c) QAS-U - CBS; d) QAS-U - TMTD

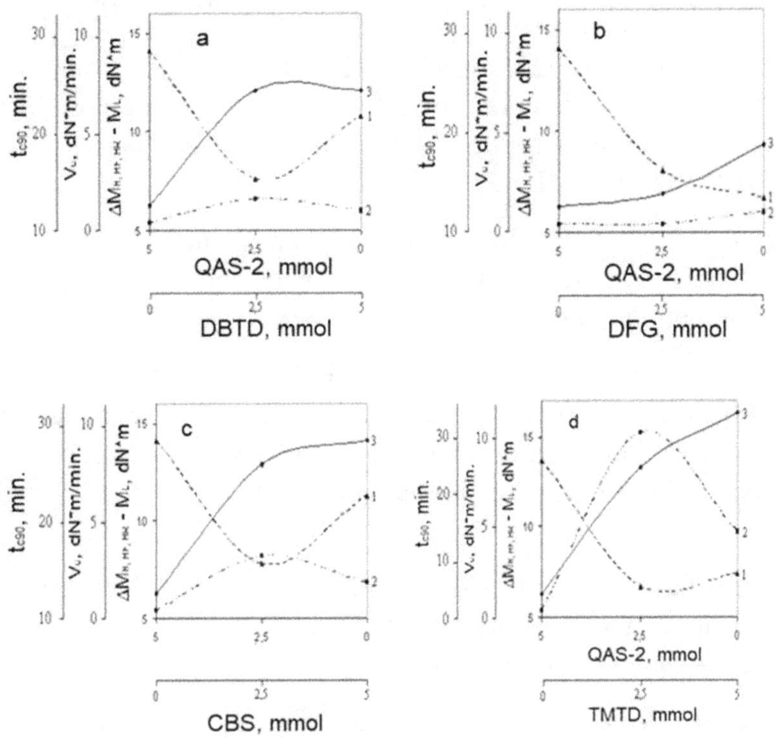

REFERENCES

Avdeev, Y. G., Nenasheva, T. A., Luchkin, A. Y., Marshakov, A. I., & Kuznetsov, Y. I. (2022). Effect of Quaternary Ammonium Salts and 1,2,4-Triazole Derivatives on Hydrogen Absorption by Mild Steel in Hydrochloric Acid Solution. *Materials (Basel)*, *15*(19), 6989. doi:10.3390/ma15196989 PMID:36234330

Baki, V. A., Ke, X., Heath, A., Calabria-Holley, J., Terzi, C., & Sirin, M. (2022). The impact of mechanochemical activation on the physicochemical properties and pozzolanic reactivity of kaolinite, muscovite and montmorillonite. *Cement and Concrete Research*, *162*, 106962. doi:10.1016/j.cemconres.2022.106962 de Araújo, M. J. G., Barbosa, F. C., Fook, M. V. L., Silva, S. M. L., & Leite, I. F. (2022). Influence of Quaternary Ammonium Salt Functionalized Chitosan Additive as Sustainable Filler for High-Density Polyethylene Composites. *Materials (Basel)*, *15*(21), 7418. doi:10.3390/ma15217418 PMID:36363010

Gallo, C., Rizzo, P., & Guerra, G. (2019). Intercalation compounds of a smectite clay with an ammonium salt biocide and their possible use for conservation of cultural heritage. *Heliyon*, *5*(12), e02991. doi:10.1016/j.heliyon.2019.e02991 PMID:31879708

Greene, J. P. (2021). 9 - Elastomers and Rubbers Editor(s): Joseph P. Greene, In Plastics Design Library Automotive Plastics and Composites, 127-147. William Andrew Publishing.

Ovcharov, V. I., Burmistr, M. V., & Sukhyy, K. M. [et al] (2005). Reinforcement of thermoplastic polymers by nanosize organo–clay. Technomer 05.19: Fachtagung uber Verarbeitung und Anwendung von Polymeren [Specialist conference on the processing and application of the polymers]. Chemnitz. 2, 71–82.

Schmidt, G., Nakatani, A. I., Butler, P. D., & Han, C. C. (2002). Polymer–layered silicate nanocomposites: Preparation, properties and uses of a new class of materials. *Macromolecules*, *35*, 4725–4732. doi:10.1021/ma0115141

Sukhyy, K. M., Privalko, E. G., Privalko, V. P., & Burmistr, M. V. (2010). Thermoelasticity and Stress Relaxation Behavior of Synthetic Rubber/Organoclay Nanocomposites. John Wiley & Sons.

The Science and Technology of Rubber Fourth Edition. (2013), Editor(s): James E. Mark, Burak Erman, C. Michael Roland, Academic Press, 333

Yu, Z., Zhu, Z., & Wang, S.-Q. (2005). Synthesis and rheological properties of polystyrene/layered silicate nanocomposite. *Polymer*, *104*, 785–796.

Zhang C., Cui Y., Lin S. & Guo J. (2022). Preparation and applications of hydrophilic quaternary ammonium salt type polymeric antistatic agents. *e-Polymers, 22*(1), 370-378.

Chapter 9
Synthesis and Properties of Modified Montmorillonite

ABSTRACT

In this chapter authors present processes of obtaining montmorillonite modified by polyionenes. Intercalation of macromolecular polymer of a quaternary ammonium salt in montmorillonite intercrystalline space is exhibited to be accompanied by increasing interlayer distances from 1.08 nm to 1.67 nm. The technique of synthesis of montmorillonite modified with polyionenes is proposed. Optimal conditions for sorption of polyionenes molecules by montmorillonite are stated as follows: the concentration of aqueous dispersion of montmorillonite equals 1%, the temperature of the reaction medium is 40 °C, the montmorillonite-polyionene ratio is 3: 1, the processing time is 24 hours.

BACKGROUND

Depletion of miniaturization possibilities with traditional technologies and global trends of miniaturization lead to decreasing objects of science and technology up to nano-size (Jeevanandam et al., 2018). This impedes developing of innovative materials. Their functional characteristics are determined not only by microdomain properties but also processes on atomic and molecular level in monolayers and nanovolumes.

It can be realized only with molecular clusters and nano-sized particles. Moreover, micro- nanotechnology transition is not only quantitive but qualitative, physical parametes (fusion temperature, bond energy, interatomic

DOI: 10.4018/978-1-7998-9728-6.ch009

distances, ionic and electronic conductivity, optical and magnetic properties) are revealed non-monotonic dependence on size (Shameem et al., 2021). Thermodynamical method allows revealing control factors based on homogenic and heterogenic nucleation and studies of particles growth dynamic and structural organization.

Intercalation of ionogenic organic compounds into porous and layered nanostructures followed by post-intercalation processes is of the crucial interest. However, criteria and general principles of directed affecting the colloid-chemical properties of disperse bodies and their systems ought to apply to clay mineral modification method (Sevim, 2017). These principles are based on comprehensive investigation of structure of disperse phase, i.e. determination of their fine structure, nature, value and energy state of surface; structure formation mechanism and influence of various factors on their structural-adsorptive characteristics, interaction colloidal phase and disperse medium; structure formation in disperse system due to interaction both particles and medium.

These principles are the base for methods of obtaining a lot of clay minerals. Properties of disperse system are strongly affected by water-soluble surfactant species with various functional groups, i.e., polyionenes with quaternary nitrogen atom in backbone chain. Interaction of polyionenes and clay minerals leads to far-reaching changes of their chemical, surface, and porous structure. Increase of mineral adsorption capacity of hydrocarbons is a consequence of these processes. Moreover, it is accompanied with expansive growth of swelling and thyxotropic gelation of organic derivatives in disperse media with various polarity.

When exchange cations replaced with organic cations, expanding of crystal gratings along z-axis and internal surface activation occur Açışlı, Karaca & Gürses (2017). Highly developed microporous structure becomes accessible for adsorption of hydrocarbon molecules and inert gases. According to imbedded ion size, pores diameter and sorbent selectivity to determined substances can be controlled.

This surface hydrophobisation leads to affinity of molecular surface properties for organic media with various polarities, it being supported by increase of adsorption of some hydrocarbons (Faghihi, Abootalebi, & Shabanian, 2013) humidification and dispergation and heat emission which results from inner phase solvate interaction with solvent molecules.

Interaction of modified minerals with liquid disperse media exhibits their important properties, i.e. swelling and thyxotropic coagulation structurization which are determined by used polyionenes, modification extent, concentration and anisotropy of disperse phase properties, presence of polar additives and solvative ability of disperse medium (Faghihi, Abootalebi, & Shabanian, 2013). As a consequence, the first main task for modeling of processes dealt with conversion of materials filled with fine particles is to determine interaction energy of particles and aggregates. So, the second main task for obtaining of polymeric composites with using of nanomaterials is to arrange the energetic conditions for nanoparticles division followed by their equal distribution. These conditions can be controlled by choice of optimal ratios operating parameters-design working elements of equipment.

When modified clay minerals, it is necessary to proceed from certain criteria and general principles of directed regulation of colloid-chemical properties of dispersed bodies and their systems, in particular Nicolais & Carotenuto (2005):

- study of the mechanism of structure formation and the influence of various factors on their structural and adsorption characteristics;
- it is necessary to thoroughly study the structure of dispersed phases, which includes determining the subtle features of their structure, nature, size and energy state of the surface;
- elucidation of the mechanism of interaction of the colloidal phase with a certain dispersion medium;
- elucidation of the mechanism of structure formation in the dispersed system due to the interaction of particles with each other and with the medium.

These principles underlie the methods of obtaining many modified clay minerals.

Water-soluble surfactants containing functional groups of different nature are of great importance for regulating the properties of dispersed systems. Such substances are polyonenes, which contain Quaternary nitrogen atoms in the main chain. The interaction of polyonenes with clay minerals of layered structure causes profound changes in their chemical structure, surface, and porous structure. As a result, the adsorption capacity of minerals in relation to hydrocarbons increases significantly, the ability to swell and thixotropic gelation of organ derivatives in dispersion media of different polarity increases sharply.

The location of organic molecules on the surface of minerals can be varied. Polyionene molecules can be located in the interlayer space of minerals parallel, at an angle or normal to the surface of silicate layers; the value of the basal interlayer distance d_{001} depends on both the number of layers of adsorbed molecules and the length of their hydrocarbon radicals. For the system under analysis, the average particle length is about 100 nm. For a fully delaminated system, the average particle thickness should be equal to the thickness of the elementary particle (lamen) (Fig. 1, a), i.e. equal to 0.94 nm. In the case of a completely homogeneous structure of unmodified montorillonite (Fig. 1, b), the thickness of each individual lamen can be calculated as:

$$t_{package}^{(u)} = d_{001}^{(u)}(n-1) + t_{lamen}$$

where d_{001} is periodicity of the structure of montmorillonite, d_{001} = 0,96 nm;

n is the number of lamen in the package,
t_{lamen} is the thickness of lamen, it being equal to 0,94 nm.

In the case of modified montmorillonite in the interlayer space intercalated molecules of polymeric Quaternary ammonium salts (Fig. 1, c). In this system, the thickness of the packets can be expressed as:

$$t_{package}^{(i)} = d_{001}^{(i)}(n-1) + t_{lamen}$$

where d_{001} is periodicity of the structure of modified montmorillonite (according with X-ray analysis data d_{001} = 1,67 nm).

The average particle thickness, taking into account radiographic data, can be defined as:

$$t_{part.}^{-i} = \frac{\Sigma(1,6h - 0,86)}{N}$$

where N is a total amount of particles.

The authors Açışlı et al. (2017) found that the replacement of exchangeable cations of clay minerals with organic ones is accompanied by the displacement of crystal lattices along the z-axis and the activation of their inner surface. There is a highly developed microporous structure available for adsorption of molecules of various hydrocarbons and inert gases. Depending on the size of the embedded cation, you can adjust the pore diameter and, consequently, the selectivity of the sorbent in relation to certain substances.

Improving the selective characteristics of the sorbent also contributes to the reduction of energy inhomogeneity of the mineral surface as a result of modification, which makes it possible to distinguish between substances with similar adsorption heat and boiling point. Along with this, the hydrophilic properties of the surface of dispersed minerals undergo significant changes. Adsorbed organic cations block the centers of water adsorption (oxygen atoms on the surface), resulting in a sharp decrease in water adsorption and thermal effect when wetted with increasing degree of surface modification and chain length of the organic cation. However, in the case of over-equivalent adsorption of polymer quaternary ammonium salt (PQAS) an increase in the heat of wetting can be observed, which indicates the reversal of the lyophilic properties of the surface. This is due to the formation of the second molecular layer of polyonenes with reverse orientation.

However, the hydrophobization of minerals leads to a sharp increase in the affinity of the molecular properties of the surface in relation to organic media of different polarity. This is confirmed by increased adsorption of some hydrocarbons Faghihi et al. (2013), improved wetting and dispersion, as well as heat release as a result of interfacial solvation interaction with solvent molecules.

In the process of interaction with liquid dispersed media, the most important properties of modified minerals are revealed, i.e., swelling and their ability to form thixotropic coagulation structure, determined by the chemical nature of polyonenes, degree of modification, concentration and anisometry of dispersed phase particles, presence of polar additives.

ENERGY OF INTERACTION BETWEEN LAYERS OF PRIMARY CRYSTALLINE ELEMENTS OF MONTMORILLONITE STRUCTURE

Polymer composite materials are increasingly used in various fields of economic activity. At the same time, fine materials, including nanomaterials, have recently become widely used as fillers. This type of material can significantly, sometimes several times, improve the basic performance of the product. However, along with the obvious advantages of using different types of nanomaterials, there are significant difficulties associated with the technology of obtaining such materials. One of such points is the increase in energy consumption per unit mass of the obtained polymer composite, which

Figure 1. Scheme for determining the thickness of (a) the individual layer (lamellae) of modified montmorillonite, (b) the original package of montmorillonite, (c) the package of polyionene-modified montmorillonite.

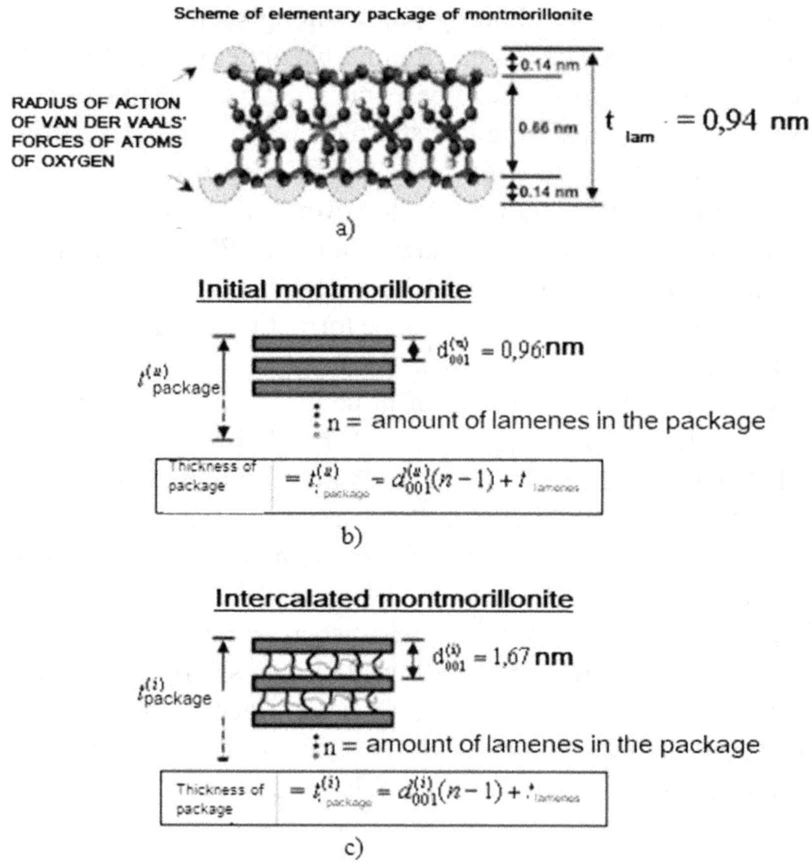

is associated with a sharp increase in the forces of interaction between the particles while reducing their size.

The interaction between small particles was first described by Van der Waals. They obtained the equation of state for real gas in the following form

$$(P + a \bullet N^2/V^2)\bullet(V - b \bullet N) = N\bullet T, \qquad (1)$$

where P is pressure, V is volume; T is temperature; N is amount of particles; a, b are parameters of a real gas (when a = b = 0 equation transfers into equation of state of a perfect gas).

Parameter b characterizes the repulsive effect of atoms at distances corresponding to atomic dimensions. As for the parameter a, the fulfillment of the relation a> 0 determines the attraction between atoms. In this case, as the parameter a increases, there is a decrease in pressure at constant values of temperature and volume.

For two atoms (molecules or particles) in certain states, the interaction potential U, which contains both the forces of attraction and the forces of ion-electrostatic repulsion, is a function of the distance H between atoms (molecules or particles). The dependences U ~ H, which are most common for real processes, are shown in Fig. 1.

Curve 1 in Fig. 2 corresponds to the case when the convergence of particles due to some external influences there is a continuous increase in energy, which determines the corresponding increase in the repulsive forces of particles. In this case, the adhesion of particles, and hence the formation of agglomerates is impossible.

For curve 2 under the action of external energy fields there is a potential minimum U_{2min} in the area of positive values of U, which can lead to the appearance of agglomerates, which, however, are metastable. The latter is due to the fact that the decay of agglomerates is accompanied by a decrease in the interaction potential.

Curve 3 has two varieties. The first variety (completely solid line) consists of three sections (I, II, III)). The nature of the first type of curve 3 is similar to curve 2. However, in this case there is an absolute minimum U_{31min} in the negative zone and therefore the state of the agglomeration is thermodynamically stable.

The second variety (with a dashed right part) consists of four sections (Ia, Ib, II, III). In this case, at a certain distance there is a second minimum U_{32min}, where thermodynamically stable agglomerates can also be formed. The behavior of curve 4 suggests that at fairly large distances the adhesion of particles can occur without external influences. However, at small distances H, repulsive forces still become predominant. Finally, curve 5 corresponds to particles for which only van der Waals forces of attraction are inherent. This dependence is inherent in most nanodisperse fillers of inorganic materials used to obtain polymer composites.

Van der Waals interaction can be caused by several reasons, among which the first mathematical description was obtained for the orientational interaction, most fully represented by Keesom, and for the induction interaction - Debye Liao, Song, Liang & Pang (2001).

Figure 2. Possible forms of potential interaction between particles.

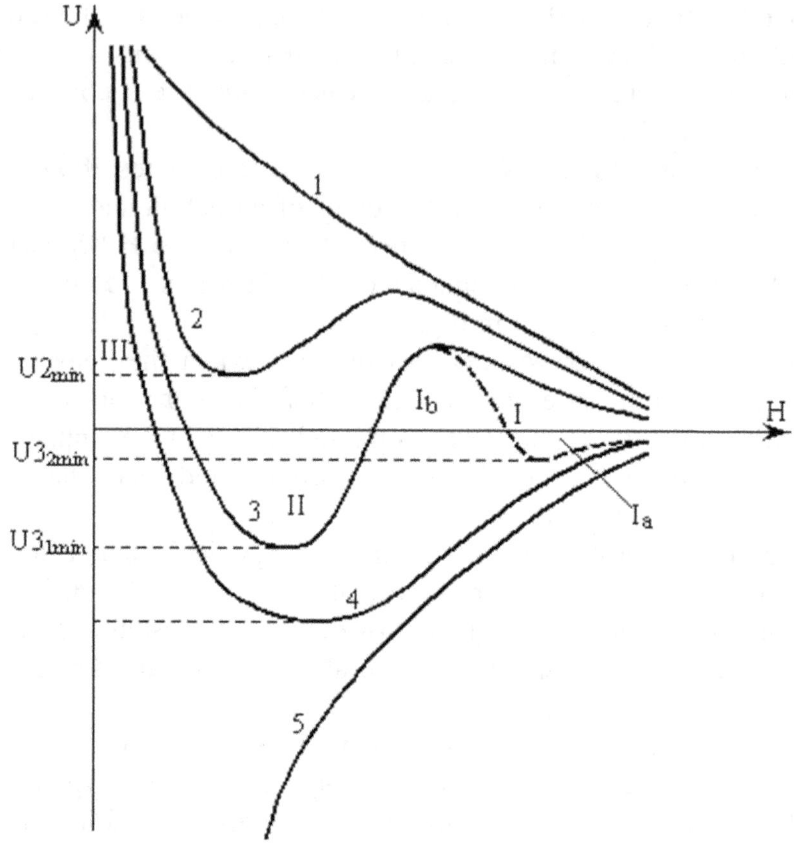

The first, rather complete mathematical apparatus for describing dispersion forces was created by F. London (1937), according to which the energy of interaction U between individual atoms and molecules at a distance H greater than their size is where C is the constant that characterizes type of atoms and is always positive.

To determine the forces of intermolecular interaction F it is necessary to use the relationship. The last two expressions are obtained in the second approximation of the perturbation theory applied to the electrostatic interaction of two dipoles. The forces of intermolecular attraction, in this case, are inversely proportional to the seventh degree of the distance between the molecules.

London's theory has significant limitations on distance. The equations obtained by him lead to incorrect results not only for very small distances between atoms, when their own wave functions overlap, but also for H >>

λ_i, where λ_i are the wavelengths of absorption or emission of atoms. In the second case, the effects of delay become significant.

For dispersive (London) forces, their characteristic feature is their additivity. In this case, the molecule induces periodic dipoles in several neighboring molecules. Then the induced dipole is attracted to the original dipole, and therefore the energy of attraction between two bodies can be considered as the sum of the energies of attraction between the respective pairs of molecules that make up these bodies.

The transition from the interaction energy of molecules to the interaction energy of condensed systems was performed by De Boer (1948) and Hamaker (1937). Assuming that the energy of interaction of two systems can be found by summing the pairwise interactions of the components of their particles and, moving from summation to integration, the energy of interaction is calculated by the formula

$$E_{12} = -C_K \cdot N_1 \cdot N_2 \cdot \int_{V_1} dV_1 \int_{V_2} H^{-6} dV_2, \tag{2}$$

where N_1, N_2 are densities of particles in each of systems; V_1, V_2 are volumes of both of systems; C_K is constant.

Hamaker derived formulas for energy between two bodies of different configurations: - two spheres of radii R_1 and R_2

$$E_{12} = -\frac{A_{12}}{6} \cdot \left\{ \frac{2 \cdot R_1 \cdot R_2}{H^2 + 2 \cdot H \cdot (R_1 + R_2)} + \frac{2 \cdot R_1 \cdot R_2}{H^2 + 2 \cdot H \cdot (R_1 + R_2) + 4 \cdot R_1 \cdot R_2} + \ln \left[\frac{H^2 + 2 \cdot H \cdot (R_1 + R_2)}{H^2 + 2 \cdot H \cdot (R_1 + R_2) + 4 \cdot R_1 \cdot R_2} \right] \right\}; \tag{3}$$

sphere and flat wall (R_2 ® ¥, R_1 = R)

$$E_{12} = -\frac{A_{12}}{6 \cdot H} \cdot \left[1 + \frac{H}{2 \cdot R + H} + \frac{H}{R} \cdot \ln \left(\frac{H}{2 \cdot R + H} \right) \right] \tag{4}$$

two flat endless walls per unit surface

$$E_{12}^{P} = -\frac{A_{12}}{12 \cdot \pi \cdot H^2} \tag{5}$$

In Eq.(3) - (5), the value of A_{12} is called the Hamaker constant (indices 1 and 2 denote the first and second bodies, respectively). As a rule, the Hamaker constant is determined either in the air (under vacuum) or in the aqueous medium (in the first case, this value is denoted as A_{1v2}, and in the second - A_{1w2}).

It should also be noted another area in the study of intermolecular interactions in condensed matter under the leadership of BV Deryaguin et al. (1985), which developed in parallel with Hamaker's theory. The main problems studied in these works are related to the aggregative stability of colloids and its violation in the coagulation process. In contrast to the previous theory, Deryagin proposed a clearly theoretically sound method for calculating the forces of ion-electrostatic repulsion. This method is based on the use of Maxwell's expressions for the components of the electric field strength tensor applied to particle surfaces and dependent on the density of surface charges. Moreover, to determine this density, assuming constancy and low potential in particle convergence, the Poisson-Boltzmann equation was used in the first stage, and in the subsequent development this equation was used to determine the charge density and hydrostatic pressure distribution between convergent particles.

In the works of Deryaguin & Landau (1941) the limit case of strongly charged particles is considered and the conditions under which the power and potential barriers separating them disappear are found. Calculations were performed for both symmetric electrolyte and electrolyte with valence types 1-2 and 2-1 (the first digit indicates the charge of the counterion). It was shown that the charge of the side ion has little effect on the stability of the sol. The charge of the counterion determines the coagulating concentration m according to the formula $m = C_0 \cdot \varepsilon^3 \cdot (k_B \cdot T)^5 / (A_{12}^2 \cdot e^6 \cdot z_1^6)$, where C_0 is a numerical coefficient that weakly depends on the valence of the side ion; ε is the dielectric constant of the medium; k_B is the Boltzmann's constant; T is the absolute temperature; e refers to the electron charge; z_1 is the ion charge.

Thus, recent work has explained the Hardy-Schulze rule about the predominant strong influence of the counterion valence on the coagulation threshold.

Further development of the theory of stability of lyophobic colloids was found in the monograph of Verwey & Overbeek (1948). It should be noted that the main points in the theory of Derjaguin and Landau, on the one hand, and the theories of Verwey and Overbeeck, on the other hand, coincide. In this regard, the general theory is called the Deryagin-Landau-Verwey-Overbeeck theory (DLFO). The merits of Verwey and Overbeeck are as follows: to obtain results for the repulsive forces of colloidal particles charged to any potential; the effect on the stability of colloids Stern correction, which takes into account the formation of a dense monolayer of adsorbed ions; theoretical consideration of the main features of peptization processes; explanation of the influence of the secondary potential well (Fig. 1) on the aggregation and thixotropy of the suspension.

Thus, the following main shortcomings can be identified for the theory under consideration, the so-called microscopic theory presented in the works of London and Hamaker and in Deryagin's early works: the applicability of the law of interaction of gas particles to condensed bodies is postulated; full additivity of electronic interactions is provided; at small distances between bodies, the summation of these interactions is replaced by volume integration. Lifshitz's macroscopic theory in which interacting bodies are considered as continuous media, lacks these shortcomings Lifshitz (1956).

The basic idea of Lifshitz's theory is that the relationship between bodies is due to the fluctuating electromagnetic field, which is always present inside any material environment and goes beyond it due to thermodynamic fluctuations. One of the clearest examples of such a field is the thermal radiation of the body.

It should be noted that even at absolute zero there are electromagnetic fluctuations, while thermal radiation is absent. In this case, the fluctuations are purely quantum. In all such cases, the thermodynamic relation has one common feature - they are all related to the non-additivity of the free energy of the body system, taking into account van der Waals forces, ie free energy is not simply proportional to the volume of the system, but depends on a certain volume also from the parameters that characterize the relative position of bodies. This type of non-additivity, which is associated with the long-term nature of the forces of Van der Waals, is a qualitatively new effect.

This non-additivity can be explained by the fact that any change in the electrical properties of the medium in a certain area leads, according to Maxwell's equations, to a change in the fluctuation field also outside this area. From the latter conclusion it follows that the part of free energy associated with electromagnetic fluctuations is not determined by the properties of matter only at this point.

Shear deformations are the main factor for dividing units into separate particles in the working areas of equipment for processing polymeric materials, for example, extrusion machines. However, in the usual form of a screw In most cases, the creation of a sufficient level of shear deformations for the separation of aggregates is an impossible task, especially for fine materials (with sizes from units to several tens of nanometers).

In Sinha Ray et al. (2003), the dependence for determining the ultimate shear stresses τ_0 in a dispersed medium was obtained. Two planes under the action of mutual shear deformation, which changes the potential energy of interaction of particles in these planes, were considered. After a certain period of time, the potential energy of the particles reaches a maximum. Thus, as a result of shear deformations over the system of particles will be carried out work A, presented as follows: $A = n \bullet P^2/a^3$, where n is the number of particles per unit volume; P is the dipole moment of the particles; a is the distance between the centers of the particles.

This work will be directed against the forces of mutual attraction. It is further proposed that the dipole potential energy of particles is converted into heat, ie into the energy of thermal motion of the molecules of the dispersion medium and the molecules that make up the particles.

Thus one cycle of such transition corresponds to an angle π/4 at which in each unit of volume the energy corresponding to work *A* will pass to heat.

As is known from the theory of plasticity Chakrabarty J. (2006)., the specific mechanical power of N_{sp}, converted into heat during the flow of the medium, can be expressed through tangential stresses τ as follows:

$$N_{sp} = \tau \bullet dVt/dy = \tau \bullet d\gamma/dy,$$

where $dVt/_d y$ is the velocity gradient in the direction perpendicular to the flow direction; dγ/dt refers to the shear rate. Equating A of the complex Nsp. Δt, we obtain:

$$n \cdot P2/a3 = N \cdot \Delta t = \tau \bullet \Delta \gamma.$$

Taking into account the equality Δγ = π/4, the latter relation is obtained taking into account the dependence n = a-3.

In Chakrabarty J. (2006), based on the last three works, the following relation is given

$$\tau 0 = 4/\pi \bullet n \bullet U0$$

where U0 is the energy of pair interaction of microobjects.

Thus, from the given materials the first main task at modeling of the processes connected with processing of the materials filled with fine particles, is definition of energy of interaction between particles and their aggregates. Hence the second main task, which for the production of polymer composites using nanomaterials is to create certain energy conditions for the separation of nanoparticles, followed by their uniform distribution. These conditions can be created by choosing the appropriate optimal relationships between the parameters of operating modes and the design of the working elements of processing equipment. In this case, if necessary, it is necessary to provide additional opportunities to create different types of energy fields that affect the recycled material - from the intensification of mechanical fields to the introduction of magnetic, electric, ultrasonic fields, etc.

PRODUCTION TECHNOLOGY, STRUCTURE AND PROPERTIES OF ORGANOMODIFIED MONTMORILLONITE

Section 3 investigated the surface activity of polyonenes and found that, in the studied series of polyonenes, it increases with hydrophobization of the structure and decreasing the concentration of ionic cents (replacement of the phenyl radical between Quaternary nitrogen atoms by dialkyl-substituted phenyl radical 1 and the binuclear phenyl radical - I-1, I-6) and increases in the series: I-5 <I- 3 <I-1 <I- 6. It was also found that with the introduction of naphthalene fragment into the structure of polyonenes surface activity is significantly reduced.

For qualitative and quantitative study of the process of modification of montmorillonite, the adsorption of polyonenes at the interface 'solution - porous adsorbent' was studied. The results of the study of adsorption activity are shown in Fig. 3. The change in adsorption values with decreasing concentration for different polyonenes indicates that the adsorption of organic cations depends on both the level of hydrophobicity of the polymer chain and the flexibility of radicals in the chain.

Adsorption proceeds with a significant increase in the free energy of the system with increasing proportion of the hydrophobic component of the surfactant. This indicates an increase in the contribution of van der Waals forces in the total adsorption energy.

As the hydrophobicity of the polymer increases and the size of the radicals decreases (steric factor), the entropy decreases. This leads to a loss of mobility of cations during their transition to the adsorbed state. Despite the significant adsorption of organic cations from the dispersed phase, the equimolar exchange of inorganic cations for organic ones almost never occurs, and the resulting organomineral derivatives always have a residual exchange capacity.

Incomplete substitution of inorganic cations for organic ones is explained by the spatial-steric factor, low diffusion rate within the crystal lattice of minerals, as well as partial overlap of the ion exchange centers of the adsorbent with adsorbate molecules.

Regarding polyonenes with naphthalene radicals, it is obvious that the adsorption activity is lower than the polymers discussed above. This is due to the decrease in the flexibility of the polymer chain due to the increase in the size of the radical. Given the low surface activity of polyonenes I-7 and I-8, they were not used to modify montmorillonite in the future.

The adsorption of polyonenes from solutions is characterized by a complex concentration dependence. This is due to conformational and structural changes in polymer solutions. The adsorption of polymers differs significantly from the adsorption of low molecular weight substances. These differences are explained not only by the macromolecular nature of the molecules adsorbed from the solution, but also by the fact that each concentration of the solution from which the adsorption is carried out corresponds to different conformations of the macromolecular ball, the degree of overlap and interpenetration adsorption.

In order to find the optimal parameters of the montmorillonite modification process, the influence of temperature, duration and ratio of components on the kinetics of the process of ion exchange of polyonenes on the surface of crystalline layers was studied.

It is established that the rate of ion exchange of polyonenes on montmorillonite increases with increasing temperature up to 30°C and then acquires constant values at a temperature of 40°C reaches the maximum level. It was also found that the maximum level of ion exchange at temperatures above 35°C is reached in 24 hours and does not depend on the chemical structure of polyones. Thus, the process of modification of montmorillonite by polyonenes should be carried out at a temperature of 35-40 °C for 24 hours.

The influence of the ratio of montmorillonite and polyionene in the aqueous dispersion on the efficiency of modification has been studied (Fig. 4). As can be seen from the above data, the ratio of components in the reaction medium significantly affects the speed of the modification process and its efficiency - the amount of precipitated polyonene Sukhyy K.M. et al. (2018). As can be

Figure 3. Isothermes of adsorption and ion exchange of polyionenes on montmorillonite.

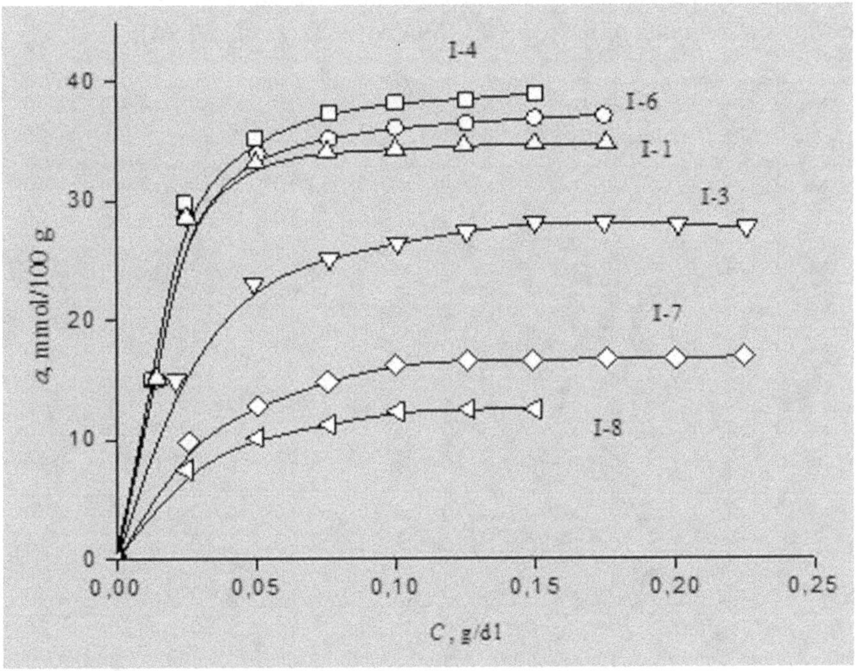

seen from Fig. 4 correlation between the ratio of polyonene / montmorillonite in aqueous dispersion and the ratio of polyonene/montmorillonite in modified montmorillonite at a polyonene content of 20% increases sharply and becomes constant the amount of polyonene on the surface of lamellae of montmorillonite. Polyionene I-4 with aliphatic and alkylaromatic radicals, which is adsorbed on montmorillonite lamellae in the amount of 15%, is the most effective. Polyonenes with binuclear phenolic radicals I-6 and I-1 are characterized by a slightly lower level of adsorption (12-13%). The lowest level of adsorption, in the studied number of polyonenes, is characteristic of polyoneones I-3 and I-5 with phenyl and dialkylphenyl radicals between quaternary nitrogen atoms. Polyonenes with aliphatic radicals were not used to modify montmorillonite due to low thermal stability, which is lower than the processing temperature of most thermoplastic polymers.

Thus, optimal, in terms of the amount of adsorbed polyonene is the mode of modification of montmorillonite at a temperature of 35-40°C for 24 hours, at a ratio of montmorillonite / polyionene in an aqueous dispersion of 3: 1.

X-ray scattering data allow us to estimate the degree of orderliness of montmorillonite in relation to montmorillonite, provide information on the

Figure 4. Dependences of polyonene content in modified montmorillonite on the montmorillonite / polyionene ratio during modification (τ = 24 h; t = 40°C). The figure shows the names of polyonenes (Sukhyy K.M. et al. (2018)).

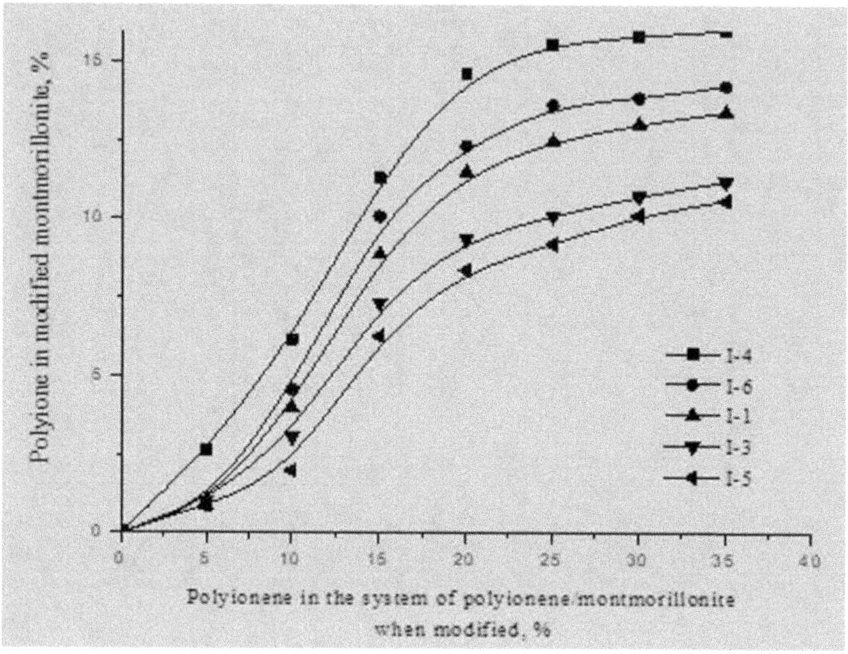

periodicity in the location of crystalline layers (lamellae) of montmorillonite, which allows determining the distance between crystalline montmorillonite layers.

Fig. 5 shows WAXS profiles of the original bentonite, "wet" montmorillonite (containing 20% moisture) and montmorillonite modified I-6 (12%). Bentonite is a hydrophilic material and can easily absorb up to 20% of air moisture. Sorbed moisture is placed in the interlayer layers, reducing the level between the batch interaction and, accordingly, increasing the interplanar distances and creating conditions for the exchange of batch metal cations for organic cations.

For the original bentonite, which contains approximately 8% of mineral impurities and does not contain sorbed moisture, there is a weak peak in the region $2\theta = 8.6°$, which corresponds to the Bragg period of the location of heterogeneous regions of 1.03 nm ($d = \lambda / 2 \sin(\theta)$), formally, to the thickness of the interlayer space. The Bragg period for pure bentonite corresponds to the literature data for montmorillonite Na-form.

For wet monmorillonite (according to the developed technology bentonite was cleaned of excess impurities and was a montmorillonite fraction) containing 20% moisture, the interplanar distance increases to 1.46 nm. This changes not only the position of the wide-angle maximum, but also the shape of this maximum (intensity and half-width), which is directly related to the degree of perfection of the macrolattice (paracrystalline ordering). It should be noted that these processes are reversible. So, during drying, the interplanar distance in montmorillonite is reduced to 1.03 nm. Replacement of interlayer exchange cations with polymer cations of polyonene I-6 is accompanied by an increase in the interplanar distance to 1.72 nm.

The increase in the interplanar distance in montmorillonite when modified with polyonenes indicates the introduction of fragments of macromolecules in the interlayer space of montmorillonite. That is, the interlayer metal cations are replaced by polyionene cations which leads to a weakening of the interlayer bond.

In order to determine the effectiveness of the studied polyonenes in the modification of montmorillonite, WAXS studies were performed, the results of which are given in Fig. 6. According to Fig. 6, the use of polyonenes with bionuclear phenolic radicals I-6 and I-1 to modify montmorillonite leads to a more significant increase in the interplanar distance of layers in montmorillonite compared to polyonenes with phenyl and dialkylphenyl radicals (I-3 and I-5). This results from the smaller size of the latter. But also greater order and homogeneity of the composition of montmorillonite modified with polyonenes I-3 and I-5, based on the shape of diffractograms. WAXS profiles of montmorillonite modified with polyionenes I-3 and I-5 are characterized by narrower diffraction maxima and almost identical half-widths of the left and right parts of the peaks. It can be concluded that polyionenes I-3 and I-5 are more evenly introduced into the interlayer galleries of montmorillonite. Polyionene I-4 with aliphatic and alkylaromatic radicals between quaternary nitrogen atoms showed the greatest efficiency in terms of increasing the interplanar distance in montmorillonite. This results from the greater flexibility of the polychain macrochains of this chemical structure, and consequently the more efficient introduction of montmorillonite into the interlayer galleries.

Synthesized organomontmorillonite was studied by the method of differential thermal analysis. The initial montmorillonite on thermograms is characterized by a double thermal endoeffect at 130°C and 200°C, which correspond to the loss of adsorbed water, and an endoeffect at 540°C, which is associated with dehydroxylation of the mineral.

Figure 5. WAXS profiles of initial bentonite, wet montmorillonite (20%) and montmorillonite modified with I-6 (Sukhyy K.M. et al. (2018)).

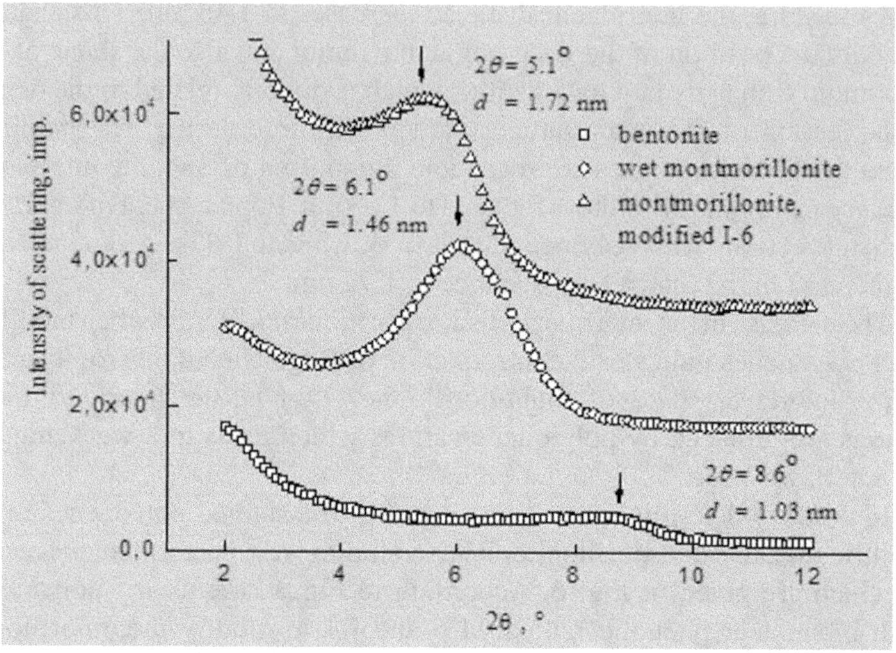

Figure 6. WAXS profiles of montmorillonites modified with polyionens (codes of polyionens are given in figure).

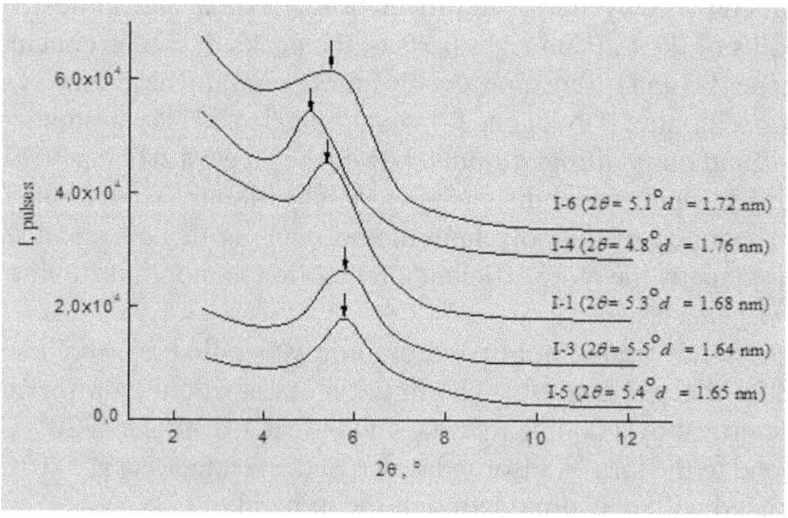

Figure 7. DTA curves of montmorillonite (1) modified with polyonene I-6 montmorillonite (2) and polyonene I-6 (3).

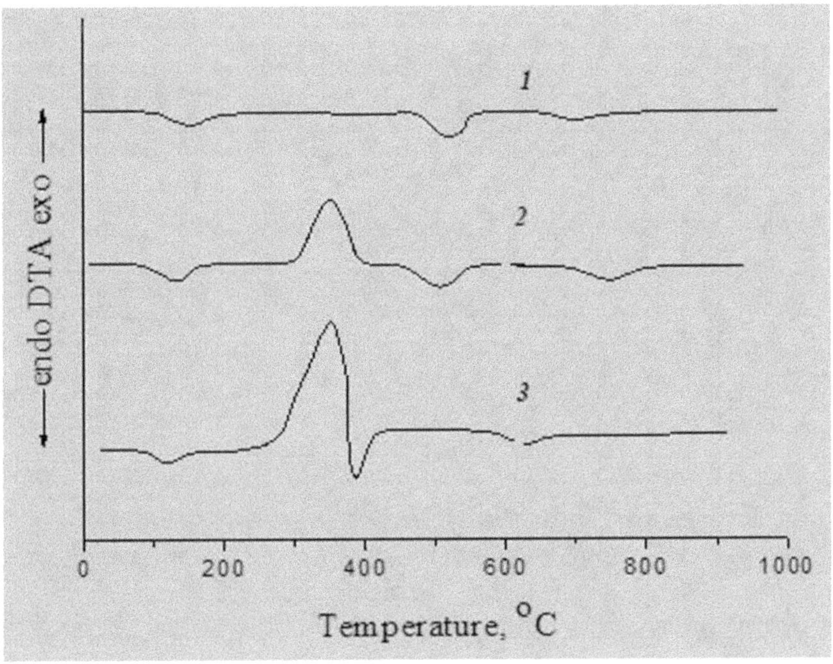

DTA curves of the original montmorillonite, montmorillonite modified with polyonene I-6 and polyionene I-6 are given in Fig.7. An exothermic peak is observed on the curve of modified montmorillonite at a temperature of 270°C, which corresponds to the decomposition of polyonene. From the comparison of the corresponding exotherms, as well as the curves of mass loss of polyonenes (Fig. 8), the content of the studied polyonenes in montmorillonite was determined (the corresponding data are shown in Fig. 4).

IR-spectroscopy is known to provide useful information about the mechanism of adsorption and the state of the adsorbent and adsorbate. For this purpose, spectroscopic studies of montmorillonite modified with the studied polyionenes were performed.

When compared the spectra of the original and modified montmorillonite, it can be noted that a number of new bands characteristic of individual polyonenes appear in the spectra of modified montmorillonite (Sukhyy et al. (2018)).

Thus, the bands in the spectra of modified montmorillonite at 2930 and 2850 cm^{-1} correspond to the valence fluctuations of the N-C bond. The new

Figure 8. Mass loss curves of montmorillonite (1) and modified polyonene I-6 montmorillonite (2).

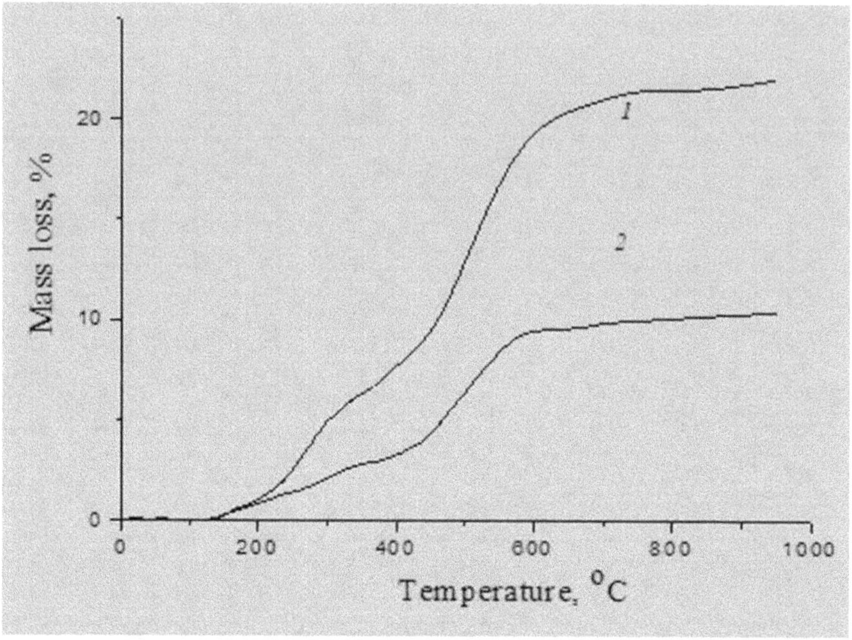

bands in the region of 1380 and 1480 cm^{-1} correspond to oscillations of the CH_2 and CH_3 groups associated with Quaternary nitrogen atoms, respectively.

No changes occurred in the region of valence vibrations of structural hydroxyl groups (bands 3690, 3670, 3620 cm^{-1}). However, an interesting situation is observed in the field of Si-O valence oscillations. The band 1120 cm^{-1}, which corresponds to the valence oscillations of Si-O with perpendicular change of the dipole moment relative to the plane of the packets, does not change during the transition from original to modified montmorillonite, and the band at 970 cm^{-1} in the original montmorillonite with transition to mode is bifurcated by two bands 1030 and 1010 cm^{-1}, which correspond to the valence oscillations of the Si-O bond with a parallel change in the dipole moment relative to the plane of the packets.

Since the octahedral and tetrahedral layers of montmorillonite have a common network of oxygen atoms, the nature of cations undoubtedly affects the position of the bands of valence Si-O oscillations Bibi et al. (2007). According with Zulfiqar & Sarwar (2008) it is noted that with increasing defect of the crystal structure of montmorillonite, which is manifested in the presence of broken bonds on the faces of crystals and defects in octahedral grids, the band of Si-O oscillations with dipole change parallel to the plane of the layers becomes more diffuse and shifts towards low frequencies. That is, the modification leads to the formation of a more perfect structure of minerals.

Based on the analysis of the nature of the interaction of amines with the surface of clay minerals, the adsorption of amines between acidic OH-groups of the surface and nitrogen atoms of amines was suggested to form a hydrogen bond Bibi et al. (2007). It is clear that in our case it would be possible to assume the existence of such interactions. However, it is known that the Quaternary nitrogen atom is not capable of forming a hydrogen bond due to the lack of an unshared pair of electrons Bibi et al. (2007).

In this case, we found the reason for grafting organic cations PQAS on the surface of montmorillonite due to the fixation of organic cations in exchange positions during ion exchange adsorption Eq.(6) and adsorption by acidic silanol groups (broken bonds on the faces of crystals) of organic cations. Thus, we can assume that the modification of montmorillonite by polyonenes leads to the formation of a more perfect structure by adsorption by silanol groups of acidic nature (broken bonds on the faces of crystals) of organic cations Eq.(7):

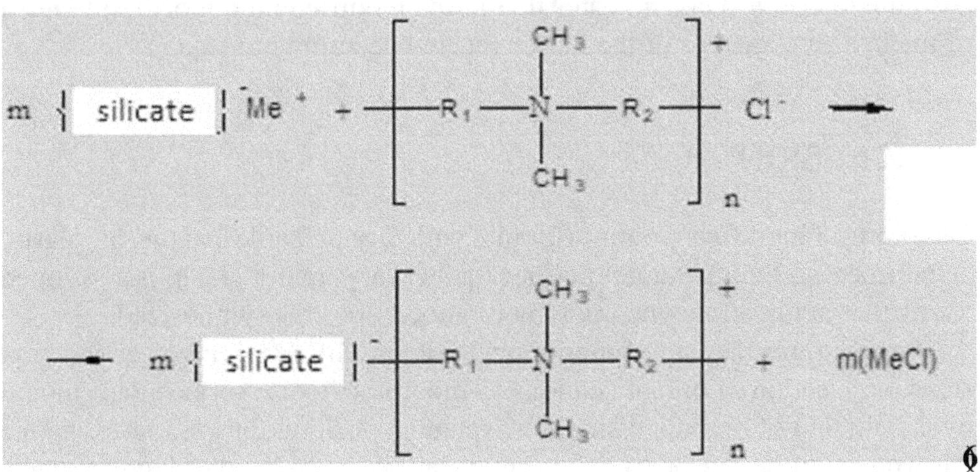

(6)

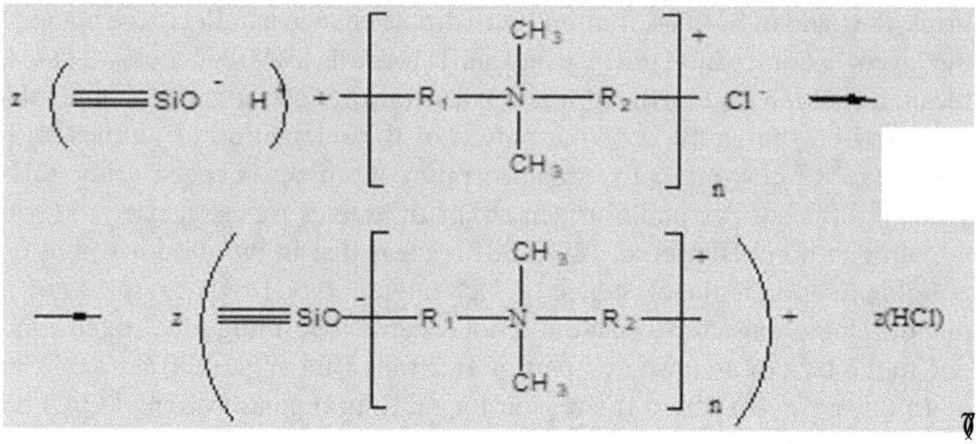

(7)

A comparison of the 770 and 730 cm^{-1} bands for the original and modified montmorillonite also supports this conclusion. Thus, Burmistr et al. (2003) attribute these bands to the deformation oscillations of the hydroxyl groups of the lattice with dipole polarization perpendicular to the plane of the layers. As can be seen (Fig. 6), for modified montmorillonite these bands are much less intense than for the original montmorillonite and are almost equal in absolute intensity. This indicates that in the modified montmorillonite most of the hydroxyl groups of the lattice reacted according to Eq.(7).

CONCLUSION

Obtaining of modified montmorillonite with thermostable and thermoplastic at the processing temperatures polymer quaternary ammonia salts is developed for further application when polymer nanocomposites synthesized.

Proposed mechanism of montmorillonite modification involves the next steps: connection of organic cations to montmorillonite surface determined by attachment of organic cations to exchange position during ion-exchange adsorption and adsorption of organic cations with acid sylanol groups, i.e. torn bonds on crystal faces. So, these processes result in more perfect structure by organic cations adsorption with acid sylanol groups (torn bonds on crystal faces).

Optimal conditions of polyionene molecules sorption with montmorillonite are stated to be: the concentration of aqueous dispersion of montmorillonite is 1%, the temperature of the reaction medium is 40 °C, the montmorillonite-polyionene ratio is 3:1, and the processing time is 24 hours.

REFERENCES

Açışlı, Ö., Karaca, S., & Gürses, A. (2017). Investigation of the alkyl chain lengths of surfactants on their adsorption by montmorillonite (Mt) from aqueous solutions. *Applied Clay Science*, *146*, 90–99. doi:10.1016/j.clay.2016.12.009

Bibi, N., Sarwar, M. I., Ishaq, M., & Ahmad, Z. (2007). Mechanical and thermal properties of nano-composites of poly (vinyl chloride) and co-poly (vinyl chloride-vinyl alcohol-vinyl acetate) with montmorillonite. *Polymers & Polymer Composites*, *15*(4), 313–319. doi:10.1177/096739110701500407

Burmistr, M. V., Ovcharov, V. I., Shilov, V. V., Sukhyy, K. M., Gomza, Yu. P., Grischenko, V. K., & Ovcharenko, V. G. (2003). Synthesis, structure and mechanical properties of nanocomposites based on organoclay and polymeric matrix different nature, *Technomer 03. 18. Fachtagung uber Verarbeitung und Anwendung von Polymeren, Chemnitz*, *2*, 65–71.

Chakrabarty, J. (2006). *Theory of Plasticity* (3rd ed.). Elsevier.

De Boer, J. (1948). Quantum theory of condensed permanent gases I the law of corresponding states. *Physica*, *14*(2-3), 139–148. doi:10.1016/0031-8914(48)90032-9

Derjaguin, B., & Landau, L. (1941). Theory of the stability of strongly charged lyophobic sols and of the adhesion of strongly charged particles in solutions of electrolytes. *Acta Physico Chimica URSS*, *14*, 633.

Derjaguin, B. V., Churaev, N. V., & Muller, V. M. (1985). *Poverkhnostnye sily* [Surface forces]. Nauka.

Faghihi, K., Abootalebi, A. S., & Shabanian, M. (2013). New clay-reinforced polyamide nanocomposite based on 4-phenylenediacrylic acid: Synthesis and properties. *Journal of Saudi Chemical Society*, *17*(2), 191–197. doi:10.1016/j.jscs.2011.03.007

Hamaker, H. C. (1937). The London – van der Waals attraction between spherical particles. *Physica*, *4*(10), 1058–1072. doi:10.1016/S0031-8914(37)80203-7

Jeevanandam, J., Barhoum, A., Chan, Y., Dufresne, A., & Danquah, M. (2018). Review on nanoparticles and nanostructured materials: History, sources, toxicity and regulations. *Beilstein Journal of Nanotechnology*, *9*, 1050–1074. doi:10.3762/bjnano.9.98 PMID:29719757

Liao, B., Song, M., Liang, H., & Pang, Y. (2001). Polymer–layered silicate nanocomposites. A study of poly(ethylene oxide)/Na+–montmorillonite nanocomposites as polyelectrolytes and polyethylene–block–poly(ethyleneglycol) copolymer/Na+–montmorillonite nanocomposites as fillers for reinforcement of polyethylene. *Polymer*, *42*(25), 10007–10011. doi:10.1016/S0032-3861(01)00563-8

Lifshitz, E. M. (1956). The theory of molecular Attractive Forces between Solids. *Soviet Physics*, *2*(1), 73–83.

London, F. (1937). The general theory of molecular forces. *Transactions of the Faraday Society*, *33*, 8–26. doi:10.1039/tf9373300008b

Nicolais, L., & Carotenuto, G. (2005). *Metal-Polymer Nanocomposites*. John Wiley & Sons.

Sevim, İ. (2017). Intercalation of vermiculite in presence of surfactants. *Applied Clay Science*, *146*, 7–13. doi:10.1016/j.clay.2017.05.030

Shameem, M. M., Sasikanth, S. M., Annamalai, R., & Raman, R. G. (2021). A brief review on polymer nanocomposites and its applications. *Materials Today: Proceedings*, *45*(Part 2), 2536–2539. doi:10.1016/j.matpr.2020.11.254

Sinha Ray, S., Yamada, K., Okamoto, M., Ogami, A., & Ueda, K. (2003). New polylactide/layered silicate nanocomposites. High performance biodegradable materials. *Chemistry of Materials*, *15*(7), 1456–1465. doi:10.1021/cm020953r

Sukhyy, K. M., Belyanovskaya, E. A., Sukhyy, M. P., Tomilo, V. I., & Lytovchenko, R. D. (2018). Obtaining technology for nanocomposites based on polyamide-6 and organomodified montmorillonite. *Journal of Chemistry and Technologies*, *26*(1), 9–19. doi:10.15421/081802

Verwey, E., & Overbeek, T. (1948). *Theory of the Stability of Lyophobic Colloids*. Elsevier.

Zulfiqar, S., & Sarwar, M. I. (2008). Inclusion of aramid chains into the layered silicates through solution intercalation route. *Journal of Inclusion Phenomena and Macrocyclic Chemistry*, *62*(3-4), 353–361. doi:10.100710847-008-9478-5

Chapter 10
Physico-Mechanical Properties and Morphology of Nanocomposites

ABSTRACT

The technology of nanocomposites based on polyamide-6 (PA-6) and modified montmorillonite is developed. The technological parameters of extrusion and injection molding, and their influence on the formation of nanoscale inorganic phase of the polyamide are investigated. Morphology and thermal behavior of polyamide-6, composites based on PA-6 and modified montmorillonite are studied by WAXS, XRD, DTA and DSC data. New γ-phase formation is revealed in composite polyamide-6 – 1 – 2% of modified montmorillonite. Optimal concentrations of modified montmorillonite in the composites stated as 1 – 2% are shown to correspond with composites improved properties as compared to initial polyamide-6. It is explained by increasing crystallinity degree, which resulted from acting of modified montmorillonite particles as nucleation heterogenetic agents.

BACKGROUND

Development and application of polymeric materials require searching for rational solutions to give new properties to bulk polymers (Ray, 2013). Design of polymeric nanocomposites based on engineering polymers is a relevant objective.

DOI: 10.4018/978-1-7998-9728-6.ch010

Physical and mechanical characteristics are crucial factors as changes accompanied transitions from macro- to nanostructure investigated (Chen, Khobaib, & Curliss, 2003). When nanocomposites based on various polymeric matrixes, substantial increasing of physical and mechanical properties of composites is sufficiently observed as a natural and predicted result with relatively little concentration of clay mineral. Such changes of polymeric composites properties are peculiar to macro-nanostructure transition (Zhong, Zhu, & Wang, 2005).

Layered silicates as nanoscale heterophase require to obtain polymeric nanocomposites. Layered silicates, in particular montmorillonite, have a specific matrix structure causing their uniquel properties. For example, montmorillonite primary crystal elements (layers) can be expanded in water media. This process allows for sizable organic cation intercalation into interlaminar space resulted from replacements of metal cations. Subsequently, such modified layered silicates can be delaminated up to elementary nanosized layers during overlaying with polymeric matrix.

Physical and mechanical properties are affected by the next factors: size and shape of filler particles, filler particles orientation and their relative positions in the matrix, filling ratio, filler-binder interaction, polymeric matrix characteristics (Yebra-Rodriguez et al., 2011).

Numerous reports describing the preparation and characterization of polymer-based clay involved reinforcing a polymer with modiðed clay (ceramic type ðller) verify a considerable improving the degree of homogeneity and adhesion between the organic (polymer) and inorganic (clay) components when reactive organoclay applied (Zulðqar & Sarwar, 2009). It results in greatly enhanced properties of the hybrid materials, i.e. mechanical (Burnside & Giannelis, 1995), thermal (Paul et al., 2003) and barrier (Messersmith & Giannelis, 1994).

Also, consideration must be given to extrusion and pressure casting technical parameters influence over physical and mechanical properties when products obtained.

PRODUCTION TECHNOLOGY, MORPHOLOGY AND PHYSICAL AND MECHANICAL PROPERTIES OF NANOCOMPOSITES BASED ON THERMOPLASTICS AND ORGANOMODIFIED MONTMORILLONITE

Physico-mechanical characteristics of nanocomposites are one of the most significant in the study of changes that occur during the transition from macro to nanostructure.

In obtaining nanocomposites based on polymer matrices of different nature and clay minerals, a significant increase in the physical and mechanical properties of composites with a relatively low content of clay mineral is quite naturally and predictably observed. Such a change in the properties of polymer composites is inherent in the transition from macro to nanoscale structure.

As mentioned above, the physical and mechanical properties of polymer nanocomposites are influenced by the following factors:

- the size and shape of the filler particles;
- the orientation of the filler particles and their relative position in the matrix;
- degree of filling;
- interaction between filler and binder;
- characteristics of the polymer matrix.

Moreover, the influence on the physical and mechanical properties of the technological parameters of injection molding in the production of products should be taken into account.

In the process of transportation and storage of bulk materials, including nanomaterials, they tend to form aggregates of different configurations with different volumes. This is due to the interaction between particles due to forces of different nature, among which for particles with a smooth surface the main forces can be considered Van der Waals. Thus, for two spherical particles the energy of the van der Waals interaction can be found from as

$$E_{12} = -C_K \cdot N_1 \cdot N_2 \cdot \int_{V_1} dV_1 \int_{V_2} H^{-6} dV_2, \tag{1}$$

where N_1, N_2 are densities of particles in each of systems; V_1, V_2 are volumes of both of systems; C_K is constant.

Table 1. Values of E_v at r_1 = 40 nm and $A_H = 8,9 \times 10^{-20}$ J

r_2	E_v, kJ/m³						
	h, nm						
	0.3	0.6	1.2	2.4	4.8	9.6	19.2
10	11257.5	1796.70	268.26	36.04	4.12	0.37	0.02
20	14961.7	2464.28	386.91	56.21	7.22	0.76	0.06
40	14972.4	2512.23	406.57	62.02	8.60	1.02	0.10
80	11028.9	1870.21	308.04	48.36	7.03	0.90	0.10
160	6200.3	1057.54	175.90	28.07	4.19	0.56	0.06

For most materials, the value of the Hamaker constant is in the range $(0.2 \div 30) \times 10^{-20}$ J Grigoriev, Shilko et al. (2017). To estimate the force of the environment on the system of two interacting particles due to the van der Waals energy, we use the specific energy E_v in the form

$$E_v = E_{1,2}/V_e, \tag{2}$$

where V_e is a special volume of nanosystem normalized to spherical form $V_e = \pi \cdot d_e^3 / 6$, ($d_e$ is the equivalent diameter of the normalized sphere).

For two spherical particles, the equivalent diameter can be determined from the expression

$$d_e = \sqrt{(r_1 + r_2) \cdot h}\,.$$

In the Table 1 the value for E_v is given in absolute value at different ratios of one of the radii and the distance between the surfaces of the spheres. The results presented in the Table indicate that the leading role for Ev is played by the distance between the surfaces of the spheres. However, the change in radius does not give such a significant gradient for the specific energy.

As a rule, the nanoaggregate contains a large number of nanoparticles, which, in addition, may differ in shape. Then for the generalized equivalent diameter deo the following dependence can be written Strawhecker & Manias (2000).

$$d_{eo} = \sum_{i,j} \sqrt{\left(K_{fi} \cdot r_i + K_{fj} \cdot z_j\right) \cdot h_{i,j}} \Big/ \sum_{i,k} \left(1 + 0,5 \cdot i_{ri} \cdot K_{hk}\right), \tag{3}$$

Figure 1. Basic schemes of connection of four particles: a –linear; b – flat (four-square); c – flat (lozenge); d – 3D (tetrahedron)

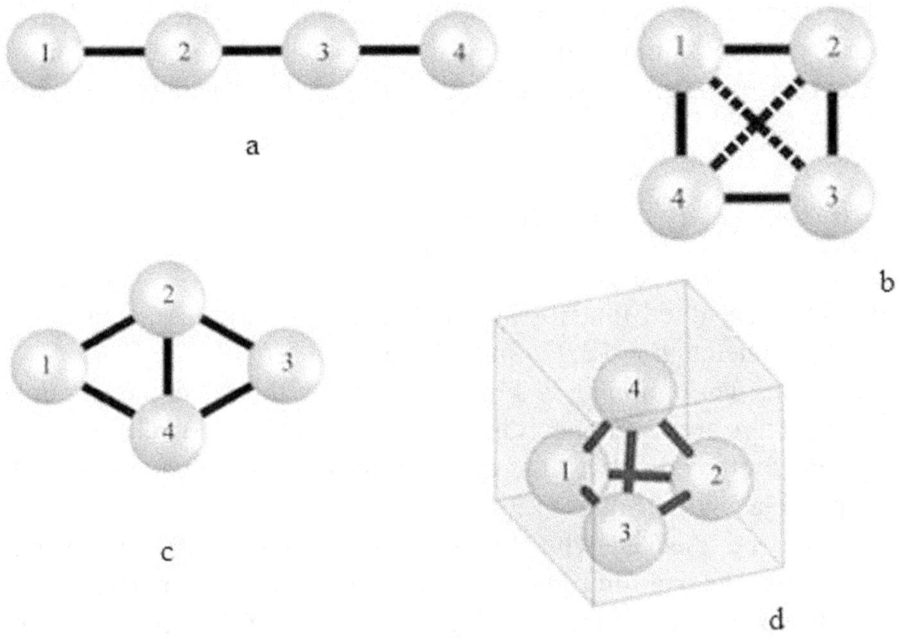

where K_{fi}, K_{fj} are factors of shape of particles;

i_{ri} is the number of closed circuits containing the same number of connections;
K_{hk} is the number of connections for one circuit.

For a system of particles, the value of specific energy will be determined by expression

$$E_{vo} = 6/\pi \cdot \min\left(\sum\nolimits_{i,j} E_{i,j}\right)/d_{eo}^{3}. \tag{4}$$

Let's consider four-particle systems. There may be several connection schemes. The basic schemes are shown in Fig. 1. For the scheme in Fig. 1 the system has three connections. We will assume that the particles have a spherical shape, i.e., and closed circuits are absent. If for all elements the sizes are identical, both radii, and distances between particles, the last equation will be the next

$$d_{eo}^{a} = 3 \cdot \sqrt{2 \cdot r_0 \cdot h} = 3 \cdot d_e. \tag{5}$$

For example, as $r_1=r_2=r_i=r_0=60$ nm and $h_{1,2}=h_{2,3}=h_{3,4}=h_{i,j}=3$ nm, we obtain d_{eo} = 56.921 nm, and when Eq. (1) used, for one connection $E_{1,2}$ = -1.086×10^{-22} kJ. Then the special energy (by absolute value) is determined for this case as:

$$E_{vo}^{a} = 6/\pi \cdot E_{1,2} / \left(d_{eo}^{a}\right)^3 = 1{,}124\,\text{kJ/m}^3.$$

It should be noted that for system from two particles according with adopted parameters d_e= 18,974 nm; E_v= 30,359 kJ/m³.

Let's consider option b in Fig. 1, again for the accepted parameters. Generally, there are six connections. Moreover, the four bonds correspond to Eq.(2), and for the other two (1–3 and 2–4) the distance between the surfaces of the particles will change.

It should be checked whether these elements will not be lower-order connections. Determining the distance between these particles as the hypotenuse of a right triangle with legs a = 2×r_0 + h, followed by subtraction of the particle diameter, we obtain: $h_{1,3}$ = $h_{2,4}$ = h_1 = 53.948 nm. Then we have $d_{e1.3}$ = $d_{e2.4}$ = 80.46 nm; $E_{1,3}$ = $E_{2,4}$ = $-6{,}808\cdot10^{-25}$ kJ.

According with obtained results, diagonal bonds are smaller than peripheral ones by three orders of magnitude, so they can be neglected. The equivalent diameter and specific energy for option b in Fig. 1, taking into account the fact that each particle has two identical connections, will be found from the ratios, according with the accepted parameters

$$d_{eo}^{l} = 4 \cdot \sqrt{2 \cdot r_0 \cdot h} / (1 + 0{,}5 \cdot 1 \cdot 4) = 25{,}298\,\text{nm};$$

$$E_{vo}^{l} = 6/\pi \cdot 2 \cdot E_{1,2} / \left(d_{eo}^{l}\right)^3 = 25{,}615\,\text{kJ/m}^3.$$

For the scheme in Fig. 1c, we have two closed loops with a minimum number of connections equal to two (for the first and third element), and the total number of connections corresponds to five, i.e. the expressions will look like:

$$d_{eo}^2 = 5 \cdot \sqrt{2 \cdot r_0 \cdot h} / (1 + 0,5 \cdot 2 \cdot 3) = 23,717\,\text{nm};$$

$$E_{vo}^2 = 6 / \pi \cdot 2 \cdot E_{1,2} / \left(d_{eo}^2\right)^3 = 31,087\,\text{kJ/m}^3.$$

According with the last results, for the separation of particles according to the scheme in Fig. 1c, it is necessary to spend a little more energy than for the simplest case, namely, for a two-particle system.

And finally, for the scheme in Fig. 1, d four closed circuits occurs with a total number of bonds - six. In this case, each particle has three connections. Then for the equivalent diameter and specific energy we obtain the following values:

$$d_{eo}^3 = 6 \cdot \sqrt{2 \cdot r_0 \cdot h} / (1 + 0,5 \cdot 4 \cdot 3) = 16,26\,\text{nm};$$

$$E_{vo}^3 = 6 / \pi \cdot 3 \cdot E_{1,2} / \left(d_{eo}^3\right)^3 = 144,63\,\text{kJ/m}^3.$$

The obtained result indicates that for nanoparticles with the same parameters the energy of their separation increases significantly during the transition from a linear structure to a flat one and then to a bulk structure, which is associated with an increase in the number of bonds and a change in equivalent diameter.

Under the conditions r_0 = 60 nm, h = 3 nm and $A_{12} = 8.9 \cdot 10^{-20}$ J we have $E_{1.2} = -1.244 \cdot 10^{-23}$ kJ, which is an order of magnitude less than for two spherical particles.

In the case of two planes, the interaction energy can be calculated by formula (3). When two square plates with side dimensions b = 2×r interact, other things being equal, we obtain $E_{1,2} = -3,777 \cdot 10^{-21}$ kJ from Eq.(3), i.e. almost an order of magnitude larger than between two identical spherical particles of corresponding radius . In addition, flat surfaces can also have protrusions and depressions of different configurations, which can also affect the amount of energy of interaction of particles with each other.

For different types of equipment for processing polymeric materials, such as extruders of different types, the main factor in the separation of nanoaggregates is the shear stresses that occur in the working volume. In general, shear stresses τ can be given as follows: $\tau = \mu_c \cdot \dot{\gamma}^n$, where μ_c is the average coefficient of consistency.

The average value of the shear rate can be determined from the following dependence: $\dot{\gamma} = V_c / l_p$ where l_p is a characteristic geometric parameter, and V_c is the linear velocity of the rotating element.

For a single-worm unit, there are two working volumes with shear deformations: the first is a helical channel with a depth of $l_p = H$; the second is a radial gap with height $l_p = \delta r$. The linear velocity, in this case, will be determined from the expression: $V_c = \pi \cdot Dc\, n0$ (Dc is an inner diameter of the material cylinder; n0 is the speed of the working body).

To estimate possible shear stresses we will accept such parameters of the unit and material: $\mu_c = 9.89\times10^3$ Pa×s^n; $n = 0.513$; $n_0 = 2.5\times s^{-1}$; $D_c = 32$ mm; $\delta_r = 0{,}12$ mm; $H = 3$ mm. Then it can be obtained: $\dot{\gamma}_\delta = 2094.395\times s^{-1}$; $\dot{\gamma}_= = 83.776\times s^{-1}$; $\tau_\delta = 0{,}5$ MPa; $\tau_= = 0.096$ MPa.

If we compare the results with the values in the table, it can be seen that with these extruder parameters and material characteristics, all systems can be separated in the radial gap, starting from the 3rd column (ie at h = 1.2 nm and above). As for the shear stresses occurring in the helical channel, the degree of separation can be achieved at h = 2.4 nm.

The research was conducted in the following areas:

1. Investigation of the influence of thermomechanical parameters in the process of worm-disk extrusion on the degree of delamination of the layered structure of montmorillonite.
2. Determination of the possible range of regulation of the properties of the polymer matrix due to the introduction of organomodified montmorillonite
3. Creation of a mathematical model of the process of worm-disk extrusion, which describes the viscoelastic properties of polymer melts and allows to determine technological parameters that affect the thermomechanical transformations of layered silicates in the polymer matrix, as well as extrapolate found in laboratory processing modes with a given set of properties on equipment of a certain capacity

The rheological equations of state of the Newtonian fluid used in common mathematical models of worm-disk extrusion processes do not adequately describe the behavior of the polymer melt in the presence of laminated layered silicates. Therefore, in the mathematical model it is necessary to use the equation of state of highly elastic White-Metzner fluid, because it describes the elastic properties of polymer composites, the formation of normal stresses, and the Beisenberg effect. The design features and conditions of the extrusion process make it possible to use the following simplifications: the movement of material in the disk zone of the extruder is a flat, forced, constant flow of continuous, viscoelastic fluid between two disks in isothermal conditions.

The magnitude of the shear component was calculated from the ratio (McNeil & Grimsditch, 1993):

$$\tau_{12} = K \cdot (W \cdot r / B)^{1+1/n} \tag{6}$$

where K is the value of the coefficient of consistency,

B is the size of the gap between the disk and the housing
W is the speed of rotation of the disk,
r - disk.

The value of normal stresses is found from the expression McNeil & Grimsditch (1993):

$$\tau_{22} = \left(K^2 / G \cdot n\right) \cdot \left(1 / (1+n)\right) \cdot (W \cdot R / B)^{2n} \tag{7}$$

where R is the radius of the disk.

Volume costs:

$$Q = \left(\frac{\pi \cdot r^3}{64}\right) \cdot \left(\frac{n}{3n-1}\right) \cdot \left(\frac{W \cdot r}{B}\right)^2 \cdot \left(\frac{H}{4G \cdot n}\right)^{1/n} \cdot \left(\frac{2r}{L}\right)^{1/n} \tag{8}$$

The residence time of the polymer melt in the zone of intense thermomechanical influence:

$$t = \frac{2\pi \cdot B}{Q} \cdot \int_r^R r dr = \frac{\pi \cdot B \cdot R^2}{Q} \cdot \left(1 + \frac{r^2}{R^2}\right) \tag{9}$$

Mathematical model of the worm-disk extrusion process allows determining the following parameters of the polymer melt flow in the disk assembly: distribution of deformation velocities along the longitudinal section of the disk gap, time of the composite in the zone of intense thermomechanical strain stress tensor, which allows calculating the power parameters of the delamination process.

The mathematical model was used to calculate the optimal technological modes and initial data for the design of process equipment of a certain productivity by the value of unobserved process parameters invariant to the geometric dimensions of the apparatus (strain rate, shear strain and composite time in the extruder disk zone).

Optimization of the technological mode of processing in the process of worm-disk extrusion was performed by combining methods of mathematical and physical modeling of the process. To conduct physical modeling of the behavior of the material during processing, the influence of process parameters on the properties of the material being processed and products from it in laboratory conditions was experimentally investigated. According to these data, the technological mode of obtaining products with improved properties was determined.

Figure 2 - 4 presents the physical and mechanical properties of the studied composites based on polyamide-6. There is a clear pattern between structure and strength characteristics.

When 1% of unmodified montmorillonite is added to the polyamide, the tensile strength increases by almost 30% compared to the original polyamide, while the toughness and elongation are reduced by 10% and 24%, respectively. With an increase in the content of the original montmorillonite in polyamide to 2%, the increase in tensile strength is already 20%, and the decrease in toughness and elongation - 25 and 36%, respectively. With a montmorillonite content of polyamide of 5%, the increase in tensile strength becomes even less pronounced and is 6%, and the toughness and elongation are reduced quite significantly - by 33% and 63%, respectively. Thus, it becomes obvious the inexpediency of increasing the concentration of montmorillonite over 2% in polyamide-6 + montmorillonite composites. Moreover, it is obvious that the change of processing modes - variation of shear stresses in the extrusion zone does not cause the destruction of the crystal structure of unmodified montmorillonite.

The use of modified montmorillonite brings the physical and mechanical properties of the composite based on polyamide-6 to a new level. This is due to the destruction of the structure of montmorillonite and the formation in the polymer matrix of nanoscale discrete phase, accompanied by an increase of 2-3 orders of magnitude of the interaction between montmorillonite and polymer matrix and the formation of interfacial transition layers that determine the strength properties of the polymer composition. Due to the high values of

Figure 2. Histograms of tensile strength of composites based on polyamide-6 with different content of montmorillonite. The figure shows the modifiers used.

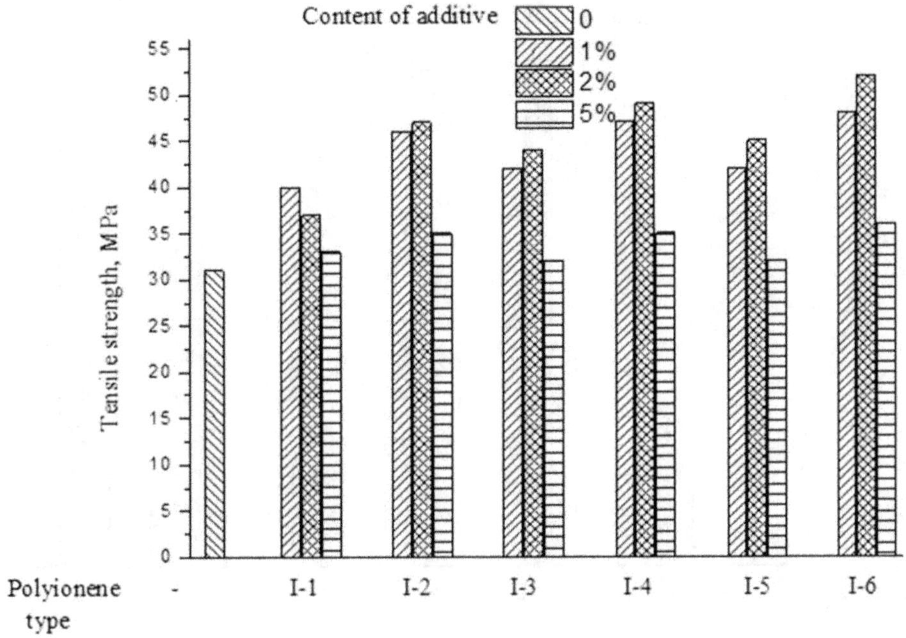

surface energy of individual montmorillonite plates, the level of interaction between nanoparticles (lamellae) of montmorillonite and chains of the polymer matrix significantly increases, which leads to the formation of a qualitatively new structure and accompanied by increased strength.

Thus, with the introduction of 1% of modified montmorillonite in polyamide, the tensile strength increases by 35-60% compared to the original polyamide (depending on the polyionene used, which will be discussed below). Impact strength increases by 100-120% and elongation by 20-25%. When the concentration of modified montmorillonite in polyamide increases to 2%, the tensile strength and elongation remain almost unchanged. Impact strength increases by 120-150%. Increasing the amount of modified montmorillonite in the composition to 5% reduces the elongation by 20-40%. At the same time, the tensile strength increases by only 10-15%, and the toughness - by 45-65%. Thus, the optimal content is 2 wt.% Modified montmorillonite in the composition based on polyamide-6.

Figure 3. Histograms of impact strength of composites based on polyamide-6 with different content of montmorillonite. The figure shows the modifiers used.

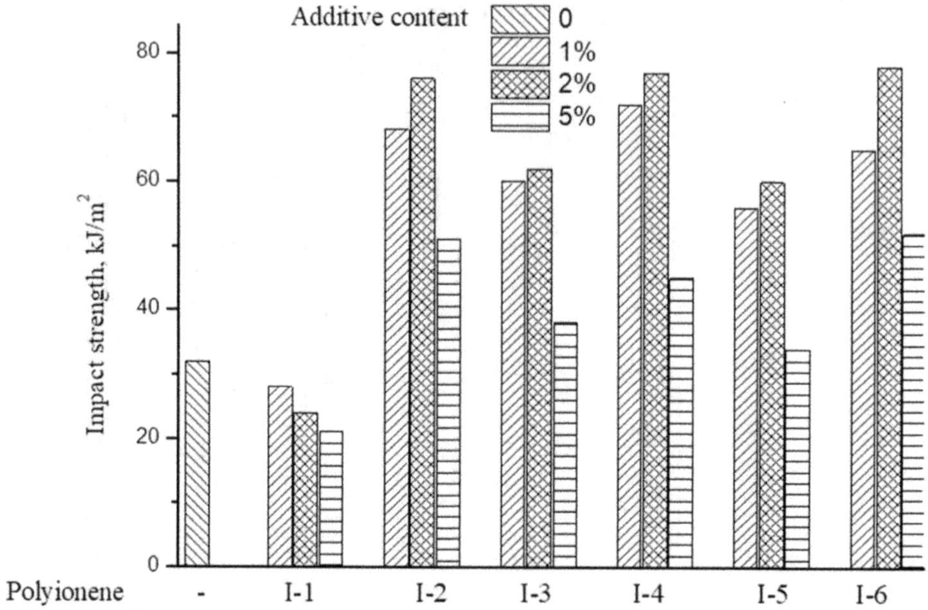

A similar tendency to increase the physical and mechanical properties is observed in compositions based on polystyrene (Fig. 5-7), but is less pronounced. Thus, at a concentration of modified montmorillonite in polystyrene of 1% and 2% tensile strength increases by 18-22% and 20-32%, respectively.

The toughness increases by 8-15% and 20-32%, respectively, and the elongation by 180-210% and 190-220%. It should be noted that for samples of the composition based on polystyrene with 2% modified montmorillonite, even the formation of "neck" during stretching is observed, which indicates significant structural changes. The increase in the concentration of modified montmorillonite in polystyrene to 5% was accompanied by a decrease in toughness by 16-28% and elongation by 8-12%. The tensile strength of polystyrene-based composites at this concentration of modified montmorillonite remained almost unchanged.

Montmorillonite modified with polyonenes I-6, I-1, and I-4 with aliphatic, alkylaromatic and binuclear phenolic radicals are the most effective in terms of improving the physical and mechanical properties by phenyl and dialkylphenyl radicals between Quaternary nitrogen atoms.

Figure 4. Histograms of elongation at break of composites based on polyamide-6 with different content of montmorillonite. The figure shows the modifiers used.

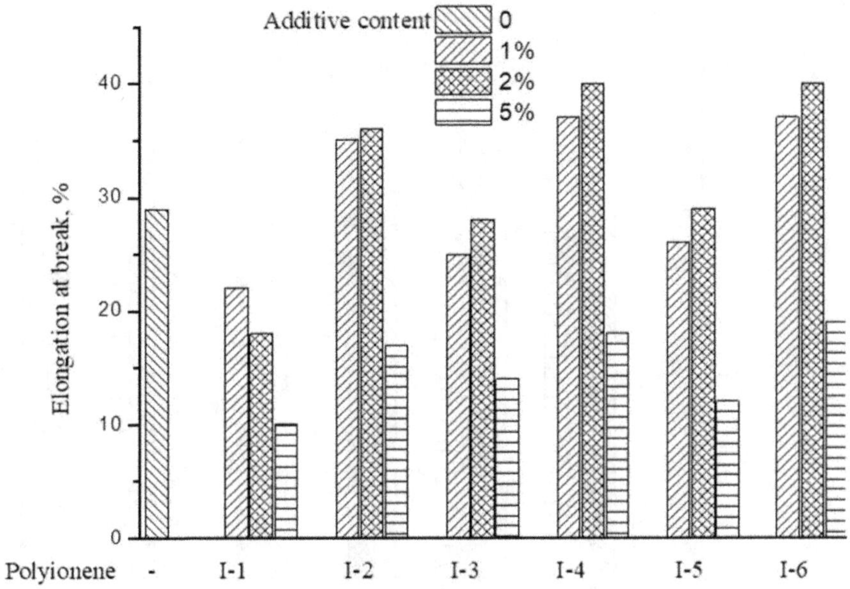

During the physical modeling of the process, the influence of the technological mode of processing on the worm-disk extruder ED-2.2 (Speed of rotation of the worm - 50-250 rpm, diameter of the worm - 45 mm, was investigated) (L / D = 8), disk diameter - 120 mm, disk gap 3–8 mm) for the formation of inorganic nanophase in the polymer matrix during extrusion.

The effectiveness of the effect was evaluated by changing the rheological properties of the material melt, during processing at different temperature - speed regimes, because the change in viscosity reflects the transformations occurring in the material.

It is known that when the melt of the polymer matrix and organomodified layered silicates is deformed, transformations occur that are associated with the destruction of the layered structure with the formation of a nanophase in the polymer matrix. This significantly changes the physical and mechanical properties of the formed composite in comparison with the original polymer matrix. This leads to the use of the degree of delamination, as the main criterion for the quality of the extrusion process.

The effect of total shear deformation and nanocomposite residence time in the zone of intense thermomechanical influence on the properties of polymer-inorganic composites was investigated by changing the processing

Figure 5. Histograms of tensile strength of polystyrene-based composites with different contents of montmorillonite. The figure shows the modifiers used.

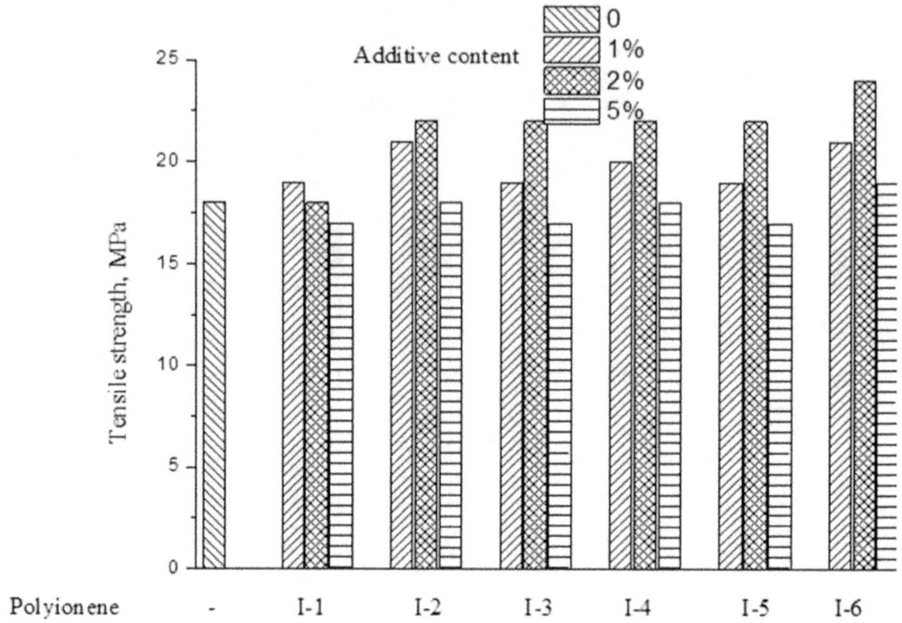

temperature at a constant shear rate and varying the shear rate in a wide range at constant processing temperature. The dependence of tensile strength and toughness of composites PA-6 + 2% montmorillonite modified I-6 on the shear deformation in the disk zone and extrusion temperature are presented in Fig. 8 and 9, respectively. The dependence of tensile strength and toughness of composites PA-6 + 2% of unmodified montmorillonite on shear deformation in the disk zone and extrusion temperature are presented in Fig. 10 and 11 respectively.

For the original polyamide, the physical and mechanical properties do not depend on the technological parameters of extrusion. The range of changes in physical and mechanical parameters is 10-15%. The extreme tensile strength behavior is apparently related to the change in the molecular weight of the polyamide in the deformation zone. For polyamide filled with unmodified montmorillonite, the change in physical and mechanical properties is typical for trivial filled polymer systems, i.e., with increasing tensile strength, the toughness decreases sharply.

Figure 6. Histograms of impact strength of polystyrene-based composites with different montmorillonite content. The figure shows the modifiers used.

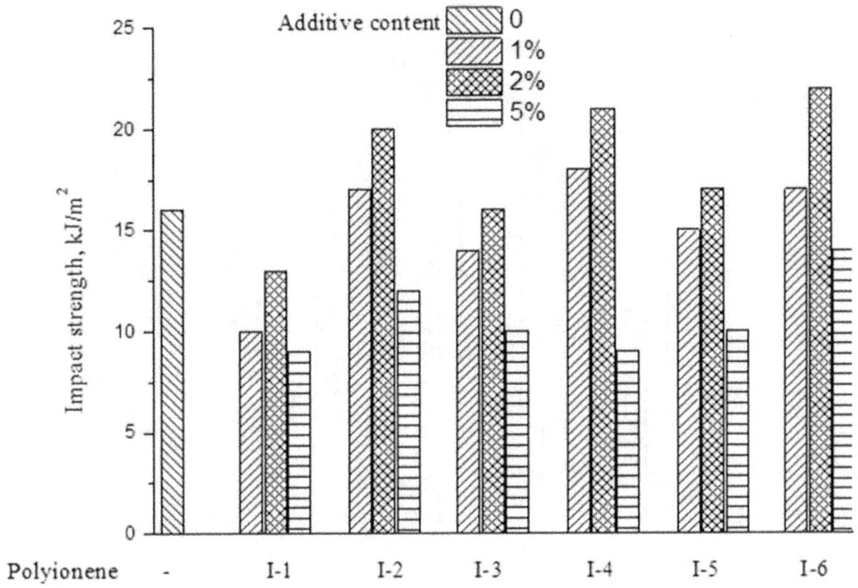

The use of modified montmorillonite brings the physical and mechanical properties of the composite based on polyamide-6 to a new level. This is due to the destruction of the structure of montmorillonite and the formation in the polymer matrix of nanoscale discrete phase. This is accompanied by an increase of 2-3 orders of magnitude in the area of interaction between montmorillonite and the polymer matrix and the formation of interfacial transition layers, which determine the strength properties of the polymer composition as a whole.

Due to the high values of surface energy of individual montmorillonite plates, the level of interaction between nanoparticles (lamellae) of montmorillonite and chains of the polymer matrix significantly increases, which leads to the formation of a qualitatively new structure and accompanied by increased strength. It is noted that the samples obtained at a processing temperature of 240 and 250 °C, have a higher viscosity than the viscosity of the original, not subjected to processing samples.

When the temperature rises above 260 °C, there is a decrease in viscosity to values below the viscosity of the original PA-6.

Figure 7. Histograms of elongation at break of composites based on polystyrene with different content of montmorillonite. The figure shows the modifiers used.

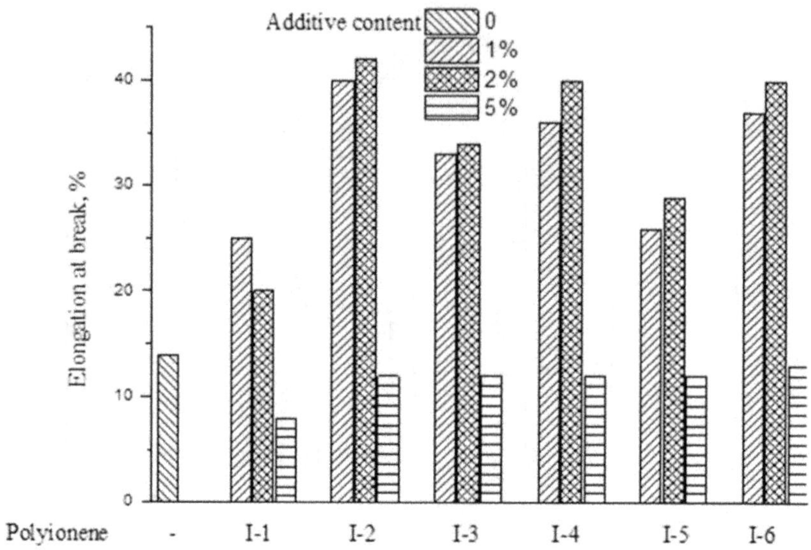

The strength changes little with increasing temperature to 260-270 °C, although slightly higher values have samples obtained at a temperature of 260 °C.

In the study of the properties of composites that were processed in the range of shear rates of 100-400 s^{-1}, due to the design capabilities of the extruder, and the processing temperature of 250 °C, the following is established. In the melts of composites filled with 2% of organomodified montmorillonite, at a shear rate of 100 and 200 s^{-1}, the effective viscosity, respectively, 32 and 25% higher than the viscosity of the original polyamide. The samples of nanocomposites processed at a shear rate of 200 s^{-1} have the greatest strength.

As the shear rate increases to 300 s^{-1}, the melt viscosity decreases by approximately 40%. At the same time, there is a shift in the curve of the relaxation spectrum towards shorter relaxation times, which was not observed before. Changing the rate of delamination processes leads to some reduction in the level of physical and mechanical properties.

In samples processed at a shear rate of 400 s^{-1} is observed. a sharp decrease in the effective viscosity of 2 times. At the same time there was a further

Figure 8. 3D plot of tensile strength vs. extrusion temperature and shear velocity of composite PA-6+2% organomodified montmorillonite.

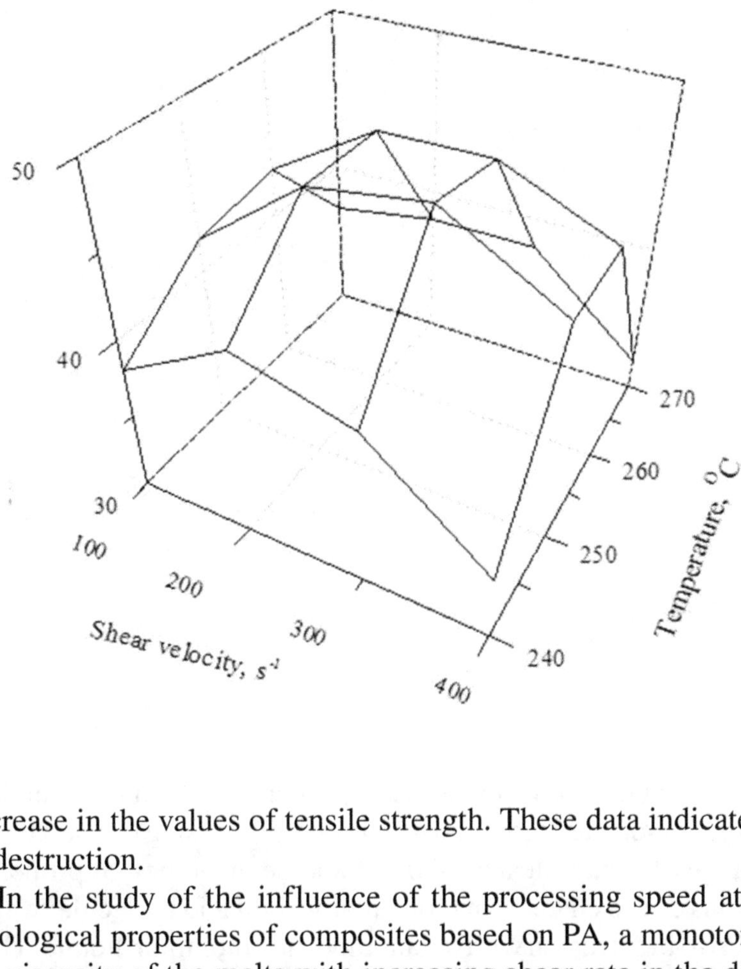

decrease in the values of tensile strength. These data indicate the processes of destruction.

In the study of the influence of the processing speed at 260°C on the rheological properties of composites based on PA, a monotonic decrease in the viscosity of the melts with increasing shear rate in the disk zone of the extruder. The strength of the samples deteriorates markedly. Thus, raising the processing temperature above 250°C in the entire range of shear rates has a negative effect on the properties of the material and accelerates the destructive processes.

The studies indicate a wide possibility of regulating the rheological and physico-mechanical properties of composites based on PA-6 and organomodified montmorillonite by changing the parameters of thermomechanical influence on the processes of delamination of the layered structure of montmorillonite.

Figure 9. 3D plot of impact strength vs. extrusion temperature and shear velocity of composite PA-6+2% organomodified montmorillonite.

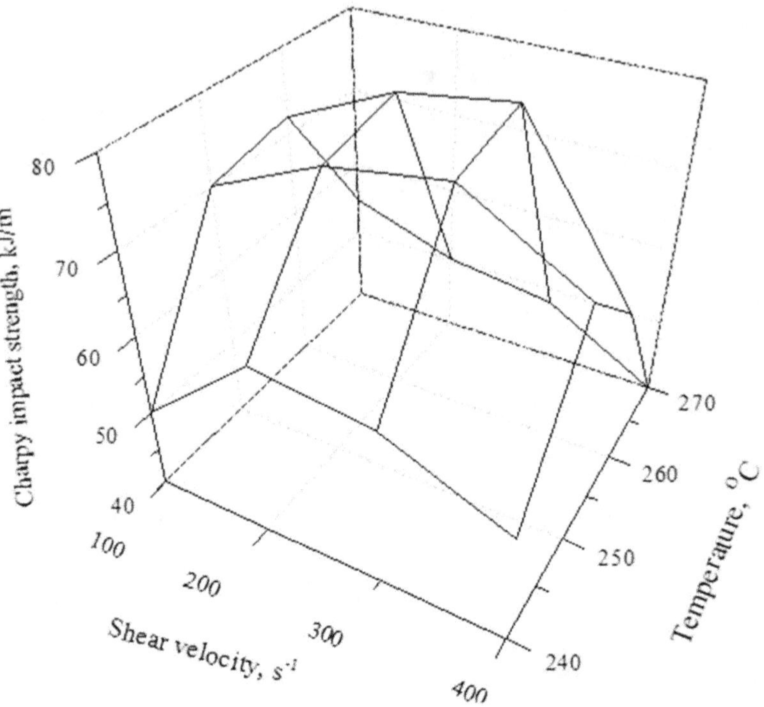

There is a significant improvement in the properties of products made of PA-6 and 2% organomodified montmorillonite at a processing temperature of 250°C in the range of shear rates 200-300 s^{-1} and deterioration of properties in other conditions due to changes in molecular weight and molecular weight distribution of PA-6 in extrusion with organomodified montmorillonite. These changes affect the rheological properties of the melt material.

The data obtained indicate that the appropriate treatment in the disk gap leads to a significant increase in the mechanical properties of products made of polymer nanocomposites.

This effect can be explained by two main reasons. First, shear velocities in the range 102 - 104 s^{-1} can be created in the disk gap, and at different time intervals, which leads to the division of nanoagregates into separate elements that have free radicals that allow forming new bonds. Second, these elementary nanoparticles are evenly distributed throughout the polymer matrix, creating a more branched contact area between the nanoparticles and the polymer melt.

Figure 10. 3D plot of tensile strength vs. extrusion temperature and shear velocity of composite PA-6+2% unmodified montmorillonite.

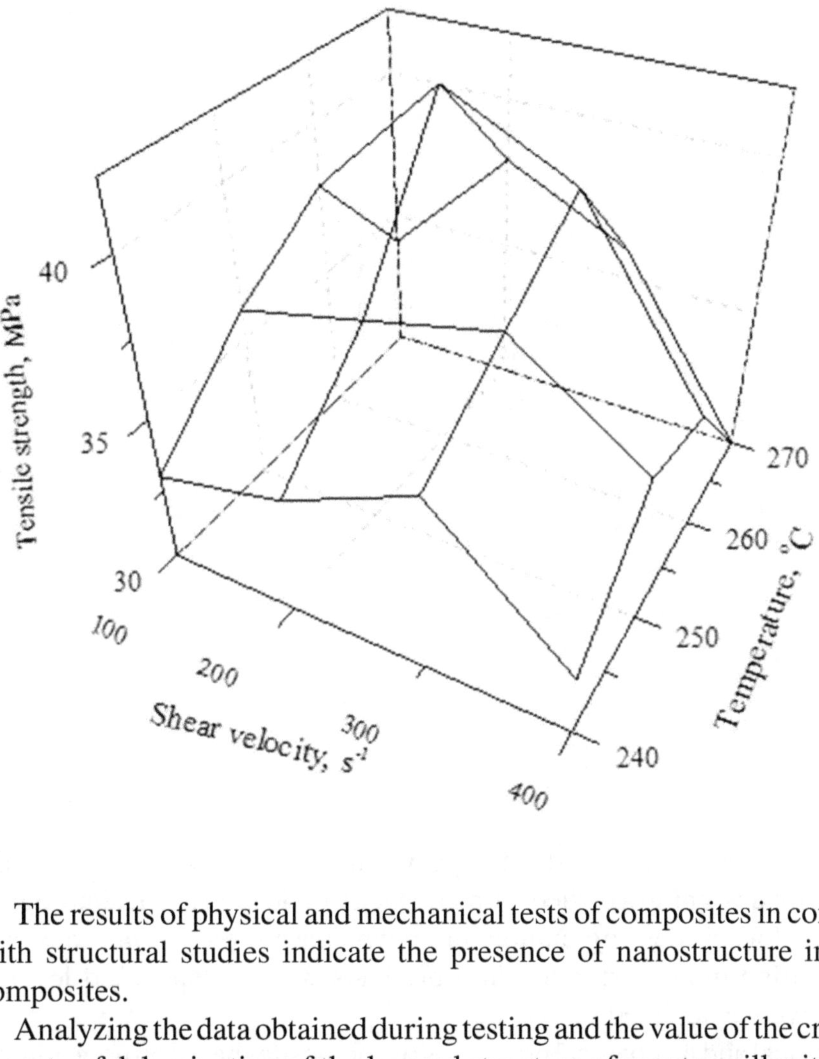

The results of physical and mechanical tests of composites in combination with structural studies indicate the presence of nanostructure in polymer composites.

Analyzing the data obtained during testing and the value of the criterion for the rate of delamination of the layered structure of montmorillonite, the data of physical and mechanical properties, comparing the values obtained during testing of samples of composites processed at different technological modes in the composite when changing, thermomechanical effect. The optimal mode of obtaining polymer nanocomposites on the used extruder in products with improved physical and mechanical characteristics is determined: processing temperature 250-260°C, disk rotation speed 100-110 rpm, which provides level, shear rate 200-300 s^{-1} and time stay in the disk zone 8-9 s, complete shear deformation 7500- 8500.

Figure 11. 3D plot of impact strength vs. extrusion temperature and shear velocity of composite PA-6+2% unmodified montmorillonite.

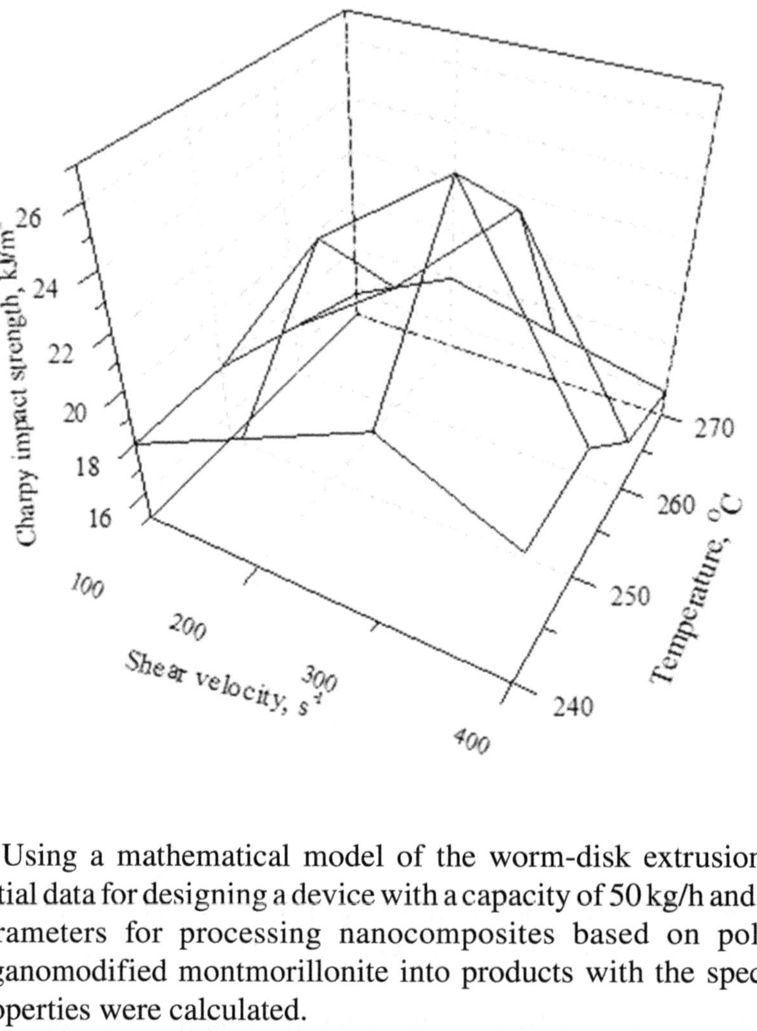

Using a mathematical model of the worm-disk extrusion process, the initial data for designing a device with a capacity of 50 kg/h and technological parameters for processing nanocomposites based on polyamides and organomodified montmorillonite into products with the specified level of properties were calculated.

The physical and mechanical properties of composites based on polyamide-6 and polystyrene using montmorillonite, which is modified with polyionene I-6, as well as the original montmorillonite during processing by injection molding were investigated.

In order to select the optimal technological parameters of the injection molding process, we studied the dependence of mechanical and thermophysical properties of composites on the casting temperature and holding time under pressure. The results of studies of the properties of composites from the temperature of injection molding are given in Table 2.

Table 2. Properties of nanocomposites based on polyamide-6 depending on the temperature of injection molding (concentration of montmorillonite modified I-6 - 2%.)

Properties	Dosing zone temperature, °C				
	240	250	260	270	280
Impact strength, kJ/m²	75	77	78	73	72
Tensile strength, MPa	47	48	51	47	43
Relative elongation at break, %	35	37	37	35	33

The dependence of the toughness of the nanocomposite based on polyamide-6 on the casting temperature is observed at a maximum temperature of 260°C. The toughness of the composite increases from 75 kJ/m^2 at 240°C to 78 kJ/m^2 at 260°C, and then decreases to 72 kJ/m^2 ‘at 280°C. Tensile strength increases from 47 MPa at 240 ° C to 51 MPa at 260°C; then it is reduced to 43 MPa at 280°C. The relative elongation of the material slowly increases from 35% at 240°C to 37% at 260°C, and at 280°C decreases to 33%. Thus, the processing temperature has little effect on the relative elongation of the samples during tension, and only in some cases - with increasing temperature from 240°C to 270°C - it increases significantly, and at 280°C - decreases to 33%. The optimum processing temperature of the composite is 260°C. The material obtained at this temperature has the best properties in comparison with composites that were obtained at other temperatures, namely: high value of elongation at break is 37%; the greatest tensile strength is 51 MPa; the highest impact strength is 78 kJ/m^2. Table 3 shows the effect of exposure time under pressure on the properties of nanocomposites based on polyamide-6.

Table 3. Influence of exposure time under pressure on the properties of nanocomposites based on polyamide-6 (concentration of montmorillonite modified I-6 - 2%.)

Properties	Exposure time under pressure, s			
	5	10	20	40
Impact strength, kJ/m²	75	78	78	78
Tensile strength, MPa	57	51	51	51
Relative elongation at break, %	35	37	37	37

Based on the data in Table 3, it can be stated that the exposure time under pressure has little effect on the physical and mechanical properties of composites.

Table 4. Thermal properties of composites based on polyamide-6 and 2% montmorillonite

Properties	PA-6	PA-6+2% MM, polyionene					
		-	I-1	I-3	I-4	I-5	I-6
Degradation start temperature, °C	361	379	380	382	375	385	384
Temperature of 50% mass loss, °C	460	485	581	587	562	573	611

The heat resistance of the obtained nanocomposites in comparison with the original polyamide-6 and polystyrene was analyzed. Tables 4 and 5 show the data of thermogravimetric analysis for composites based on polyamide-6 and polystyrene, with the optimal, in terms of improving the physical and mechanical characteristics, the concentration of modified montmorillonite - 2%. For the composite with unmodified montmorillonite, the heat resistance increases by only 18°C compared to the original polyamide. Analysis of mass loss curves for polyamide nanocomposites shows a change in the nature of the destruction process. The heat resistance of the composite varies insignificantly - 20-25 °C, but the temperature of 50% weight loss increases by 100-150°C for polyamide with modified montmorillonite. For polyamide with modified montmorillonite, two processes that occur during destruction can be traced. In the temperature range 380-550°C, the mass decreases slowly, and after 550 ° C avalanche, at a higher speed compared to the original polyamide.

For polystyrene composites with modified montmorillonite is characterized by an increase in heat resistance by 25-30°C compared to the original polystyrene, the temperature of 50% weight loss for the composite with modified montmorillonite and the original polystyrene differ by 30-50°C. The rate of destruction of the original polystyrene is higher compared to polystyrene, which contains modified montmorillonite.

Table 5. Thermal properties of composites based on polystyrene (PS) and 2% montmorillonite

Properties	PS	PS+2% Mm, polyionene					
		-	I-1	I-3	I-4	I-5	I-6
Degradation start temperature, °C	283	295	311	315	312	314	312
Temperature of 50% mass loss, °C	372	385	418	422	405	416	421

Figure 12. WAXS profiles of the original polystyrene (1), a nanocomposite containing 1 (2) and 2 (4)% of montmorillonite modified I-6 compared to scattering by a mechanical mixture containing 2 wt.% Montmorillonite modified I-6 (3).

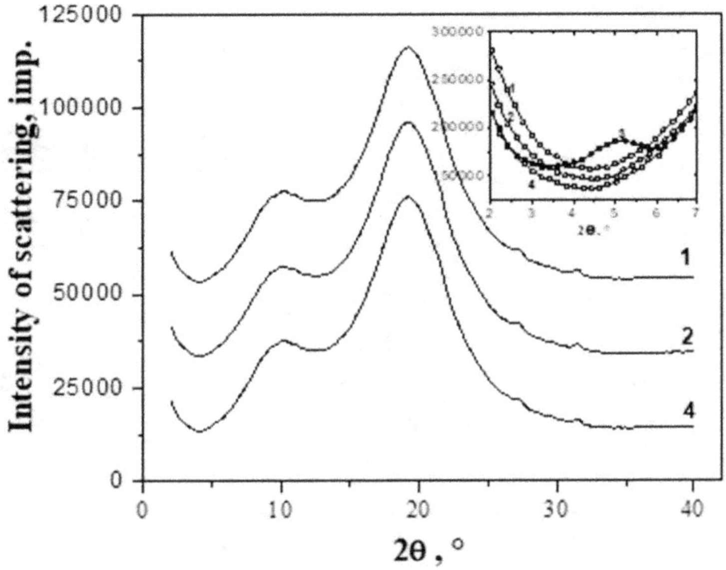

The increase in heat resistance of the studied nanocomposites based on polyamide and polystyrene is associated with changes in barrier properties. Nanosized montmorillonite particles with a thickness of about 1.5 nm and a length and width of several μm, prevent the diffusion of oxygen molecules and thus slow down the catalytic destruction of the polymer Burmistr et al. (2005).

WAXS profiles of composites obtained by extrusion in comparison with the scattering profiles of mechanical mixtures of polymer powders with 2% montmorillonite, modified I-6 are given in Fig. 12 and 13. A common pattern for diffractograms of nanocomposites based on polystyrene (Fig. 12) is the absence of changes in profiles, relative intensity and angular positions of the characteristic maximum for the original polystyrene. This is evidence of the preservation of the features of the near-order polymer matrix. On the scattering curves of mechanical mixtures of polystyrene and 2% of modified montmorillonite, a maximum is observed in the region 2θ = 5.7°.

Figure 13. WAXS profiles of the original PA-6 (4), a nanocomposite containing 5 (1) and 2 (2) and 1 (3)% of montmorillonite modified I-6 (1) compared to scattering by a mechanical mixture containing 2% of montmorillonite modified I-6 (5).

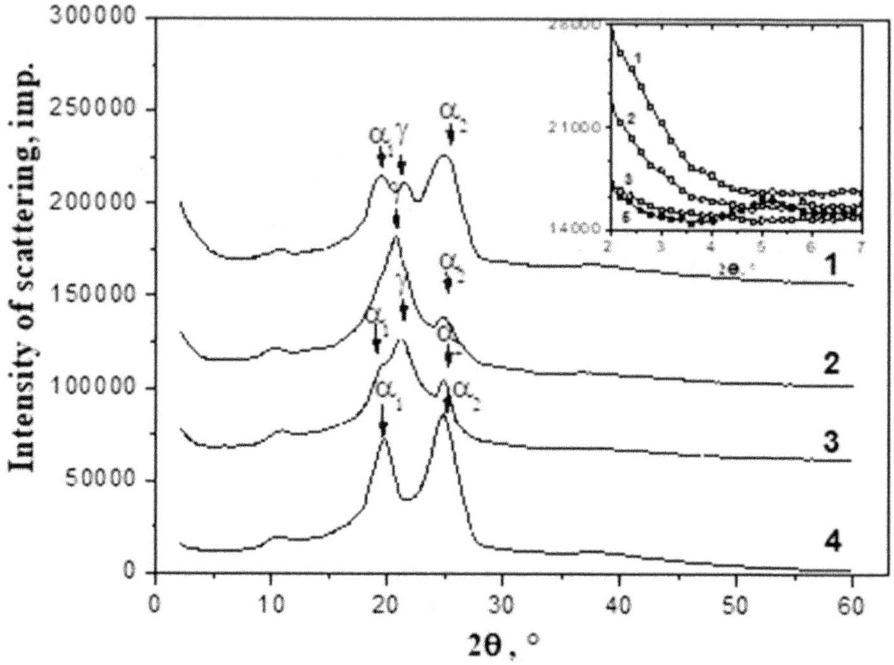

On the WAXS profiles of nanocomposites in this area of angles, a diffusion take-off towards the zero scattering angle is observed, the level of which increases with increasing content in the composition of modified montmorillonite. Extrusion leads to the destruction of the ordered layered structure of the modified montmorillonite and the distribution of individual layers (lamellae) in the polymer matrix.

In the partially crystalline polyamide-6 more significant structural changes are observed. In the area of small angles for nanocomposite based on PA-6 (Fig. 13) we observe the absence of the maximum characteristic of modified montmorillonite, which together with its presence in the mechanical mixture of PA-6 and 2% montmorillonite modified I-6 indicates the destruction of the layered structure modified montmorillonite and formation in PA-6 of nanoscale inorganic phase.

Two stable crystalline modifications are possible for PA-6, which are identified as α- and γ-phases. Two intense maxima at $2\theta \cong 200$ (d = 0.44

nm) and 2θ ≅ 250 (d = 0.37 nm) are observed on the diffraction curve of the α-phase PA-6 (Fig. 13), and on the diffraction pattern of the γ-phase PA -6 only one peak in the region 2θ = 210 (d = 0.41 nm). In Fig. 12 diffraction patterns of the original PA-6 and compositions based on PA-6 and 1.5.5% of montmorillonite modified I-6 are given.

Characteristic features of the α-phase are the formation of hydrogen bonds between molecules "directed" in opposite directions and the shift of associates by 3/14 of the identity period. Molecules in this case take a completely rectified conformation. The γ-phase is characterized by the formation of hydrogen bonds between parallel chains "directed" in one direction. This leads to a somewhat twisted conformation of macromolecules and, consequently, to a decrease in identity parameters compared to the α-phase.

The original PA-6 is characterized by α-modification of the crystal lattice. For nanocomposites based on PA-6 and modified montmorillonite, the formation of a new γ-phase is observed. The introduction of 1% and 2% of modified montmorillonite in PA-6 significantly reduces the diffraction intensity of the α-phase peaks along with the appearance of a new peak in the area of angles 2θ = 210, characteristic of the crystalline γ-phase PA-6. When increasing the concentration of modified montmorillonite to 5%, the content of γ-phase in the composite becomes insignificant.

It is obvious that delamination of the crystal structure of modified montmorillonite is accompanied by a change in the modes of crystallization of composites and the formation of a new γ-crystalline phase by a heterogeneous mechanism of nucleation.

The presence of crystalline γ-form in the studied nanocomposites is also confirmed by experimental data of differential scanning calorimetry. Fig. 14 shows the DSC curves of the original PA-6 and the studied nanocomposites.

For the original polyamide, there is only one melting peak at a temperature of 220°C, which corresponds to the melting of the crystalline α-phase PA-6. Nanocomposites on DSC curves are characterized by the presence of two melting peaks: high-temperature - at 220°C, which corresponds to the melting of the crystalline α-form, and low-temperature - in the region 210 ÷ 215°C, which corresponds to the melting of the crystalline γ-form PA-6. The maximum intensity of the melting peak of the γ-form is observed for a nanocomposite with 2% of modified montmorillonite, which is fully correlated with the results of wide-angle radiography. Since both α- and γ-phases are observed for all composites, an average value of 190 J / g was chosen for the melting enthalpy Sukhyy K.M. et al. (2018).

Figure 14. DSC curves of the original PA-6 (4), nanocomposites PA-6 + 1% (3), PA-6 + 2% (2), PA-6 + 5% (1) montmorillonite modified I-6 .

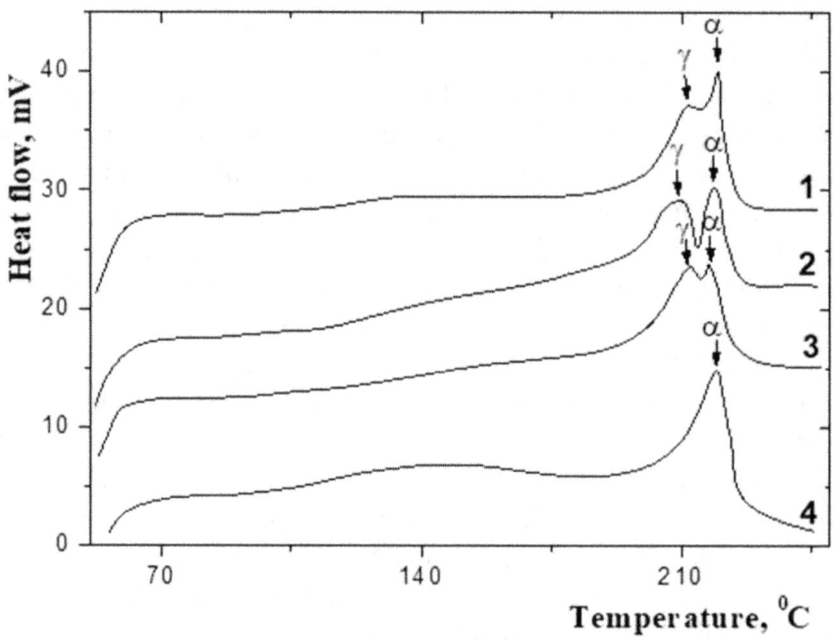

Fig. 15 presents the dependence of the degree of crystallinity of the sample on the content of modified montmorillonite in composites based on PA-6.

The degree of crystallinity for nanocomposites is higher than the original PA-6, i.e. the particles of modified PQAS montmorillonite act as heterogeneous nucleating agents.

CONCLUSION

Nanocomposites based on polyamide-6 and montmorillonite modified with polyionene have been obtained with screw-disk extrusion. Qualitative difference of morphology and thermal behavior of PA-6, composites based on modified montmorillonite is shown by WAXS, XRD, DTA and DSC data. Destruction of modified montmorillonite layered structure and nanosized inorganic phase in polyamide-6 is confirmed.

These processes have been explained by formation of crystalline γ-phase of polyamide in the presence of 1 – 2% of modified polyamide in contrast

Figure 15. The degree of crystallinity of the sample vs. the content of modified montmorillonite in composites based on PA-6 and montmorillonite modified I-6.

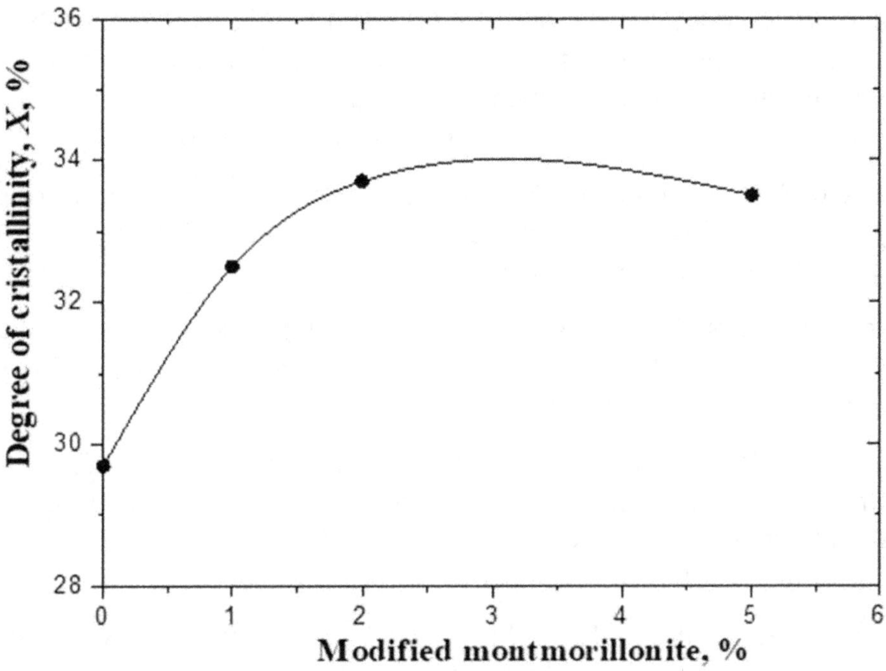

with initial polyamide-6 and mechanical mixtures of polyamide and montmorillonite. Optimal concentrations of modified montmorillonite in the composites are stated to be 1 – 2%. It is established by increasing crystallinity degree, that being explained by acting of modified montmorillonite particles as nucleation heterogenetic agents. The technology of nanocomposites based on polyamide-6 and organo-montmorillonite is developed. The technological parameters of extrusion and injection molding, and their influence on the formation of nanoscale inorganic phase of the polyamide were studied.

REFERENCES

Burmistr, M. V., Sukhyy, K. M., Shilov, V. V., Pissis, P., Spanoudaki, A., Sukha, I. V., Tomilo, V. I., & Gomza, Y. P. (2005). Synthesis, structure, thermal and mechanical properties of nanocomposites based on linear polymers and layered silicates modified by polymeric quaternary ammonium salts (ionenes). *Polymer*, *46*(26), 12226–12232. doi:10.1016/j.polymer.2005.10.094

Burnside, S. D., & Giannelis, E. P. (1995). Synthesis and properties of new poly (dimethylsiloxane) nanocomposites. *Chemistry of Materials*, *7*(9), 1597–1600. doi:10.1021/cm00057a001

Chen, Ch., Khobaib, M., & Curliss, D. (2003). Epoxy layered–silicate nanocomposites. *Progress in Organic Coatings*, *47*(3-4), 376–383. doi:10.1016/S0300-9440(03)00130-9

Dultz, S., & Rytwo, G. (2005). Effects of different organic cations on the electrokinetic surface charge from organo-montmorillonites - consequences for the adsorption properties. *Annual Meeting of Swiss, Austrian and German Clay Group - DTTG*, (*vol. 11*, pp. 6-14). DTTG Reports.

Grigoriev, A. S., Shilko, E. V., Skripnyak, V. A., & Psakhie, S. G. (2017) An approach to determining the parameters of kinetic strength theory based dynamic model of brittle solids mechanical behavior. In *Proceedings of the International Conference on Advanced Materials with Hierarchical Structure for New Technologies and Reliable Structures*, *1909*(1). doi:10.1063/1.5013745

McNeil, L. E., & Grimsditch, M. J. (1993). Structural and mechanical properties of monomineral clays. *Phys. Condens. Matter.*, *5*(11), 1681–1690. doi:10.1088/0953-8984/5/11/008

Messersmith, P. B., & Giannelis, E. P. (1994). Synthesis and characterization of layered silicate-epoxy nanocomposites. *Chemistry of Materials*, *6*(10), 1719–1725. doi:10.1021/cm00046a026

Paul, M. A., Alexandre, M., Degée, P., Henrist, C., Rulmont, A., & Dubois, P. (2003). New nanocomposite materials based on plasticized poly (L-lactide) and organo-modified montmorillonites: Thermal and morphological study. *Polymer*, *44*(2), 443–450. doi:10.1016/S0032-3861(02)00778-4

Ray, S. S. (2013) Clay-Containing Polymer Nanocomposites: From Fundamentals to Real Applications. Elsevier, 416.

Strawhecker, K. E., & Manias, E. (2000). Structure and properties of poly(vinylalcohol)/Na+–montmorillonite nanocomposites. *Chemistry of Materials*, *12*(10), 2943–2949. doi:10.1021/cm000506g

Sukhyy, K. M., Belyanovskaya, E. A., Sukhyy, M. P., & Tomilo, V. I. (2018). Modification technology of montmorillonite by polyionenes. *Journal of Chemistry and Technologies.*, *26*(1), 1–8. doi:10.15421/081801

Yebra-Rodriguez, A., Alvarez-Lloret, P., Cardell, C. A., & Rodriguez-Navarro, B. (2011). Influence of processing conditions on the optical and crystallographic properties of injection molded polyamide-6 and polyamide-6/montmorillonite nanocomposites. *Applied Clay Science*, *51*(4), 414–418. doi:10.1016/j.clay.2010.12.031

Zhong, Y., Zhu, Z., & Wang, S. (2005). Synthesis and rheological properties of polystyrene/layered silicate nanocomposite. *Polymer*, *104*(9), 785–796. doi:10.1016/j.polymer.2005.02.014

Zulðqar, S., & Sarwar, M. I. (2009). Synthesis and Characterization of Aromatic–Aliphatic Polyamide Nanocomposite Films Incorporating a Thermally Stable Organoclay. *Nanoscale Research Letters*, *4*(5), 391–399. doi:10.100711671-009-9258-1 PMID:20596518

Chapter 11
Analysis of Properties of Nanocomposites Based on Polyamide-6 and Organomodified Layered Silicates by Models of Elastic Modulus of Composites

ABSTRACT

In the present chapter, authors analyzed properties of nanocomposites based on polyamide-6 and organomodified layered silicates according to the theories of Halpin-Tsai and Mori-Tanaka. The morphology of nanocomposites under consideration was analyzed by radiographic analysis and transmission electron microscopy. The degree of delamination of the structure of layered silicates in the polymer matrix is shown to be a crucial factor to improve the physical and mechanical characteristics of the nanocomposites. A high level of orientation of elementary particles of layered silicates is stated when shear forces applied in comparison with fibers both due to individual features and due to a more favorable geometric shape. Correlation of the simulation and experimental data is shown.

DOI: 10.4018/978-1-7998-9728-6.ch011

BACKGROUND

As mentioned above, the studied nanocomposites are of great interest because they show high levels of reinforcement at very low levels of filling in contrast to traditional composites. A comparative estimation of the growth of the modulus of elasticity (normalized to the modulus of elasticity of the polymer matrix) is given in Fig. 1 for the studied nanocomposites and composites filled with glass fibers depending on the degree of filling. Increase the modulus of elasticity of the composite material in 2 times compared to the original polymer matrix requires 3 times more glass fibers as compared to organomodified layered silicates (20% vs. 6.5%).

Layered silicates consist of thin ~ 1-1.5 nm layers, with a length and width of such layers of several microns. Thus, for the elementary layer, the ratio of linear dimensions and thickness can exceed 1000. In addition, the plates of layered silicates, which have extremely high modulus of elasticity and high specific surface area, can make a significant contribution to the formation of polymer composite properties. An important issue is the possibility of explaining the properties of the studied nanocomposites using the theory of elasticity of polymer composites.

An attempt is made to model the properties of the studied nanocomposites based on models of modulus of elasticity analysis of composite materials to identify the importance of factors such as individual influence of polymer matrix and filler, geometry and orientation of filler, interaction of polymer matrix and filler. In addition, as will be discussed below, there are many additional factors that have an indirect effect on the formation of a set of properties of polymer composites. The main purpose of this work is to better understand the high efficiency of reinforcement, which is observed in well-laminated polymer-organosilicate nanocomposite in comparison with traditional polymer composite materials.

ANALYSIS OF MODULUS OF ELASITITY

Theories of analysis of the modulus of elasticity of Halpin-Tsai and Mori-Tanaka composite materials were used for modeling He et al. (2019). Empirical dependences were analyzed in comparison with direct experimental data studying morphology, physico-mechanical and relaxation properties for the studied nanocomposites based on polyamide-6 and layered silicates modified I-6 Burmistr et al (2005) and polyamide-6 filled with glass fibers.

For several decades, theories of predicting the properties of composite materials have been developed, the main factors of which are the properties of raw materials and the morphology of composites. The assumption that is characteristic of almost all theories is that the ingredients are considered independently of each other, while the purpose of the theory is to predict the set of properties for the composite material as a whole. The most important initial parameters are the properties of the original polymer matrix and filler, the volume content of the filler, the ratio between the linear dimensions of the filler, its orientation. The relationship between linear dimensions (shape factor) is the main factor responsible for the implementation of a high level of reinforcement in nanocomposites using organomodified layered silicates. Halpin and Tsai developed the well-known composite theory of inelastic unidirectional polymer composites based on the filler shape factor Mittal, V. This theory is based on the work of Hermans (1967) and Hill (1964), who considered single fibers that are "filled" in the polymer matrix. Halpin and Tsai brought these results into a simple analytical form and adapted them for different filler configurations, including dispersed.

According to this theory, the modulus of the composite can be defined as:

$$\frac{E}{E_{<}} = \frac{1+\zeta\eta\varphi_{ad}}{1-\eta\varphi_{ad}}, \tag{1}$$

where E and E_M are Young's modulus of composite and matrix, respectively,

ζ is a scale factor which depends on geometry and distribution of filler,

φa_d – volume fraction of filler,

$$\eta = \frac{E_{=}/E_{<} - 1}{E_{=}/E_{<} + \zeta}, \tag{2}$$

$E_н$ – Young's modulus of filler.

For discontinuous fibers, the scale factor in the longitudinal direction is $\zeta = 2$ (a / b) Hermans (1967)., where a and b are the length and thickness of the fiber. The scale factor in the perpendicular direction for the fibers is

Figure 1. Comparison of strengthening of PA-6 with organomodified layered silicate – 1 and glass fiber - 2 Sukhyy et al. (2007).

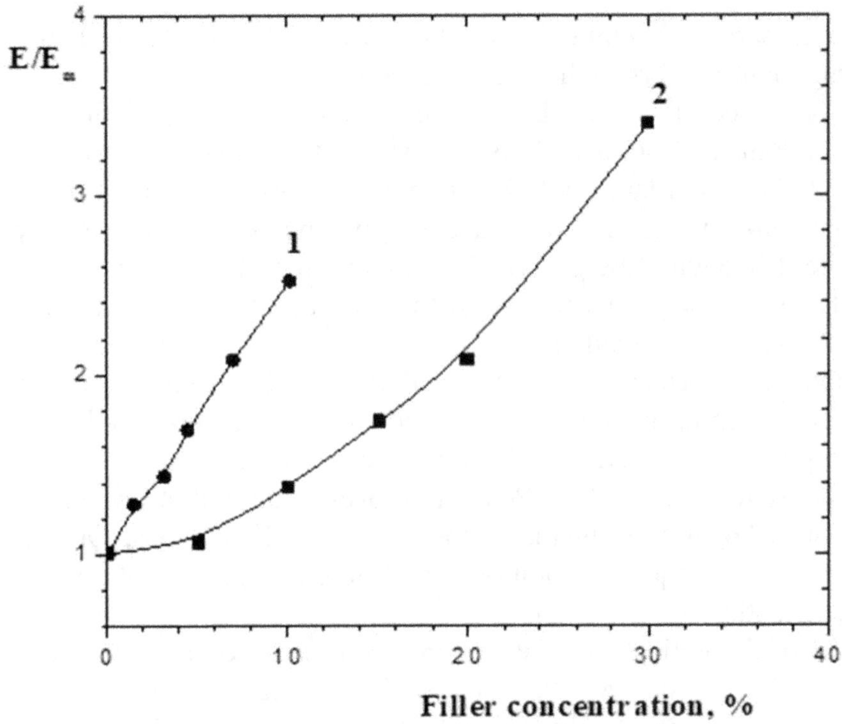

approximately Hermans (1967). Based on the Halpin-Tsai theory, at ζ ® 0 the value of the modulus of the composite can be described as:

$$\frac{1}{E} = \frac{\varphi_{=}}{E_{=}} + \frac{1-\varphi_{=}}{E_{<}},$$

And when $\zeta \to \infty$

$$E = \varphi a_{dE} a_d + (1 - \varphi ad)_E M$$

$$\frac{E_{11}}{E_M} = \frac{A}{A + \varphi_H \left(A_1 + 2\upsilon_M A_2 \right)} \quad (3)$$

$$\frac{E_{22}}{E_M} = \frac{2A}{2A + \varphi_H\left[-2\upsilon_0 A_3 + (1-\upsilon_M)A_4 + (1+\upsilon_M)A_5 A\right]} \quad (4)$$

where $\phi_н$ is parts by volume of the filler; A_1, A_2, A_3, A_4, A_5 are Eshelby's tenzor function; υ refers to the Poisson's ratio

Details of theoretical calculations are given in Tandon & Weng (1984).

The modulus of fibrous and disc-shaped particles in different directions is calculated in accordance with Mori-Tanaka theory (Fig. 2a).

The Halpin-Tsai and Mori-Tanaka theory will be used to evaluate polymer composites filled with fiberglass and organomodified layered silicates. They were selected, among other things, by the level of citations in the literature Oshinski, Keskkula & Paul (1996).

Composite materials based on PA-6 filled with glass fibers were selected as objects for comparison with the studied nanocomposites based on PA-6. Such composites were made by injection molding (injection zone temperature 270°C, mold temperature 80°C). Preparation of organomodified layered silicate was performed by modifying the layered silicate with polymeric Quaternary ammonium salt in aqueous solution for 24 hours at a temperature of 40°C and constant stirring.

Details of theoretical calculations are given in Tandon & Weng (1984). In Fig. 2 (b) in accordance with Mori-Tanaka theory, the modulus of fibrous and disc-shaped particles in different directions is calculated. The Halpin-Tsai and Mori-Tanaka theory will be used to evaluate polymer composites filled with fiberglass and organomodified layered silicates. They were selected, among other things, by the level of citations in the literature Tandon & Weng (1984). Composite materials based on PA-6 filled with glass fibers were selected as objects for comparison with the studied nanocomposites based on PA-6 Kanapitsas, A. & Pissis, P. & Kotsilkova, R. (2002). Such composites were made by injection molding (injection zone temperature 270°C, mold temperature 80°C) van Es et al. (2001).

Preparation of organomodified layered silicate was performed by modifying the layered silicate with polymeric Quaternary ammonium salt in aqueous solution for 24 hours at a temperature of 40°C and constant stirring.

The manufacturing procedure of the studied nanocomposites is as follows: organomodified layered silicate was mixed with a polymer matrix in a worm-disk extruder. The obtained extrudate underwent the stage of granulation. Then, standard injection specimens were obtained by injection molding. The Young's modules of the studied nanocomposites together with the

Figure 2. Physical images in coordinate systems corresponding to the equations of theories of Halkin-Tsai (a) and Mori-Tanaka (b), which are used to calculate the level of reinforcement of composites filled with fibers and disc-shaped particles.

(a)

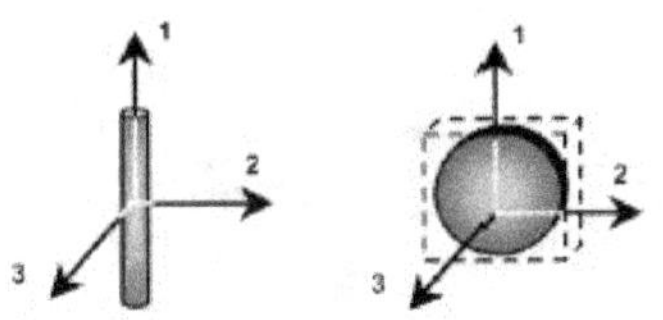

Direction of applied impact	Module of composites according to Halpin-Tsai's theory		Scale factor $\zeta^{(a)}$	
	Fibers	Layers	Fibers	Layers
1	$E_{11} = E_{\parallel}$	$E_{11} = E_{\parallel}$	$\zeta = 2(l/d)$	$\zeta = 2(l/t)$
2	$E_{22} = E_{\perp}$	$E_{22} = E_{\parallel}$	$\zeta = 2$	$\zeta = 2(l/t)$
3	$E_{33} = E_{\perp}$	$E_{33} = E_{\perp}$	$\zeta = 2$	$\zeta = 2$

fiber disk-shaped particles

(b)

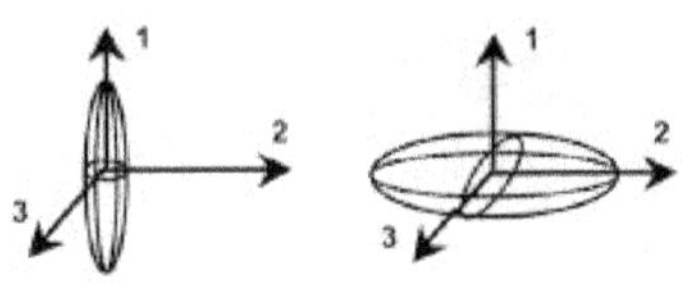

fiber disk-shaped particles

Direction of applied impact	Module of composites according to Mori-Tanaka's theory	
	Fibers	Layers
1	$E_{11} = E_{\parallel}$ (Eq.5)	$E_{11} = E_{\perp}$(Eq.6)
2	$E_{22} = E_{\perp}$(Eq.6)	$E_{22} = E_{\parallel}$(Eq.5)
3	$E_{33} = E_{\perp}$(Eq.6)	$E_{33} = E_{\parallel}$(Eq.5)

Table 1. Young's modules of nanocomposites PA-6 + organomodified layered silicates and composites PA-6 + glass fibers

Filler content, wt. %	Filler content by volume, %	Young module, GPa
PA-6 system + glass fibers		
0	0	2.84
5	2.3	2.96
10	4.8	3.88
15	7.3	4.93
20	10.1	5.89
30	16.1	9.6
PA-6 system + organomodified layered silicates		
0	0	2.75
1	0.4	3.31
2	0.8	3.92
5	2.0	4.64

corresponding characteristics of composites that are filled with glass fibers are presented in Table. 1.

Figure 3. Examples of complications in determining the form factor of fillers in polymer nanocomposites.

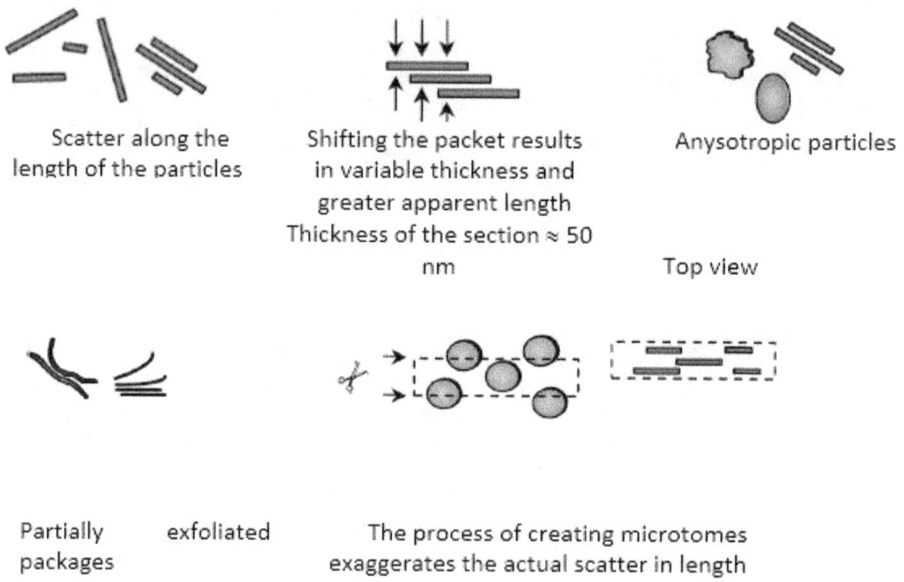

MORPHOLOGY OF NANOCOMPOSITES BASED ON POLYAMIDE-6 AND ORGANOMODIFIED LAYERED SILICATES

The morphology of composites based on polyamide-6 and glass fibers has been studied by transmission electron microscopy (TEM). The average diameter of the glass fibers was 13 μm, the average fiber length was 262 μm. Accordingly, the ratio of length to diameter (shape factor) is about 20 units.

The morphology of nanocomposites based on PA-6 and organomodified layered silicates was analyzed on the basis of radiographic analysis and transmission electron microscopy. Unlike fiberglass, determining the shape factor for exfoliated layered silicates is much more difficult. The various obstacles that arise are given in Fig. 3.

The scatter of microtomes of nanocomposite samples also shows a large particle size distribution. Reliable TEM image analysis requires high resolution and contrast. In fact, TEM images consist of different shades of gray, which is primarily due to the different thickness of the samples for analysis.

In addition, in the case of nanocomposites, the contrast is also the result of differences in the molecular packaging of the composites. TEM image of a nanocomposite sample with 2% organomodified layered silicate is given in Fig. 4. To facilitate the interpretation of TEM data, the original TEM image is converted to black and white. The file was then imported into Scion Image 1.63 Mac OS Sukhyy et al. (2007), where each filler particle is assigned a numeric value and its length is calculated.

The data of Fig. 4 were used to assess the particle shape factor of the nanocomposite filler in accordance with the expression:

$$\$\,\$_{\text{lay.sil.}} = \frac{\bar{l}}{\bar{t}}$$

where $\bar{l}$ refers to an average length of particles, $\bar{t}$ is an average thickeness of particles.

The histogram of the particle size distribution of organomodified layered silicate based on the analysis of TEM data is given in Fig. 5.

For the system under analysis, the average particle length is about 100 nm. For a fully delaminated system, the average particle thickness should be equal to the thickness of the layered silicate elementary plate, i.e. equal to 0.94 nm. As mentioned earlier, the direct calculation of the particle thickness of incompletely delaminated layered silicate presents significant difficulties due to the various possible variants of the particle structure in the composite (see Fig. 4).

Substituting the average number of layers of particles in the nanocomposite, determined according to TEM in Eq. (5), we find the average particle thickness of organomodified layered silicate, which is equal to » 2 nm.

The shape factor will be:

$$\Phi\Phi = \frac{\bar{l}}{\bar{t}} = \frac{100}{2} = 50$$

Based on generalized experimental data for nanocomposites based on PA-6 and organomodified layered silicates, literature data for composites based on PA-6 and fiberglass, as well as data on physical parameters of all starting

Figure 4. Example of semi-automatic analysis of TEM images to determine (a) the length of the distributed particles and (b) the number of layers in the package.

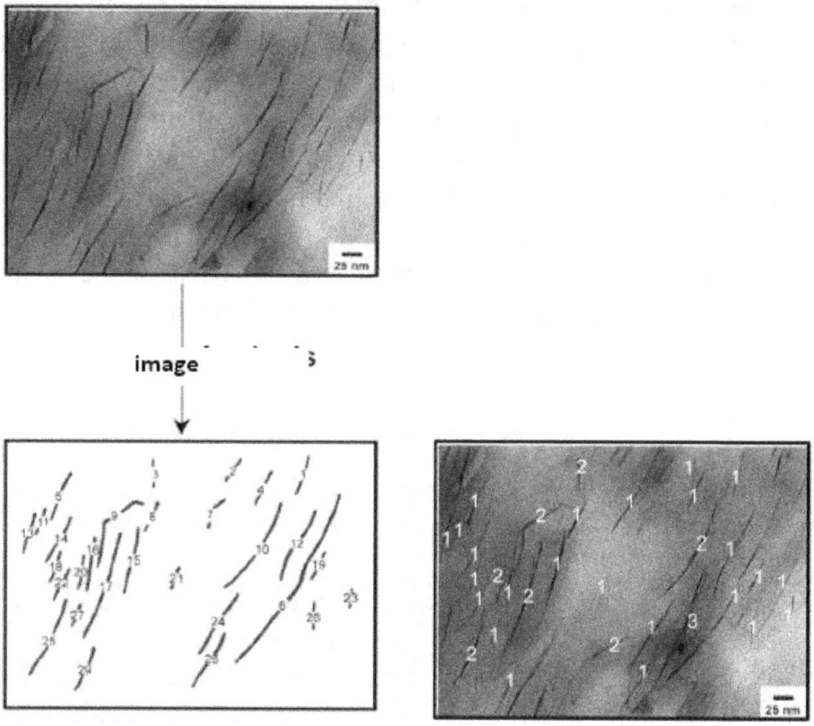

ingredients, which are given in Table. 2, it is possible to conduct modeling in accordance with the theories of Halpin-Tsai and Mori-Tanaka.

Young's modulus for PA-6, determined experimentally, corresponds to the literature data García–Lópezet al. (2005).

Young's modulus for layered silicate was determined by ultrasound Alexandrov & Ryshova (1961), Upadhyay & Singh (2014) and Brillouin scattering McNeil & Grimsditch (1993). The values obtained by these methods are in good agreement. The density value for montmorillonite was calculated based on the mass of the Na-form montmorillonite elementary plate with a surface area of 0.515x8.9 nm^2 Alexandrov & Ryshova (1961) with a thickness of 0.94 nm (Fig. 1 (a), Chapter 17).

Fig. 6 presents the dependences of the Young's modulus of the analyzed composites, normalized to the Young's modulus PA-6 on the content and form factor of the filler, calculated in accordance with the theories of Halpin-Tsai and Mori-Tanaka. With increasing form factor for composites of two types

Figure 5. Histogram of organomodified layered silicate: the number of layers in the package for nanocomposites based on PA-6.

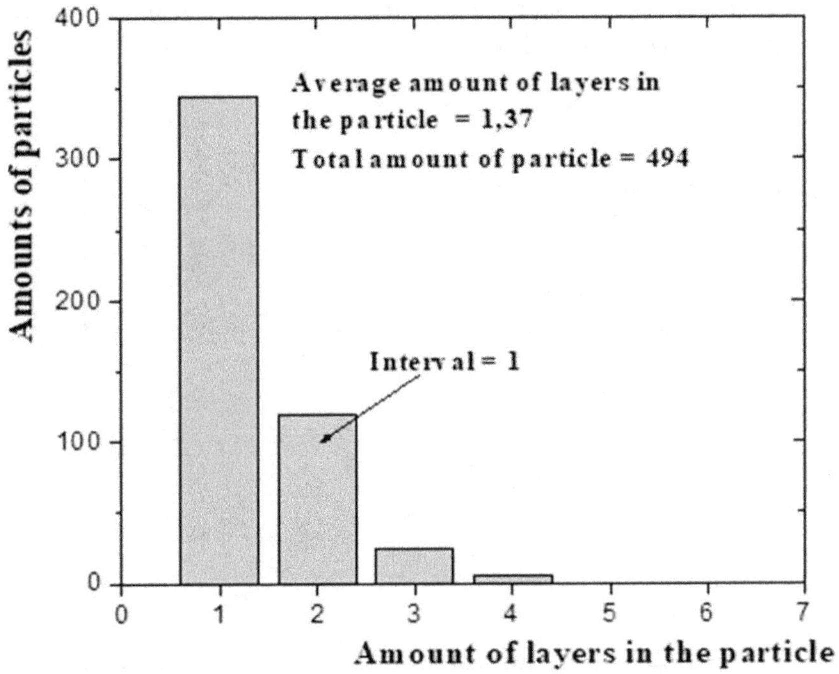

from 20 to 100 there is a significant increase in the modulus of elasticity, but for nanocomposites this effect is more pronounced due to the higher value of the modulus of elasticity of layered silicates. The influence of the value of the filler modulus on the strengthening of the composite is also clearly seen when comparing composites with fillers with equal shape factor.

Table 2. Physical properties of the ingredients of the composites analyzed.

Material	Young's modulus, GPa	Density, g/cm³	Poisson's ratio, ν
system PA-6 + glass fibers			
PA-6	2.82	1.14	0.35
Glass fiber	72.4	2.54	0,20
system PA-6 + organomodified layered silicates			
PA-6	2.75	1.14	0.35
Organomodified layered silicate	178	2.83	0.20

Figure 6. Longitudinal reinforcement curves of composites based on PA-6 and layered silicate with disc-shaped particles and glass fibers. The curves were predicted by the Halpin-Tsai equations. Solid lines correspond to composites that are reinforced with disc-shaped particles of layered silicate, and dashed lines correspond to composites that are reinforced with glass fibers.

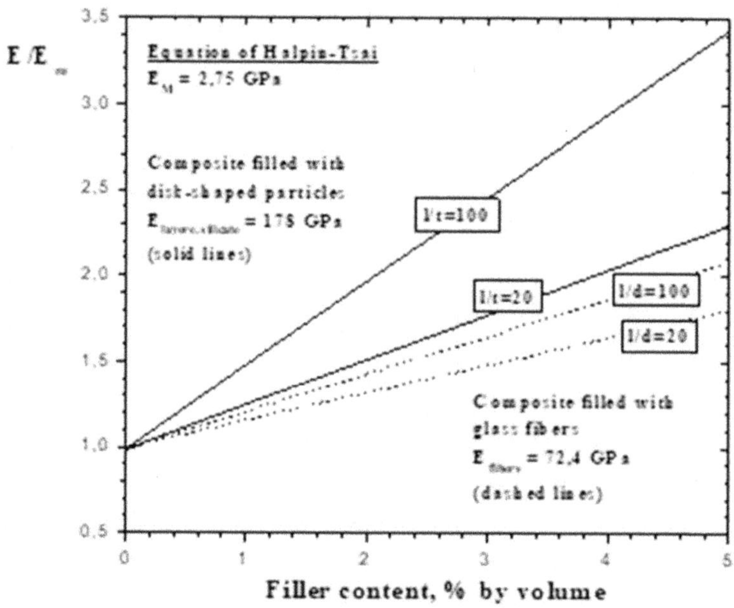

For nanocomposites, there is a significant increase in the Young's modulus compared to glass-filled composites, despite the fact that the Young's modulus of layered silicates exceeds the Young's modulus of glass fibers only 2.5 times. The combination of high modulus and high form factor is accompanied by even more significant differences in the degree of reinforcement for the compared composites. It is important to note that in the Halpin-Tsai theory, the location of both fibers and disc-shaped particles of layered silicates is taken in a direction parallel to the study, i.e. unidirectional - oriented, which gives a "head start" to fiber-filled composites.

Mori-Tanaka theory gives somewhat different tendencies of strengthening of the analyzed composites (Fig. 7). At the form factor 20 for composites filled with glass fibers, higher values of the modulus of elasticity in comparison with nanocomposites are predicted. These differences can be explained by different approaches in the analysis of the geometry of the fillers. Thus, the equations for calculating longitudinal hardening in the Halpin-Tsai theory are identical in shape for fibers and disc-shaped particles of layered silicates, while in the Mori-Tanaka theory they are calculated differently. In the form factor 20, the unidirectional strengthening of composites by fibers is more significant in contrast to the disc-shaped particles of layered silicates.

In contrast to fibers, disc-shaped particles of rectangular cross-section in the direction of force application have areas with different shape factors that decrease from the center to the edges (see Fig. 2). Thus, the effective length of the disk-shaped particle is less than its diameter, and the disk segments on the left and right in relation to the direction of the applied load are stress concentrators. This effect becomes less pronounced with an increase in the shape factor of the disc-shaped particles. In this case, as shown in Fig. 6, in the framework of Mori-Tanaka theory, the degree of reinforcement of composites in the case of disc-shaped particles depends more on the shape factor than for fibers. At a form factor of 100, the level of gain for nanocomposites is more significant than for composites filled with fibers.

Thus, in general, both theories emphasize the significant influence of the form factor on the degree of reinforcement in composites. Trends presented in Fig. 6 and 7, show a significant potential for strengthening nanocomposites with increasing form factor by stratification of layered silicate particles.

In addition to the factor of filler shape discussed above, the orientation of the dispersed phase in the composite has a significant effect on the degree of reinforcement of composites. Deviation from unidirectional hardening can lead to a decrease in the degree of hardening calculated for the ideal case, and this effect may be manifested to a greater or lesser extent depending on the geometric shape of the filler.

According to the geometric shape of the disc-shaped particles, it is obvious that they can provide an equal degree of gain in two directions, while fibrous fillers can provide a reinforcing effect only in the direction of the fibers. In the case of complete anisotropy of the location of the filler particles, the disc-shaped particles provide a much higher level of gain than the fibers, which can be described by the following expressions:

$$E^{fibers} = 0{,}184E_{\parallel} + 0{,}816E_{\perp} \tag{6}$$

$$E^{disks} = 0{,}49E_{\parallel} + 0{,}51E_{\perp} \tag{7}$$

Figure 7. Longitudinal reinforcement curves of composites based on PA-6 and layered silicate with disc-shaped particles and glass fibers. The curves were predicted by the Mori-Tanaka theory. Solid lines correspond to composites reinforced with disc-shaped layers of layered silicate, and dashed lines correspond to composites reinforced with glass fibers.

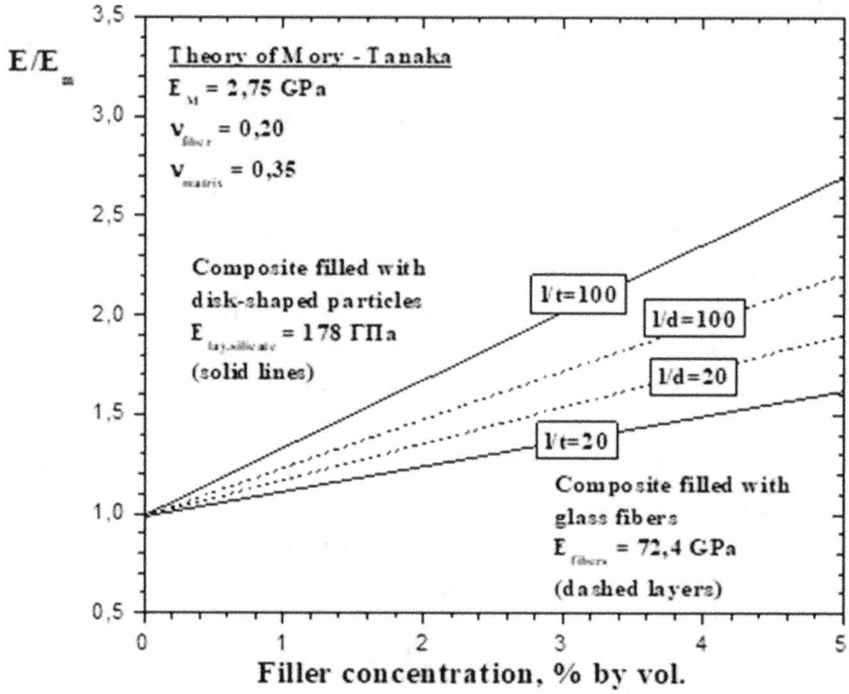

where $E_{\parallel}$ and $E_{\perp}$ refer to the elastic modulus in the directions which are parallel and normal to the main axis of filler.

The effect of filler orientation on the degree of reinforcement of composites in the framework of Halpin-Tsai and Mori-Tanaka theory, respectively, using Eq.(6) and (7) is given in Fig. 8 and 9. To simplify the modulus of glass fibers and particles of layered silicates are taken equal. For the random arrangement of disc-shaped particles in the composite, a much higher level of gain is observed (even if the initial modulus of elasticity is equal) in comparison with composites filled with fiber, regardless of the theory used. The curves for anisotropic samples have a much smaller angle of inclination compared to oriented samples.

Predictions performed in the framework of the analyzed theories indicate that the initial modulus of the filler, the form factor and its orientation in the polymer matrix have a significant impact on the modulus of elasticity of the composite material.

The results of the analysis allow understanding to what extent the stress concentrators (fiber ends for fibrous fillers and segments in the direction parallel to the applied load for disc-shaped particles) limit the realization of the target properties of composites.

A key issue in the formation of polymer nanocomposites using layered silicates is to achieve a high level of delamination of the latter in the polymer matrix. However, in most cases such delamination is thermodynamically unfavorable, and most of the work in this direction is accompanied, at best, by the intercalation of the organic phase in the layered structure of silicates.

The effect of filler orientation on the degree of reinforcement of composites in the framework of Halpin-Tsai and Mori-Tanaka theory, respectively, using Eq.(6) and (7) is shown in Fig. 8 and 9. To simplify the modulus of glass fibers and particles of layered silicates are taken equal. For the random arrangement of disc-shaped particles in the composite, a much higher level of gain is observed (even if the initial modulus of elasticity is equal) in comparison with composites filled with fiber, regardless of the theory used. The curves for anisotropic samples have a much smaller angle of inclination compared to oriented samples.

Predictions performed in the framework of the analyzed theories indicate that the initial modulus of the filler, the form factor and its orientation in the polymer matrix have a significant impact on the modulus of elasticity of the composite material.

The results of the analysis allow to understand to what extent the stress concentrators (fiber ends for fibrous fillers and segments in the direction parallel to the applied load for disc-shaped particles) limit the realization of the target properties of composites.

A key issue in the formation of polymer nanocomposites using layered silicates is to achieve a high level of delamination of the latter in the polymer matrix. However, in most cases such delamination is thermodynamically unfavorable, and most of the work in this direction is accompanied, at best, by the intercalation of the organic phase in the layered structure of silicates.

In this case, the modulus of elasticity of the layered silicate package can be estimated using the mixture rule Tam et al. (2012):

$$E_{package} = \varphi l_{ay.silcate.} El_{ay.silicate} + \varphi ga_{l.\,E} ga_{l.} \qquad (8)$$

Figure 8. Influence of filler concentration (disks and fibers) on the gain of PA-6 according to the Halpin-Tsai equation.

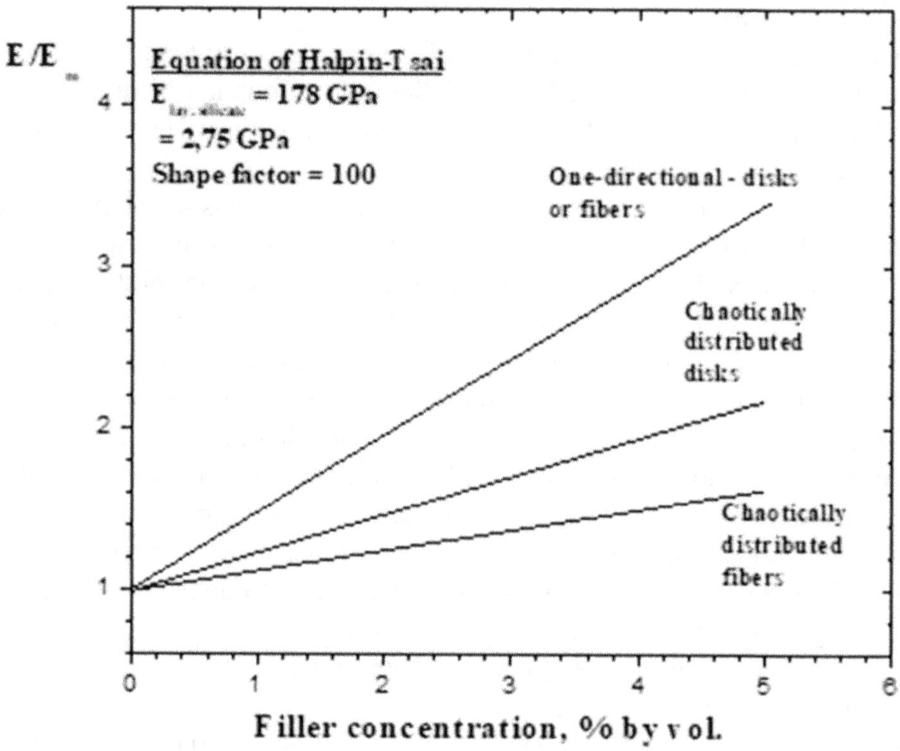

where $\varphi la_{y.silicate.i}$s parts by volume of silicate layers; $Ela_{y.silicate.i}$s elastic modulus of silicate layer; $\varphi gal_{.is}$ parts by volume of galleries; Ė gal $_{.Is}$ elastic modulus of the material in the gallery which is a priori is far below then modulus of silicate layer.

Parts by volume of galleries between layers $\varphi gal_{.is}$ determined as:

$$\varphi_{gal.} = \frac{(n-1)\left(d_{001} - t_{lay}\right)}{d_{001}(n-1) + t_{lay}} \quad (9)$$

where n is amount of layers in packages, d_{001} is periodicity of silicate layers, t_{lay} is thickness of silicate layer.

Figure 9. The influence of the concentration of filler (disks and fibers) on the strengthening of PA-6 according to Mori-Tanaka theory.

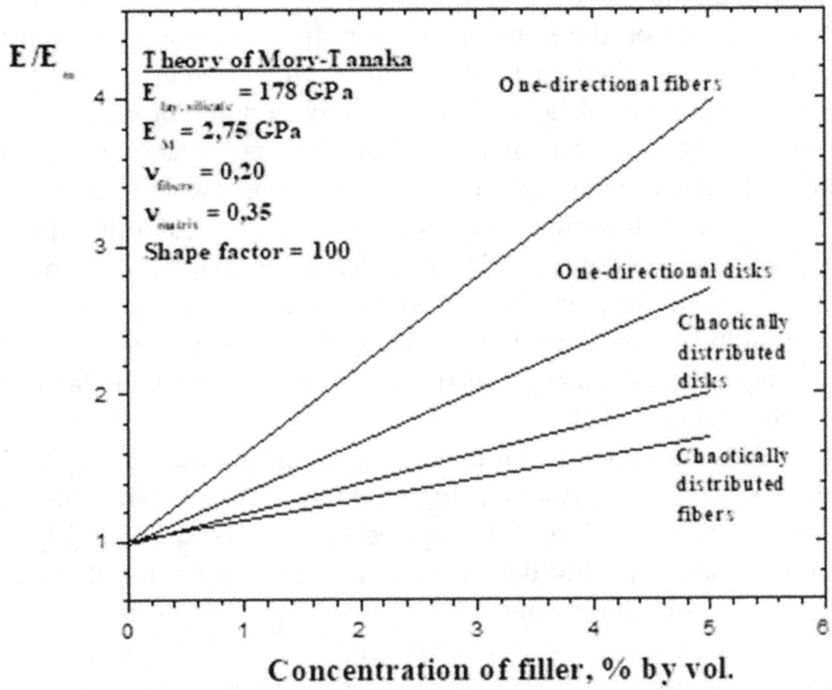

Figure 10. Image of the layered structure of incompletely exfoliated particles: (a) simplified representation and (b) real structure variants.

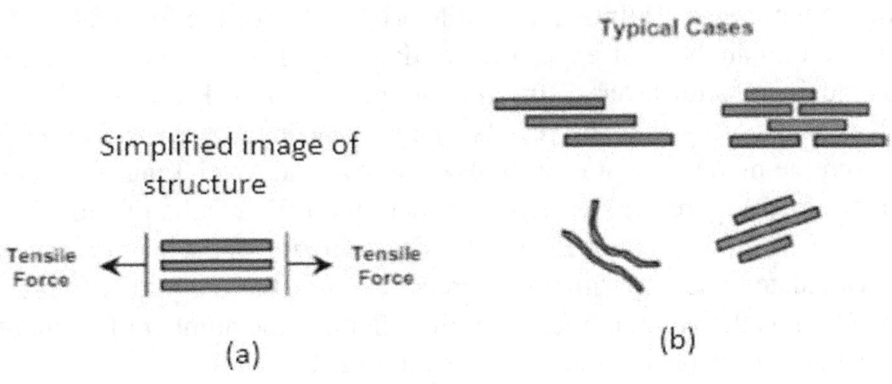

As noted earlier, for the original layered silicates, the periodicity in the location of the layers is $d_{001} = 0.96$ nm, and for organomodified - $d_{001} = 1.67$ nm. The modulus of elasticity of the elementary layer is assumed to be equal to 178 GPa in both cases.

Fig. 11 presents the dependences of the modulus of elasticity of the packages of layered silicates on the number of layers in the package in logarithmic coordinates for the original and modified forms, which are calculated by Eq.(8).

For the initial form of layered silicate, the modulus of elasticity of the package is weakly dependent on the number of silicate layers in it. In contrast, for modified forms, when the volume of the interlayer galleries and the volume occupied by the silicate layers are almost identical, this dependence is significant. The modulus of elasticity decreases sharply with increasing number of layers in the package and asymptotically approaches the value of 93 GPa. It should be noted that the characteristic that is analyzed - the number of silicate layers in the package - is inextricably linked with the factor of the form of the filler (Fig. 12).

Data of X-ray analysis and transmission electron microscopy (TEM) of nanocomposites based on PA-6 and organomodified layered silicates indicate a high level of delamination of the layered structure of silicates (Fig. 4).

Moreover, based on TEM data, nanocomposite samples are characterized by a high level of orientation of the filler phase in the casting direction. These results are in accordance with numerous applied and theoretical data on the study of the rheology of such systems (Ray & Okamoto, 2003). Thus, in Murugan, Phiri, and Mookkandi (2022), using low-angle X-ray scattering, it was shown that when applying shear forces in an aqueous solution of polyethylene oxide, saponite particles are strongly oriented in the direction of shear.

The experimental values of the modulus of elasticity of the studied nanocomposites and modeling data, which was carried out in accordance with the theories of Halpin-Tsai and Mori-Tanaka for the form factor FF = 57 (based on analysis of experimental data) and FF = 97 is the maximum possible form factor layered structure are given in Fig. 13.

It is noteworthy that the results of the experiment are exactly halfway between the modeling data for the Halpin-Tsai and Mori-Tanaka equations. It ought to compare the experimental data with the results of modeling in terms of gain, depending on the level of delamination of the layered structure. The calculated and experimental dependences of the modulus of elasticity depending on the volume fraction of the filler and the number of elementary layers in the package of filler are given in Fig. 14.

Figure 11. Influence of the number of layers in the package on the module for layered silicates.

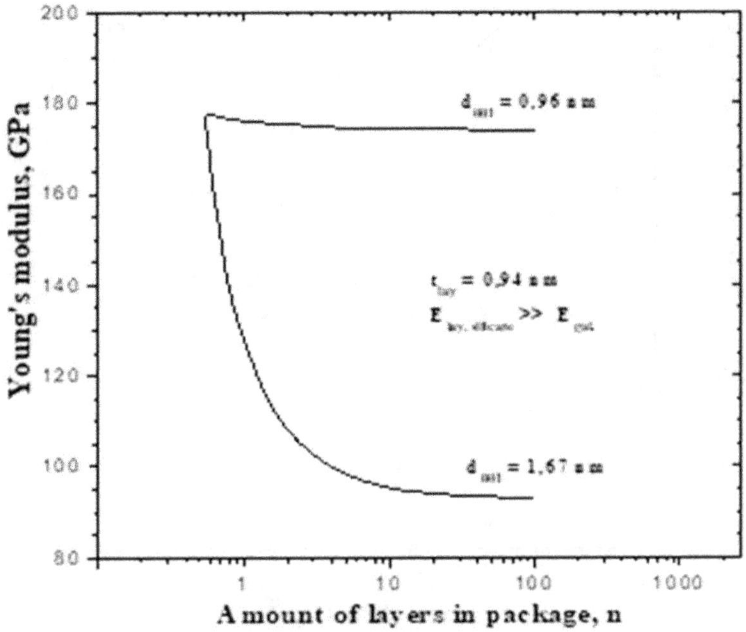

The experimental data are between the curves calculated from the Halpin-Tsai's equations, which correspond to one and two silicate layers in the elementary particle, which is very close to the experimentally calculated value of 1.4. The experimental data correspond to a fully delaminated sample in the analysis of the Mori-Tanaka theory (Fig. 15).

Thus, the theories of polymer composite materials Halpin-Tsai and Mori-Tanaka are used to better understand the observed improvement of physical and mechanical properties of nanocomposites based on PA-6 and organomodified layered silicates and explain their greater potential compared to traditional materials, in particular, covenient materials filled with fibers.

Such theoretical comparisons indicate a higher viability of organomodified layered silicates compared to fibers, due to the possibility of achieving a greater form factor, which causes a primary impact on the degree of strengthening and, moreover, for elementary particles of layered silicates there is a higher level of orientation. with fibers both due to individual characteristics and

Figure 12. Influence of the number of layers in the package on the form factor for layered silicates.

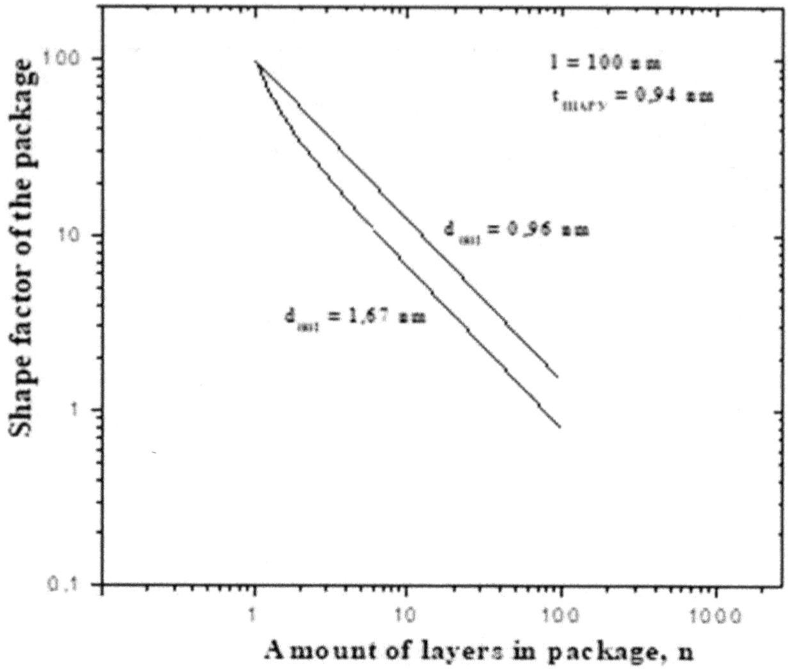

due to a more favorable geometric shape. The simulation data are in good agreement with the experimental data.

CONCLUSION

The present chapter is devoted to analysis properties of nanocomposites based on polyamide-6 and organomodified layered layer silicates using models of elasticity by theories of Halpin-Tsai and Mori-Tanaka. The main factors responsible for improving the physical and mechanical characteristics of the nanocomposites under consideration are the degree of delamination of the structure of layered silicates in the polymer matrix are shown. For elementary particles of layered silicates, a high level of orientation is observed when shear forces are applied in comparison with fibers both due to individual

Figure 13. Dependences of Young's modules of PA-6-based nanocomposites normalized to Young's PA-6 modulus on the content of layered silicate experimental data (Table 1)).

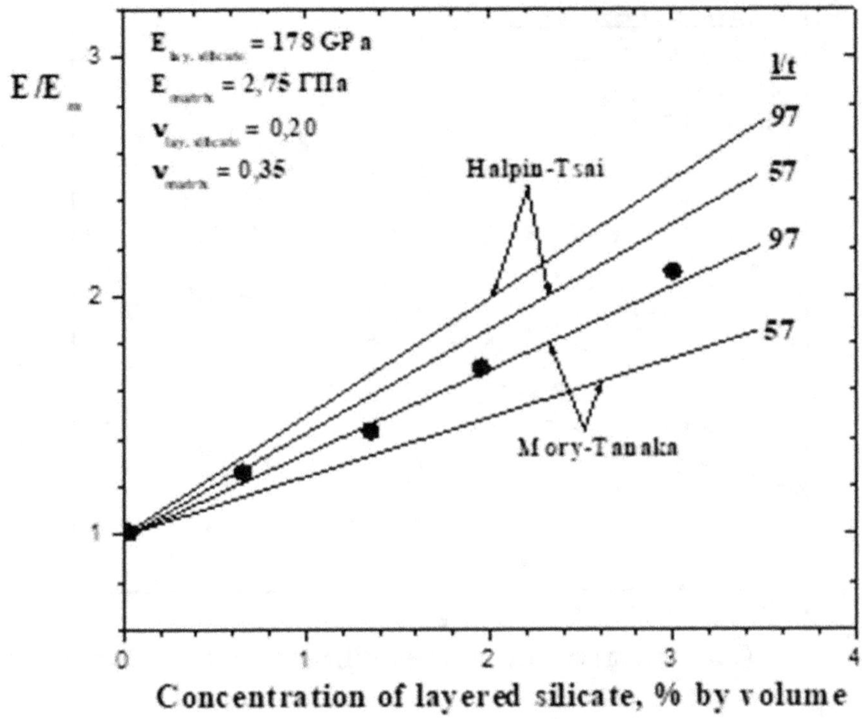

features and due to a more favorable geometric shape. The simulation data are shown to be in good agreement with the experimental data.

Figure 14. Dependences of Young's modules of nanocomposites based on PA-6, normalized to Young's PA-6 modulus vs. the concentration of layered silicate (Table 1)).

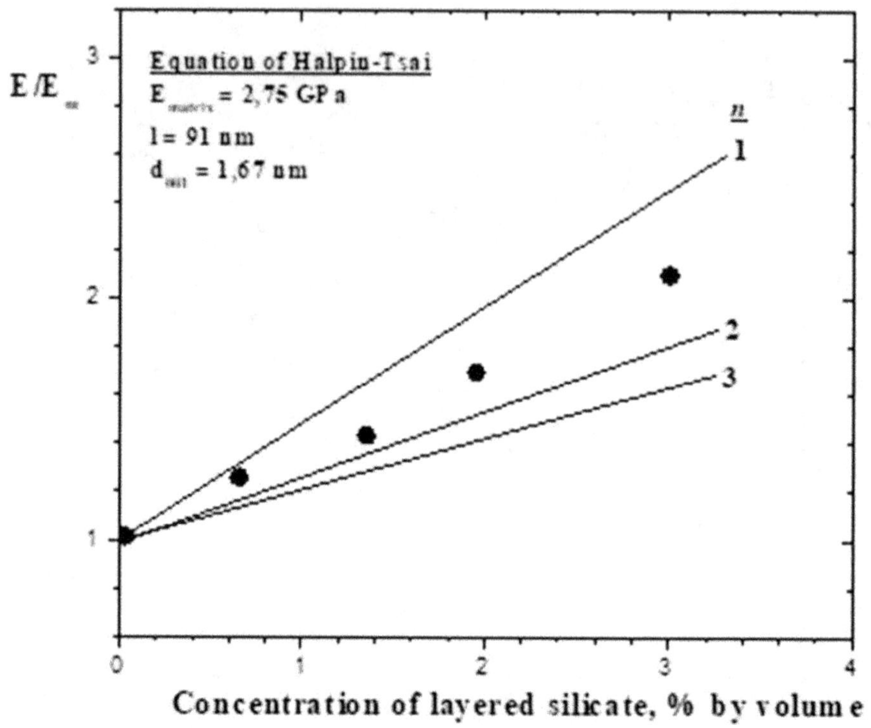

Figure 15. Dependences of Young's modules of PA-6-based nanocomposites normalized to Young's PA-6 modulus on layered silicate content (lines show modeling results for composites with the number of elementary layers in package n according to Mori-Tanaka model Table 1), experimental data are given by points.).

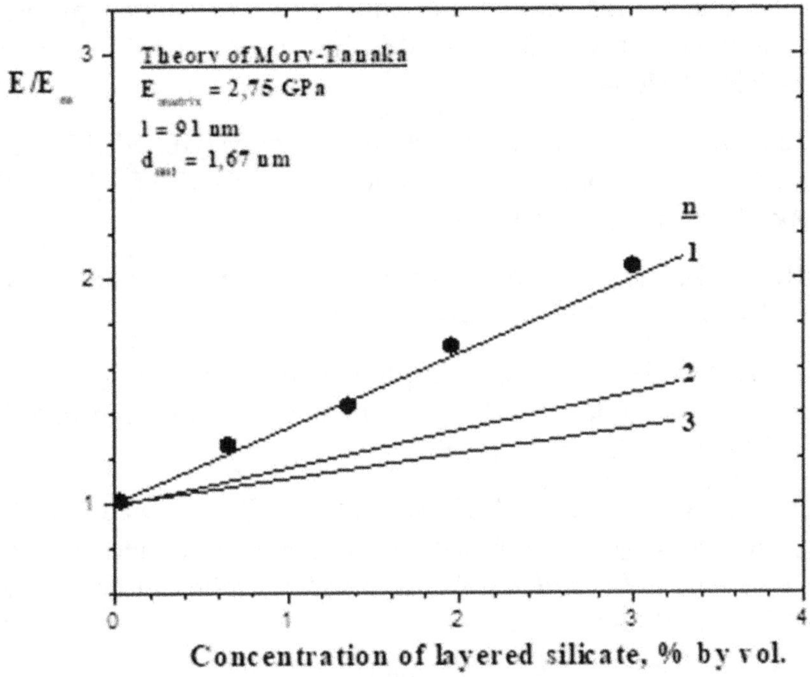

REFERENCES

Alexandrov, K. S., & Ryshova, T. V. (1961). The Elastic Properties of Rock-Forming Minerals. II: Layered Silicates. *Bulletin USSR Academy of science. Geophysics Series 9, 12*, 1165–1168.

Burmistr, M. V., Sukhyy, K. M., Shilov, V. V., Pissis, P., Spanoudaki, A., Sukha, I. V., Tomilo, V. I., & Gomza, Y. P. (2005). Synthesis, structure, thermal and mechanical properties of nanocomposites based on linear polymers and layered silicates modified by polymeric quaternary ammonium salts (ionenes). *Polymer, 46*(26), 12226–12232. doi:10.1016/j.polymer.2005.10.094

García–López, D., Gobernado–Mitre, I., Fernández, J. F., Merino, J. C., & Pastor, J. M. (2005). Influence of clay modification process in PA6–layered silicate nanocomposite properties. *Polymer*, *5*(8), 23–25. doi:10.1016/j.polymer.2005.01.038

He, S., Carolan, D., Fergusson, A., & Taylor, A. (2019). Toughening epoxy syntactic foams with milled carbon fibers: Mechanical properties and toughening mechanisms. *Materials & Design*, *169*, 107654. doi:10.1016/j.matdes.2019.107654

Hermans, J. (1967). Exfoliated high–impact polystyrene/MMT composites prepared using anchored cationic radical initiator–MMT hybrid. *Proc Kon Ned Akad v Wetensch B*, *65*, 1–9.

Hill, R. (1964). Photodegradation of polypropylene nanocomposites. *Journal of the Mechanics and Physics of Solids*, *12*, 119.

Hui, C. Y., & Shia, D. (1998). Morphology development in layered silicate thermoset nanocomposites. *Polymer Engineering and Science*, *5*, 774–782. doi:10.1002/pen.10243

Kanapitsas, A., Pissis, P., & Kotsilkova, R. (2002). Dielectric studies of molecular mobility and phase morphology in polymer–layered silicate nanocomposites. *Journal of Non-Crystalline Solids*, *305*(1-3), 204–211. doi:10.1016/S0022-3093(02)01107-9

McNeil, L. E., & Grimsditch, M. (1993). Elastic Moduli of Muscovite Mica. *Journal of Physics Condensed Matter*, *5*(11), 1681–1690. doi:10.1088/0953-8984/5/11/008

Mittal, V. (2009). Polymer Layered Silicate Nanocomposites: A Review. *Materials (Basel)*, *2*(3), 992–1057. doi:10.3390/ma2030992

Murugan, R., Phiri, C., & Mookkandi, V. P. S. (2022). Approaches, attributes and applications of matrix nanocomposites -a review Approaches, attributes and applications of matrix nanocomposites -a review. *Cogent Engineering*, *9*, 2152650. doi:10.1080/23311916.2022.2152650

Oshinski, A. J., Keskkula, H., & Paul, D. R. (1996). Dielectric studies of molecular mobility and phase morphology in polymer–layered silicate nanocomposites. *Polymer*, *22*, 4891–4907. doi:10.1016/0032-3861(96)00373-4

Ray, S., & Okamoto, M. (2003). Polymer/Layered Silicate Nanocomposites: A Review from Preparation to Processing. *Progress in Polymer Science -. Progress in Polymer Science, 28.* doi:10.1016/j.progpolymsci.2003.08.002

Sukhyy, K. M., Tomilo, V. I., Burmistr, O. M., Burmistr, M. V. (2007). Modelyuvannya vlastyvostej nanokompozytiv na osnovi poliamidu–6 i organomodyfikavanykh sharuvatykh sylikativ [Modeling of the properties of nanocomposites based on polyamide-sh and organomodified layered silicates.]. *Voprosy khimii i khimichnoi tekhnologii [Question of chemistry and chemical technology]. 6,* 76–95.

Tam, D. K. Y., Ruan, S., Gao, P., & Yu, T. (2012). 10 - High-performance ballistic protection using polymer nanocomposites. In *Woodhead Publishing Series in Textiles, Advances in Military Textiles and Personal Equipment* (pp. 213–237). Woodhead Publishing. doi:10.1533/9780857095572.2.213

Tandon, G. P., & Weng, G. J. (1984). Preparation, mechanical and rheological properties of a thermoplastic polyolefin (TPO)/organoclay nanocomposite with reference to the effect of maleic anhydride modified polypropylene as a compatibilizer. *Polymer Composites, 4,* 327–333. doi:10.1002/pc.750050413

Upadhyay, A., & Singh, R. (2014). Use of Artificial Neural Network and Theoretical Modeling to Predict the Effective Elastic Modulus of Composites with Ellipsoidal Inclusions. *Open Access Library Journal, 1*(07), 1–14. doi:10.4236/oalib.1100903

van Es, M., Xiqiao, F., van Turnhout, J., & van der Giessen, E. (2001). Comparing polymer–clay nanocomposites with conventional composites using composite modeling. Specialty polymer additives: principles and applications, 245 –260.

Conclusion

Physico-chemical principles of highly efficient and competitive technologies for polymer-inorganic nanostructured composites based on amorphous silica, layered silicates and polyonenes for use in creating active environments of electrochemical devices and composites for structural purposes have been developed. The results of systematic research give grounds to draw the following conclusions.

Technological principles of control of sol-gel synthesis processes of a new class of organo-inorganic nanostructured materials based on tetraethoxysilane and polyonenes doped with orthophosphoric acid with improved functional properties to create active environments of electrochemical devices have been developed. Improvement of chemical compositions is based on the results of complex studies of morphology, physicochemical properties, transfer processes and conductivity.

Polyionenes are shown to increase the rate of hydrolytic polycondensation of tetraethoxysilane. In this case, gelation occurs the faster the greater the molecular weight of the polyionene. The presence of polyionenes with aromatic heterocycles with two nitrogen atoms in sol-gel systems based on tetraethoxysilane and H_3PO_4 is found to affect the increase in proton conductivity of nanocomposites by 1-2 orders to ~ 10^{-2} S/cm in the temperature range 0 – 120°C; which allows to reduce by 30% the optimal amount of orthophosphoric acid in the ash (from 1.5 to 1.0 mole of H_3PO_4 per mole of $Si(OEt)_4$),

By the methods of X-ray phase analysis, polyionenes introduced into phosphorosilicate sols in the amount of $8 \cdot 10^{-4}$ mole polyionene per mole $Si(OEt)_4$ are revealed to prevent crystallization processes occurring in unmodified phosphorosilicate xerogels, with the formation of phosphates and silicons which prevents the processes associated with the transfer.

The inorganic and organo-inorganic xerogels synthesized by the sol-gel method are stated to be aggregates with a fractal type of structural organization, i.e. complex multilevel nanocomposites that are structured in the form of complex hierarchically aggregated systems. At the lower level, fractal particles are formed, which, in turn, are elements for the formation of aggregates of a higher level. It was shown that for systems with high proton conductivity at all three structural levels the formation of surface-fractal aggregates with a branched pore system is observed, in which phosphoric acid, the main proton-conducting component, is strongly retained. The possibility of influencing the fractal structure of silicophosphate nanocomposites by changing the conditions of sol-gel synthesis and hybridization of sol-gel systems with polyionenes was stated. Conditions of sol-gel synthesis of nanostructured organo-inorganic materials based on tetraethoxysilane, polyionenes and orthophosphoric acid have been optimized according to high proton conductivity. On the basis of the obtained materials, organo-inorganic membranes which retain high proton conductivity (up to 10^{-2} S/cm) at temperatures up to 140°C are formed.

The technology of obtaining highly efficient and competitive resistive humidity sensors based on tetraethoxysilane, polyonenes and orthophosphoric acid has been developed and introduced into production. It is established that when the values of relative humidity of the medium varied from 23.5 to 96.5%, the conductivity of alternating current increases by 2-3 orders of magnitude.

Silica adsorbents with immobilized ions were obtained by methods of adsorption of polyonenes on the surface of silica and implementation of in situ polycondensation of polyonenes in the surface layer of modified silica gel. The developed adsorption and chemical methods of ion immobilization are stated to give the surface of silica pronounced anion exchange properties, which are due to the introduction into the modifying layer of quaternary ammonium groups. The obtained silicas with immobilized ions are shown to have a sufficiently high sorption capacity and to be characterized with satisfactory kinetics of extraction of anions from dilute aqueous solutions (all studied anions are almost completely removed within 10 minutes) and thus combine the positive properties of polymeric and silica.

New composite sorbents of water "silica gel crystal hydrate" were synthesized and their sorption properties were studied. It is shown that such composite sorbents are characterized by high sorption of water at the level of 0.70-0.85 g / g. The sorption properties of such composite sorbents are in strong juxtaposition with those for massive salts. It is shown that the sorption properties of the synthesized composite sorbents 'silica gel crystal hydrate' are not a linear combination of the properties of silica gel and massive salt, the rate of sorption of water being three times higher. The hydration process passes from the kinetic regime characteristic of massive salt to the diffusion regime, in which the limiting stage is the transport of water through the pore system of the silicon-oxygen matrix. By the method of WAXS it was found that the difference between the properties of the massive salt of the crystalline hydrate and that in the composite sorbent is due to the nanostructuredness of salt in the pores of the silicon-oxygen matrix. Thermodynamic analysis of the cycle of sorption heat storage unit based on synthesized composite sorbents was performed. The density of thermal energy storage is 0.36 kWh / kg.

Physico-chemical bases of creation of nanostructured composites of structural purpose on the basis of thermoplastic polymer matrices and organomodified montmorillonite are developed. The technology of obtaining montmorillonite modified with polyonenes was developed. The process of intercalation of polyionene macromolecules into the intercrystalline space of montmorillonite consists in the replacement of interlayer metal cations by polyionene cations, which is accompanied by a 1.5-2-fold increase in interlayer distances. It is shown that when modified montmorillonite combined with polymer matrices in the process of worm-disk extrusion, delamination of montmorillonite structure and distribution in the polymer matrix in the form of separate nanolayers of structure is accompanied by an increase of 2-3 orders of area of interaction between montmorillonite and polymer matrix and the formation of interfacial transition layers which determine the strength properties of the polymer composition as a whole. With the introduction of 2% of modified montmorillonite in polyamide, the tensile strength increases by 35-60%, impact strength and relative elongation are increased by 120-150% and by 20-40%, respectively.

The optimal mode of obtaining polymer nanocomposites based on PA-6 on a worm-disk extruder in products with improved physical and mechanical characteristics was stated as follows: processing temperature 250 – 260°C, disk rotation speed 100 – 110 rpm, which provides a level shear velocities 200 – 300 s^{-1} and residence time in the disk zone 8 – 9 s, total shear deformation 7500 – 8500.

Initial data for the design of the device with a capacity of 50 kg/h and technological parameters for processing nanocomposites based on polyamides and organomodified montmorillonite into products with the specified level of properties are calculated.

The main factors responsible for improving the physical and mechanical characteristics of the nanocomposites under consideration are the degree of delamination of the structure of layered silicates in the polymer matrix are shown by models of elastity of composites. A high level of orientation of elementary particles of layered silicates is observed when shear forces applied in comparison with fibers both due to individual features and due to a more favorable geometric shape. The simulation data are in good agreement with the experimental data.

The modification of montmorillonite with quarternary ammonium salt and polymer quarternary ammonium salt is shown to result in destruction of montmorillonite structure when mixed with elastomer and more efficient dispersion in elastomeric matrix, increasing contact surface with elastomer and formation the adhesive bonds in rubber-reinforced systems as a consequence of changing the chemical composition of montmorllonite surface, its reactivity and the structure of aluminosilicate by increasing the interplanar distance.

Compilation of References

Abdollahi, A., Roghani-Mamaqani, H., Salami-Kalajahi, M., Mousavi, A., Razavi, B., & Shahi, S. (2018). Preparation of organic-inorganic hybrid nanocomposites from chemically modified epoxy and novolac resins and silica-attached carbon nanotubes by sol-gel process: Investigation of thermal degradation and stability. *Progress in Organic Coatings*, *117*, 154–155. doi:10.1016/j.porgcoat.2018.01.001

Açışlı, Ö., Karaca, S., & Gürses, A. (2017). Investigation of the alkyl chain lengths of surfactants on their adsorption by montmorillonite (Mt) from aqueous solutions. *Applied Clay Science*, *146*, 90–99. doi:10.1016/j.clay.2016.12.009

Adnan, M. M., Nylund, I.-E., Jaworski, A., Hvidsten, S., Ese, M.-H. G., Glaum, J., & Einarsrud, M.-A. (2021). The Structure, Morphology, and Complex Permittivity of Epoxy Nanodielectrics with In Situ Synthesized Surface-Functionalized SiO_2. *Polymers*, *2021*(13), 1469. doi:10.3390/polym13091469 PMID:34062857

Affdl, J. C. H., & Kardos, J. L. (1976). The Halpin-Tsai equations: A review. *Polymer Engineering and Science*, *16*(5), 344–352. doi:10.1002/pen.760160512

Agrawal, R. C., & Gupta, R. K. (1999). Superionic solid composite phase – an overview. *Journal of Materials Science*, *34*(6), 1131–1162. doi:10.1023/A:1004598902146

Ahmed, S., & Jones, F. (1990). A Review of Particulate Reinforcement Theories for Polymer Composites. *Journal of Materials Science*, *25*(12), 4933–4942. doi:10.1007/BF00580110

Al Khudhair, A., Bouchmella, K., Andrei, R. D., Mehdi, A., Mutin, P. H., & Hulea, V. (2021). One-step non-hydrolytic sol-gel synthesis of mesoporous SiO_2-Al_2O_3-NiO catalysts for ethylene oligomerization. *Microporous and Mesoporous Materials*, *322*, 111165. doi:10.1016/j.micromeso.2021.111165

Alexandre, M., & Dubois, P. (2000). Polymer-layered silicate nanocomposites: Preparation, properties and uses of a new class of materials. *Materials Science and Engineering R Reports*, *28*(1-2), 1–63. doi:10.1016/S0927-796X(00)00012-7

Alexandrov, K. S., & Ryshova, T. V. (1961). The Elastic Properties of Rock-Forming Minerals. II: Layered Silicates. *Bulletin USSR Academy of science. Geophysics Series 9*, *12*, 1165–1168.

Ali S.M., Emran K.M. (2022). Surfactant-template sol-gel protective coatings for cast iron corrosion: Effect of the surfactant type and concentration, *Journal of Molecular Liquids*, *367*, 120366.

Armelao, L., Barreca, D., & Moraru, B. (2003). A molecular approach to RuO2-based thin films: Sol–gel synthesis and characterization. *Journal of Non-Crystalline Solids*, *316*(2-3), 364–367. doi:10.1016/S0022-3093(02)01636-8

Avdeev, Y. G., Nenasheva, T. A., Luchkin, A. Y., Marshakov, A. I., & Kuznetsov, Y. I. (2022). Effect of Quaternary Ammonium Salts and 1,2,4-Triazole Derivatives on Hydrogen Absorption by Mild Steel in Hydrochloric Acid Solution. *Materials (Basel)*, *15*(19), 6989. doi:10.3390/ma15196989 PMID:36234330

Baig, N., Kammakakam, I., & Falath, W. (2021). Nanomaterials: A review of synthesis methods, properties, recent progress, and challenges. *Mater. Adv.*, *2021*(2), 1821–1871. doi:10.1039/D0MA00807A

Baki, V. A., Ke, X., Heath, A., Calabria-Holley, J., Terzi, C., & Sirin, M. (2022). The impact of mechanochemical activation on the physicochemical properties and pozzolanic reactivity of kaolinite, muscovite and montmorillonite. *Cement and Concrete Research*, *162*, 106962. doi:10.1016/j.cemconres.2022.106962

Balazs, A. C., Singh, C., Zhulina, E., & Lyatskaya, Yu. (1999). Modeling the Phase Behavior of Polymer/Clay Nanocomposites. *Accounts of Chemical Research*, *32*(8), 651–657. doi:10.1021/ar970336m

Barshilia, H. C., Lakshmi, R. V., Bera, P., Basu, B. B. J., Srinivas, G., Kumar, V. P., & Manikandanath, N. T. (2022). Transition metal nitride/oxide-based multilayer PVD coating with sol–gel derived ormosil passivation layer as an efficient solar absorber: Studies on high temperature stability and performance evaluation. *Solar Energy*, *239*, 283–293. doi:10.1016/j.solener.2022.05.016

Bazhenov, S., Li, J. X., Hiltner, A., & Baer, E. (1994). Ductility of filled polymers. *Journal of Applied Polymer Science*, *52*(2), 243–254. doi:10.1002/app.1994.070520211

Ben, F., Boury, B., Corriu, R. J. P., Delord, P., & Nobili, M. (2001) Anisotropic silica–based hybrid materials by sol–gel process. *11th International Workshop on Glasses, Ceramics, Hybrids and Nanocomposites from Gels (Sol–Gel 1981–2001): abstract*, (pp. 19–20). University of Padova.

Biazzotto, J. C., Vidoto, E. A., Nascimento, O. R., Lamamoto, Y., & Serra, O. (2002). Iron(III)-tetra-o-ureaphenylporphyrinosilica obtained by a sol–gel process: A study of EPR, surface area and catalytic activity. *Journal of Non-Crystalline Solids*, *304*(1-3), 101–104. doi:10.1016/S0022-3093(02)01010-4

Bibi, N., Sarwar, M. I., Ishaq, M., & Ahmad, Z. (2007). Mechanical and thermal properties of nano-composites of poly (vinyl chloride) and co-poly (vinyl chloride-vinyl alcohol-vinyl acetate) with montmorillonite. *Polymers & Polymer Composites*, *15*(4), 313–319. doi:10.1177/096739110701500407

Bigg, D. M. (1987). Mechanical properties of particulate filled polymers. *Polymer Composites*, *8*(2), 115–122. doi:10.1002/pc.750080208

Biron, M. (2016). 2 - Thermoplastic Specific Properties. In *Material Selection for Thermoplastic Parts. Practical and Advanced Information* (pp. 39–75). William Andrew Applied Science Publishers. doi:10.1016/B978-0-7020-6284-1.00002-7

Bokov, D., Jalil, A. T., Supat Chupradit, S., Suksatan, W., Ansari, M. J., Shewael, I. H., Valiev, G. H., & Kianfar, E. H. (2021). Nanomaterial by Sol-Gel Method: Synthesis and Application. *Advances in Materials Science and Engineering*, *2021*, 5102014. doi:10.1155/2021/5102014

Brinker, C. J., Keefer, K. D., Schaefer, D. W., & Ashley, C. S. (1982). Sol–gel transition in simple cilicates I. *Journal of Non-Crystalline Solids*, *48*(1), 47–64. doi:10.1016/0022-3093(82)90245-9

Brinker, C. J., & Scherer, G. W. (1990). *Sol–Gel Science.The Physics and Chemistry of Sol–Gel Processing*. Academic Press, Inc.

Brisard, S., Guilleminot, J., Sab, K., & Tran, V. P. (2018). Mori–Tanaka estimates of the effective elastic properties of stress-gradient composites. *International Journal of Solids and Structures*, *146*, 55–68. doi:10.1016/j.ijsolstr.2018.03.020

Broadstreet, S. W. (1955). Solution ceramic for enameling *Ceram. Age*, *66*(6), 24–27.

Buchdahl, R. (1975). Mechanical properties of polymers and composites – Vols. I and II, Lawrence E. Nielsen, Marcel Dekker, Inc. *Journal of Polymer Science. Part B: Polymer Letters*, *13*, 120–121.

Burmistr, M. V., Ovcharov, V. I., Shilov, V. V., Sukhyy, K. M., Gomza, Yu. P., Grischenko, V. K., & Ovcharenko, V. G. (2003). Synthesis, structure and mechanical properties of nanocomposites based on organoclay and polymeric matrix different nature, *Technomer 03. 18. Fachtagung uber Verarbeitung und Anwendung von Polymeren, Chemnitz*, *2*, 65–71.

Burmistr, M. V., Sukhyy, K. M., Shilov, V. V., Pissis, P., Spanoudaki, A., Sukha, I. V., Tomilo, V. I., & Gomza, Y. P. (2005). Synthesis, structure, thermal and mechanical properties of nanocomposites based on linear polymers and layered silicates modified by polymeric quaternary ammonium salts (ionenes). *Polymer*, *46*(26), 12226–12232. doi:10.1016/j.polymer.2005.10.094

Burnside, S. D., & Giannelis, E. P. (1995). Synthesis and properties of new poly (dimethylsiloxane) nanocomposites. *Chemistry of Materials*, *7*(9), 1597–1600. doi:10.1021/cm00057a001

Çankaya, N. (2021). Synthesis, characterization and thermal properties of oxo methacrylate-containing polymer/clay nanocomposites. *Politeknik Dergisi*, *24*, 1017–1022.

Chakrabarty, J. (2006). *Theory of Plasticity* (3rd ed.). Elsevier.

Chen, B., Evans, J., Greenwell, C., Boulet, P., Coveney, P., Bowden, A., & Whiting, A. (2008). A critical appraisal of polymer-clay nanocomposites. *Chemical Society Reviews*, *37*(3), 568–594. doi:10.1039/B702653F PMID:18224264

Chen, Ch., Khobaib, M., & Curliss, D. (2003). Epoxy layered–silicate nanocomposites. *Progress in Organic Coatings*, *47*(3-4), 376–383. doi:10.1016/S0300-9440(03)00130-9

Cheng, L., Cai, J., & Ke, Y. (2020). Ultrasonic-Assisted Sol–Gel Synthesis of Core–Shell Silica Particles for High-Performance Liquid Chromatography. *Journal of Inorganic and Organometallic Polymers and Materials*, *30*(3), 859–868. doi:10.100710904-019-01239-4

Chou, Ch.-Ch., & Lin, J.-J. (2004). One-Step Exfoliation of Montmorillonite via Phase Inversion of Amphiphilic Copolymer Emulsion. *Macromolecules*, *38*. doi:10.1021/ma047761x

Cui, Y., Kumar, S., Kona, B., & Houcke, D. (2015). Gas Barrier Properties of Polymer/Clay Nanocomposites. *RSC Advances*, *5*(78), 63669–63690. Advance online publication. doi:10.1039/C5RA10333A

Dahmouche, K., Atik, M., & Mello, N.C. (1998). New Li+ ion–conducting ormolytes. *Solar Energy Materials and Solar Cells*, *54*. 1–8.

de Araújo, M. J. G., Barbosa, F. C., Fook, M. V. L., Silva, S. M. L., & Leite, I. F. (2022). Influence of Quaternary Ammonium Salt Functionalized Chitosan Additive as Sustainable Filler for High-Density Polyethylene Composites. *Materials (Basel)*, *15*(21), 7418. doi:10.3390/ma15217418 PMID:36363010

De Boer, J. (1948). Quantum theory of condensed permanent gases I the law of corresponding states. *Physica*, *14*(2-3), 139–148. doi:10.1016/0031-8914(48)90032-9

Deng, Z., Wang, J., Wu, A., Shen, J., & Zhou, B. (1998). High strength SiO2 aerogel insulation. *Journal of Non-Crystalline Solids*, *225*, 101–104. doi:10.1016/S0022-3093(98)00106-9

Derjaguin, B. V., Churaev, N. V., & Muller, V. M. (1985). *Poverkhnostnye sily* [Surface forces]. Nauka.

Derjaguin, B., & Landau, L. (1941). Theory of the stability of strongly charged lyophobic sols and of the adhesion of strongly charged particles in solutions of electrolytes. *Acta Physico Chimica URSS*, *14*, 633.

Deshpande, A. V., & Kumar, U. (2002). Effect of method of preparation on photophysical properties of Rh-B impregnated sol–gel hosts. *Journal of Non-Crystalline Solids*, *306*(2), 149–152. doi:10.1016/S0022-3093(02)01054-2

Dultz, S., & Rytwo, G. (2005). Effects of different organic cations on the electrokinetic surface charge from organo-montmorillonites - consequences for the adsorption properties. *Annual Meeting of Swiss, Austrian and German Clay Group - DTTG*, (*vol. 11*, pp. 6-14). DTTG Reports.

Ebelmen (1844). Sur les éthers siliciques [On silicic ethers]. *Journ. de Pharm*, *5*(1), 262–264.

Ebelmen (1846). Recherches sur les combinaisons des acides borique et silicique avec les éthers [Research on the combinations of boris and silicic acids with ethers]. *Annales de Chemle et de Physique*, *16*(3), 129–166.

Echeverri, M., Hamad, C., & Kyu, T. (2014). Highly conductive, completely amorphous polymer electrolyte membranes fabricated through photo-polymerization of poly(ethylene glycol diacrylate) in mixtures of solid plasticizer and lithium salt. *Solid State Ionics*, *254*, 92–100. doi:10.1016/j.ssi.2013.10.050

Faghihi, K., Abootalebi, A. S., & Shabanian, M. (2013). New clay-reinforced polyamide nanocomposite based on 4-phenylenediacrylic acid: Synthesis and properties. *Journal of Saudi Chemical Society*, *17*(2), 191–197. doi:10.1016/j.jscs.2011.03.007

Fidalgo, A., & Ilharco, L. M. (2003). Thickness, morphology and structure of sol–gel hybrid films: I – the role of the precursor solution's ageing. *Journal of Sol-Gel Science and Technology*, *26*(1–3), 363–367. doi:10.1023/A:1020756728315

Franco-Urquiza, E. A. (2021). Clay-Based Polymer Nanocomposites: Essential Work of Fracture. *Polymers*, *13*(15), 2399. doi:10.3390/polym13152399 PMID:34372002

Friedel C., Crafts J.M. (1867) Research on the ethers of silic acid. *The American Journal of Science and Arts. Ser. 2. 43*(127, 128, 129), 155–171.

Friedel C., J.M. Crafts (1870) The combination of silicon with alcoholic radicals *The American Journal of Science and Arts*. Ser. 2., 49(145, 146, 147), 307–330.

Gallo, C., Rizzo, P., & Guerra, G. (2019). Intercalation compounds of a smectite clay with an ammonium salt biocide and their possible use for conservation of cultural heritage. *Heliyon*, *5*(12), e02991. doi:10.1016/j.heliyon.2019.e02991 PMID:31879708

García–López, D., Gobernado–Mitre, I., Fernández, J. F., Merino, J. C., & Pastor, J. M. (2005). Influence of clay modification process in PA6–layered silicate nanocomposite properties. *Polymer*, *5*(8), 23–25. doi:10.1016/j.polymer.2005.01.038

Gautier–Luneau, A., Denoyelle, A., Sanchez, J. Y., & Poinsignon, C. (1615–1618). Denoyelle, J.Y. Sanchez C. Poinsignon (1992) Organic–inorganic protonic polymer electrolytes as membrane for low–temperature fuel cell. *Electrochimica Acta*, *37*(9), 1615–1618. doi:10.1016/0013-4686(92)80122-3

Geiculescu, A. C., & Rack, H. J. (2002). X–ray scattering studies of polymeric zirconium species in aqueous xerogels. *Journal of Non-Crystalline Solids*, *306*(1), 30–41. doi:10.1016/S0022-3093(02)01056-6

Giannelis, E., Krishnamoorti, R., & Manias, E. (1999). Polymer-Silicate Nanocomposites: Model Systems for Confined Polymers and Polymer Brushes. *Advances in Polymer Science*, *138*, 107–147. doi:10.1007/3-540-69711-X_3

Gillett, S. L. (2002) Foresight Institute. *Nanotechnology: Clean energy and resources for future. White paper for Foresight Institute.* University of Nevada.

Gkourmpis, T., & Mitchell, G. R. (2020). The Use of Scattering Data in the Study of the Molecular Organisation of Polymers in the Non-Crystalline State. *Polymers*, *12*(12), 2917. doi:10.3390/polym12122917 PMID:33291414

Gloaguen, J.-M., & Lefebvre, J.-M. (2001). Plastic deformation behaviour of thermoplastic/clay nanocomposites. *Polymer*, *42*(13), 5841–5847. doi:10.1016/S0032-3861(00)00901-0

Gomes, A. L. M., Andrade, P. H. M., Palhares, H. G., Dumont, M. R., Soares, D. C. F., Volkringer, C., Houmard, M., & Nunes, E. H. M. (2021). Facile sol–gel synthesis of silica sorbents for the removal of organic pollutants from aqueous media. *Journal of Materials Research and Technology*, *15*, 4580–4594. doi:10.1016/j.jmrt.2021.10.069

Gomza Yu.P., Klepko V.V., Nesin S.D., Sukhyy K.M., Burmistr M.V. (2006). Ion–conducting organic–inorganic sol–gel materials as an active medium for resistive sensors, *Investigation on sensor system and technologies*, 1, 290–303.

Gordon, T., & Fakley, M. (2003). The influence of elastic modulus on adhesion to thermoplastics and thermoset materials. *International Journal of Adhesion and Adhesives*, *23*(2), 95–100. doi:10.1016/S0143-7496(02)00064-7

Gray, F. M. (1997). Polymer Electrolytes. RSC Materials Monographs, 175. London: The Royal Society of Chemistry.

Greene, J. P. (2021). 9 - Elastomers and Rubbers Editor(s): Joseph P. Greene, In Plastics Design Library Automotive Plastics and Composites, 127-147. William Andrew Publishing.

Greil, P. (1998). Near net shape manufacturing of polymer derived ceramics. *Journal of the European Ceramic Society*, *18*(13), 1905–1914. doi:10.1016/S0955-2219(98)00129-0

Grigoriev, A. S., Shilko, E. V., Skripnyak, V. A., & Psakhie, S. G. (2017) An approach to determining the parameters of kinetic strength theory based dynamic model of brittle solids mechanical behavior. In *Proceedings of the International Conference on Advanced Materials with Hierarchical Structure for New Technologies and Reliable Structures, 1909* (1). doi:10.1063/1.5013745

Guimarães, A. P., Viana, A. P. P., Lago, M. L., & Mohallem, N. D. S. (2002). The effect of thermal treatment on the properties of sol–gel palladium–silica catalysts. *Journal of Non-Crystalline Solids*, *304*(1-3), 70–75. doi:10.1016/S0022-3093(02)01006-2

Halpin, J. C., Finlayson, K. M., & Ashton, J. E. (1992) Primer on Composite Materials Analysis. Taylor and Francis, 354.

Hamaker, H. C. (1937). The London – van der Waals attraction between spherical particles. *Physica*, *4*(10), 1058–1072. doi:10.1016/S0031-8914(37)80203-7

Han, W.-T. (1999). Synthesis and linear and non-linear optical properties of (0.8PPV+0.2DMPPV)/ silica glass composites by sol–gel process. *Journal of Non-Crystalline Solids*, *259*(1-3), 107–110. doi:10.1016/S0022-3093(99)00500-1

Haruvy, Y., Ryabov, Y., Arkhipov, V., Gutinz, A., Axelrod, E., & Feldman, Y. (2002). Fast-sol-gel derived silsequioxane glasses embodying glycerol moieties: Dielectric properties and morphology. *Journal of Non-Crystalline Solids*, *305*(1–3), 226–234. doi:10.1016/S0022-3093(02)01096-7

Haudin, J.-M. (1982). Plastic Deformation of Amorphous and Semicrystalline Materials. *Les Editions de Physique*. 291-311.

Hermans, J. (1967). Exfoliated high–impact polystyrene/MMT composites prepared using anchored cationic radical initiator–MMT hybrid. *Proc Kon Ned Akad v Wetensch B, 65*, 1–9.

He, S., Carolan, D., Fergusson, A., & Taylor, A. (2019). Toughening epoxy syntactic foams with milled carbon fibers: Mechanical properties and toughening mechanisms. *Materials & Design*, *169*, 107654. doi:10.1016/j.matdes.2019.107654

Hill, R. (1964). Photodegradation of polypropylene nanocomposites. *Journal of the Mechanics and Physics of Solids*, *12*, 119.

Huang, H.-H., & Wilkes, G. L. (1987). Structure–property of behavior of a TEOS–poly(tetramethylene oxide) hybrid system prepared by the sol–gel process. I. Effect of PTMO molecular weight. *Papers Presented at the ... Meeting. American Chemical Society. Division of Polymer Chemistry*, *28*(2), 244–245.

Hui, C. Y., & Shia, D. (1998). Morphology development in layered silicate thermoset nanocomposites. *Polymer Engineering and Science*, *5*, 774–782. doi:10.1002/pen.10243

Imai, Y., Naka, K., & Chujo, Y. (1998). Reversible formation of Penetrating Polymer Network Structure in Organic–Inorganic Polymer Hybrid. *Polymer Journal*, *30*(12), 90–95. doi:10.1295/polymj.30.990

Imai, Y., Yoshida, N., Naka, K., & Chujo, Y. (1999). Thermoresponsive Organic–Inorganic Polymer Hybrids from Poly(N-isopropylacrylamide). *Polymer Journal*, *31*(3), 258–260. doi:10.1295/polymj.31.258

Ishikawa, M., Ihara, M., Morita, M., & Matsuda, Y. (1995). Electric double layer capacitors with new gel electrolytes. *Electrochimica Acta*, *40*(13–14), 2217–2222. doi:10.1016/0013-4686(95)00166-C

Iwamoto, T., & Mackenzie, J. D. (1995). Hard ormosils prepared with ultrasonic irradiation. *Journal of Sol-Gel Science and Technology*, *4*(2), 141–150. doi:10.1007/BF00491680

Jeevanandam, J., Barhoum, A., Chan, Y., Dufresne, A., & Danquah, M. (2018). Review on nanoparticles and nanostructured materials: History, sources, toxicity and regulations. *Beilstein Journal of Nanotechnology*, *9*, 1050–1074. doi:10.3762/bjnano.9.98 PMID:29719757

Kamila, R., Ridwan, M., Akhir, M. P. M., Patriati, A., & Insani, A. (2022). Synthesis of silic a particles through conventional sol-gel and sonochemistry methods and the effect of catalyst, water concentration and sample environment to the particle size. *Journal of Physics: Conference Series*, *2193*(1), 012044. doi:10.1088/1742-6596/2193/1/012044

Kanapitsas, A., Pissis, P., & Kotsilkova, R. (2002). Dielectric studies of molecular mobility and phase morphology in polymer–layered silicate nanocomposites. *Journal of Non-Crystalline Solids*, *305*(1-3), 204–211. doi:10.1016/S0022-3093(02)01107-9

Kazuki, M., Hu, Y., & Mackenzie, J. D. (1994). The effects of Ultrasonic irradiation on the Preparation and Properties of Ormosils. *Journal of Non-Crystalline Solids*, *3*, 109–116.

Klein, L. C. (1996) Processing of nanostructured sol–gel materials In A. S. Edelstein & R. S. Cammarata (eds), Nanomaterials: synthesis, properties and applications, 147–165. Bristol and Philadelphia: Institute of Physics Publishing. doi:10.1887/0750305789/b712c7

Kojima, Y., Usuki, A., Kawasumi, M., Okada, A., Kurauchi, T., Kamigaito, O., & Kaji, K. (1994). Fine structure of nylon–6–clay hybrid. *Journal of Polymer Science. Part B, Polymer Physics*, *32*(4), 625–630. doi:10.1002/polb.1994.090320404

Kononova, I., Kononov, P., Moshnikov, V., & Ignat'ev, S. (2021). Fractal-Percolation Structure Architectonics in Sol-Gel Synthesis. *International Journal of Molecular Sciences*, *22*(19), 10521. doi:10.3390/ijms221910521 PMID:34638862

Koo, J. (2006). *Polymer Nanocomposites: Processing, Characterization, And Applications: Processing, Characterization, And Applications*. McGraw Hill Professional.

Kumari, N., & Mohan, C. (2021). Basics of Clay Minerals and Their Characteristic Properties. In (Ed.), Clay and Clay Minerals. IntechOpen. doi:10.5772/intechopen.97672

Kumar, S. K., & Krishnamoorti, R. (2010). Nanocomposites: Structure, Phase Behavior, and Properties. *Annual Review of Chemical and Biomolecular Engineering*, *1*(1), 37–58. doi:10.1146/annurev-chembioeng-073009-100856 PMID:22432572

Kusmono, K., Ishak, M., Chow, W. S., Takeichi, T., & Rochmadi, R. (2009). Effect of Clay Modification on the Morphological, Mechanical, and Thermal Properties of Polyamide 6/Polypropylene/Montmorillonite Nanocomposites. *Polymer Composites*, *31*, 1156–1167. doi:10.1002/pc.20902

Laird, D. (2006). Influence of layer charge on swelling of smectites. *Applied Clay Science*, *34*(1-4), 74–87. doi:10.1016/j.clay.2006.01.009

Landel, R., & Nielsen, L. (1993). Mechanical Properties of Polymers and Composites. CRC Press, 580 doi:10.1201/b16929

Larminie, J., & Dicks, A. (2003). *Fuel Cells Systems Explained* (2nd ed.). John Wiley & Sons. doi:10.1002/9781118878330

Lewis, T. B., & Nielsen, L. E. (1970). Dynamic mechanical properties of particulate-filled composites. *Journal of Applied Polymer Science*, *14*(6), 1449–1471. doi:10.1002/app.1970.070140604

Liao, B., Song, M., Liang, H., & Pang, Y. (2001). Polymer–layered silicate nanocomposites. A study of poly(ethylene oxide)/Na+–montmorillonite nanocomposites as polyelectrolytes and polyethylene–block–poly(ethyleneglycol) copolymer/Na+–montmorillonite nanocomposites as fillers for reinforcement of polyethylene. *Polymer*, *42*(25), 10007–10011. doi:10.1016/S0032-3861(01)00563-8

Lifshitz, E. M. (1956). The theory of molecular Attractive Forces between Solids. *Soviet Physics*, *2*(1), 73–83.

Liu J., Li Y., Tan N., Hu Z., Zhou Y., Deng Q., Zhang G., Lu Z. (2023) Microstructure and properties of the solid solution ceramic coating by high speed laser cladding, *Optics & Laser Technology, 158.*

Liu, W., Yi, S., Liu, Z., Mo, Q., Xian, X., Ma, L., & Li, Y. (2022). Sol-gel silica modified recovered lignin particles and its application as blend with HDPE. *Industrial Crops and Products, 189*, 115842. doi:10.1016/j.indcrop.2022.115842

Liu, Y., Lin, Q., Chen, J., Shao, Y., Wang, Y., & Wang, J. (2022). PDMS-OH and nano-SiO2 Modified KH570-TEOS silica-sol coating and protective effect on concrete. *Colloids and Surfaces. A, Physicochemical and Engineering Aspects*, *648*, 129279. doi:10.1016/j.colsurfa.2022.129279

Li, X., Yang, P., Zhu, Z., You, Z., Zhang, W., Zhang, T., Chen, M., & Zhou, X. (2020). Sol-Gel condensation of temperature sensitive and shape stabilized phase change materials for thermal energy storage. *Thermochimica Acta*, *693*, 178758. doi:10.1016/j.tca.2020.178758

Li, Yu., Wang, S., Wang, Q., & Xing, M. (2018). A comparison study on mechanical properties of polymer composites reinforced by carbon nanotubes and graphene sheet. *Composites. Part B, Engineering*, *133*, 5–41. doi:10.1016/j.compositesb.2017.09.024

London, F. (1937). The general theory of molecular forces. *Transactions of the Faraday Society*, *33*, 8–26. doi:10.1039/tf937330008b

Loy Douglas, A. (2001). Bridget polysilsesquioxanes. Molecular engineering of hybrid organic–inorganic materials. *11th International Workshop on Glasses, Ceramics, Hybrids and Nanocomposites from Gels (Sol–Gel 1981–2001): abstract*, (pp. 20–21). University of Padova.

Lu, X., Manners, I., & Winnik, M. A. (2001). Polymer/Silica Composite Films as Luminescent Oxygen Sensors. *Macromolecules*, *34*(6), 1917–1921. doi:10.1021/ma001454j

Lu, Y., Cao, G., Kale, R. P., Prabakar, S., López, G. P., & Brinker, C. J. (1999). Microporous Silica Prepared by Organic Templating: Relationship between the Molecular Template and Pore Structure. *Chemistry of Materials*, *11*(5), 1223–1226. doi:10.1021/cm980517y

Mallikarjuna, S., & Sivaram, S. (2013). Influence of Organic Modification and Polyurethane Structure on Clay Dispersion. In: Thermoset Nanocomposites. doi:10.1002/9783527659647.ch3

Mandelbrot, B. (1982). The Fractal Geometry of Nature. *The Freeman.*

Marr, C., & Li, X. (1999). Composition and performance modelling of catalyst layer in a proton exchange membrane fuel cell. *Journal of Power Sources*, *77*(1), 17–27. doi:10.1016/S0378-7753(98)00161-X

Matejka, L., Plestil, J., & Dusek, K. (1998). Structure evolution in epoxy–silica hybrids–sol–gel process. *Non–Crystal. Solids*, *226*, 114–121.

Matsuda, A., Nakamoto, N., Tadanaga, K., Minami, T., & Tatsumisago, M. (2003). Proton conductivity at medium temperature range and chemical durability of phosphosilicate gels added with a third component. *Solid State Ionics*, *162-163*, 247–252. doi:10.1016/S0167-2738(03)00261-3

McNeil, L. E., & Grimsditch, M. J. (1993). Structural and mechanical properties of monomineral clays. *Phys. Condens. Matter.*, *5*(11), 1681–1690. doi:10.1088/0953-8984/5/11/008

Messersmith, P. B., & Giannelis, E. P. (1994). Synthesis and characterization of layered silicate-epoxy nanocomposites. *Chemistry of Materials*, *6*(10), 1719–1725. doi:10.1021/cm00046a026

Messersmith, P. B., & Giannelis, E. P. (1995). Synthesis and barrier properties of poly(ε-caprolactone)-layered silicate nanocomposites. *Journal of Polymer Science Part A*, *33*(7), 1047–1057. doi:10.1002/pola.1995.080330707

Mittal, V. (2009). Polymer Layered Silicate Nanocomposites: A Review. *Materials (Basel)*, *2*(3), 992–1057. doi:10.3390/ma2030992

Mohanty, A. K., Banerjee, S., Komber, H., & Voit, B. (2014). Imidoaryl biphenol based new fluorinated sulfonated poly(arylene ether sulfone) copolymers and their proton exchange membrane properties. *Solid State Ionics*, *254*, 82–91. doi:10.1016/j.ssi.2013.11.029

Monasse, B., Queyroy, S., & Lhost, O. (2008). Molecular Dynamics prediction of elastic and plastic deformation of semi-crystalline polyethylene. *International Journal of Material Forming*, *1*(S1), 1111–1114. doi:10.100712289-008-0174-1

Murugan, R., Phiri, C., & Mookkandi, V. P. S. (2022). Approaches, attributes and applications of matrix nanocomposites -a review Approaches, attributes and applications of matrix nanocomposites -a review. *Cogent Engineering*, *9*, 2152650. doi:10.1080/23311916.2022.2152650

Nam, C. W., & Woo, S. I. (1994). Characterization of spin-coated silicate and phosphosilicate thin films prepared by the sol-gel method. *Thin Solid Films*, *237*(1-2), 314–315. doi:10.1016/0040-6090(94)90278-X

Newman, A. C. D. (1987). *Chemistry of Clays and Clay Minerals*. Longman Scientific & Technical.

Nicolais, L., & Carotenuto, G. (2005). *Metal-Polymer Nanocomposites*. John Wiley & Sons.

Nogami, M., & Moriya, Y. (1980). Glass Formation hydrolysis of $Si(OC_2H_5)_4$ with NH_4OH and HCl solution. *Journal of Non-Crystalline Solids*, *37*(2), 191–201. doi:10.1016/0022-3093(80)90150-7

Olad, A. (2011). Polymer/Clay Nanocomposites. In *Advances in Diverse Industrial Applications of Nanocomposites*. doi:. doi:10.5772/14464

Olemskoy, A.I. Flat A.Ya. (1989). Ispolzovaniye kontseptsii fraktala v fizike kondensirovannoy sredy [Using the fractal concept in condensed matter physics]. *Успехи физических наук [Advances in the physical sciences]*, *163*(12), 1–50.

Onitiri, M., & Ubi, P. (2021). Failure modes in particle filled plastic matrix composites. *Journal of Engineering Science and Technology*, *5*, 79–96.

Osborne, H., Blohowiak, K. Y., Taylor, S. R., Hunter, C., Bierwagon, G., Carlson, B., Bernard, D., & Donley, M. S. (2001). Testing and evaluation of nonchromated coating systems for aerospace applications. *Progress in Organic Coatings*, *41*(4), 217–220. doi:10.1016/S0300-9440(01)00132-1

Oshinski, A. J., Keskkula, H., & Paul, D. R. (1996). Dielectric studies of molecular mobility and phase morphology in polymer–layered silicate nanocomposites. *Polymer*, *22*, 4891–4907. doi:10.1016/0032-3861(96)00373-4

Ou, D. L., Chevalier, P. M., Mackinnon, I. A., Eguchi, K., Boisvert, R., & Su, K. (2003). Studies on highly porous hybrid prepared by a novel fast gellation process under ambient pressure. *Journal of Sol-Gel Science and Technology*, *26*(1/3), 657–662. doi:10.1023/A:1020737618745

Ovcharov, V. I., Burmistr, M. V., & Sukhyy, K. M. [et al] (2005). Reinforcement of thermoplastic polymers by nanosize organo–clay. Technomer 05.19: Fachtagung uber Verarbeitung und Anwendung von Polymeren [Specialist conference on the processing and application of the polymers]. Chemnitz. 2, 71–82.

Pal, K. (2016). Effect of different nanofillers on mechanical and dynamic behavior of PMMA based nanocomposites. *Composites Communications*, *1*, 25–28. doi:10.1016/j.coco.2016.08.001

Pandey, P. C., Upadhyay, S., Tiwari, I., & Tripathi, V. S. (2001). An ormosil-based peroxide biosensor— A comparative study on direct electron transport from horseradish peroxidase. *Sensors and Actuators*, *72*(3), 224–226. doi:10.1016/S0925-4005(00)00669-9

Park, Y. (2001). Fractal geometry of porous materials. *Fractals*, *8*(3), 301–306.

Paul, M. A., Alexandre, M., Degée, P., Henrist, C., Rulmont, A., & Dubois, P. (2003). New nanocomposite materials based on plasticized poly (L-lactide) and organo-modified montmorillonites: Thermal and morphological study. *Polymer*, *44*(2), 443–450. doi:10.1016/S0032-3861(02)00778-4

Pell, W. G., & Conway, B. E. J. (2001). Analysis of power limitations at porous supercapacitor electrodes under cyclic voltametry modulation and DC charge. *Power Soureces*, *96*(1), 57–67. doi:10.1016/S0378-7753(00)00682-0

Penaloza, D. P. Jr. (2019). Modified clay for the synthesis of clay-based nanocomposites. *Journal of Silicate Based and Composite Materials*, *71*(1), 5–11. doi:10.14382/epitoanyag-jsbcm.2019.2

Perelomov, L., Mandzhieva, S., Minkina, T., Atroshchenko, Y., Perelomova, I., Bauer, T., Pinsky, D., & Barakhov, A. (2021). The Synthesis of Organoclays Based on Clay Minerals with Different Structural Expansion Capacities. *Minerals (Basel)*, *11*(7), 707. doi:10.3390/min11070707

Peterlin, A. (1972). Morphology and Properties of Crystalline Polymers with Fiber Structure. *Textile Research Journal*, *42*(1), 20–30. doi:10.1177/004051757204200105

Peterlin, A. (1975). Plastic deformation of polymers with fibrous structure. *Colloid & Polymer Science*, *253*(10), 809–823. doi:10.1007/BF01452401

Pinero, M., La Rosa–Fox, N. D., Erce–Montilla, R., & Esquivias, L. (2003). Small angle neutron scattering study of PbS quantum dots synthetic routes via sol–gel. *Journal of Sol-Gel Science and Technology*, *26*(1–3), 527–531. doi:10.1023/A:1020755517353

Pirmoradian M., Al-Bakhakh B.A.J., Behroozibakhsh M., & Pedram P. (2022). Repairability of aged dimethacrylate-free ORMOCER-based dental composite resins with different surface roughening methods and intermediate materials, *The Journal of Prosthetic Dentistry*. doi:. doi:10.1016/j.prosdent.2022.04.001

Qian, Z., Zhang, S., & Yang, M. (2008). Effect of Clay Modification on Photooxidation of Polyethylene/Clay Nanocomposites. *Polymers & Polymer Composites*, *16*(8), 535–546. doi:10.1177/096739110801600807

Rabinovich, L., Glezer, V., Wu, Z., & Lev, O. (2001). Naphthoquinone–silicate based gas electrodes: Chemical–electrochemical mode of operation. *Journal of Electroanalytical Chemistry (Lausanne, Switzerland)*, *504*(2), 146–149. doi:10.1016/S0022-0728(01)00406-5

Rahman, I. A., & Padavettan, V. (2012). Synthesis of Silica Nanoparticles by Sol-Gel: Size-Dependent Properties, Surface Modification, and Applications in Silica-Polymer Nanocomposites—A Review. *Journal of Nanomaterials*, *2012*, 132424. doi:10.1155/2012/132424

Ray, S. S. (2013) Clay-Containing Polymer Nanocomposites: From Fundamentals to Real Applications. Elsevier, 416.

Ray, S., & Okamoto, M. (2003). Polymer/Layered Silicate Nanocomposites: A Review from Preparation to Processing. *Progress in Polymer Science -. Progress in Polymer Science*, *28*. doi:10.1016/j.progpolymsci.2003.08.002

Rivero-Antúnez, P., Zamora-Ledezma, C., Sánchez-Bajo, F., Moreno-López, J. C., Anglaret, E., & Morales-Flórez, V. (2023). Sol–gel method and reactive SPS for novel alumina–graphene ceramic composites. *Journal of the European Ceramic Society*, *43*(3), 1064–1077. doi:10.1016/j.jeurceramsoc.2022.10.043

Roldugin, V. I. (2003). Svoystva fraktalnykh dispersnykh system [Properties of fractal disperse systems]. [in Russian]. *Uspekhi Khimii*, (11), 1028–1054.

Rolling, B., Happen, A., Funk, K., & Ingram, M. D. (1977). Dielectric studies of same conducting systems. *Physical Review Letters*, *78*, 2160–2164.

Rothon, R. N. (2003). Particulate-filled polymer composites. Shrewsbury, UK: Rapra Technology, 320

Rothon, R. N. (2017). Fillers for Polymer Applications. Springer Verlag., 211 doi:10.1007/978-3-319-28117-9

Sá Ferreira, R.A., & Carlos, R.D., V. De Zea Bermude [et al]. (2003). Room temperature visible/infrared emission and energy transfer in Nd^{3+}–based organic/inorganic hybrids. *Sol–Gel. Sciences et Techniques (Paris)*, *26*(1–3), 315–319.

Sakai, Y., Sadaoka, Y., Matsuguchi, M., & Kanakura, Y. (1991). Kinetic analysis of Cathodic Deposition. *Journal of the Electrochemical Society*, *138*, 2474. doi:10.1149/1.2085997

Santos, L. R. B., Belin, S., Brios, V., Santilli, C. V., Pulcinelli, S. H., & Larbot, A. (2003). Study of Structural Surface Modified Tin Oxide Membrane Prepared by Sol-Gel Route Sintered at 400°C. *Journal of Sol-Gel Science and Technology*, *26*(1/3), 171–173. doi:10.1023/A:1020778411051

Sato, Y., & Furukawa, J. (1963). A Molecular Theory of Filler Reinforcement Based upon the Conception of Internal Deformation (A Rough Approximation of the Internal Deformation). *Rubber Chemistry and Technology*, *36*(4), 1081–1106. doi:10.5254/1.3539632

Schaefer, D. W., & Keefer, K. D. (1984). Fractal geometry of silica condensation polymers. *Physical Review Letters*, *53*(14), 1383–1386. doi:10.1103/PhysRevLett.53.1383

Schaefer, D. W., Keefer, K. D., Aubert, J. H., & Rand, P. B. (1986). *Science of Ceramic Chemical Processing* (L. L. Hench & D. R. Ulrich, Eds.). Wiley.

Schaefer, D. W., Martin, J. E., & Keefer, K. D. (1985). *Physics of Finely Divided Matter*. Springer–Verlag.

Schmidt, G., Nakatani, A. I., Butler, P. D., & Han, C. C. (2002). Polymer–layered silicate nanocomposites: Preparation, properties and uses of a new class of materials. *Macromolecules*, *35*, 4725–4732. doi:10.1021/ma0115141

Schöllhorn, R. (1996). Intercalation Systems as Nanostructured Functional Materials. *Chemistry of Materials*, *8*(8), 1747–1757. doi:10.1021/cm9601421

Schubert, U. (2003). Silica–based and transition metal–based inorganic–organic hybrids materials – a comparison *Sol–Gel. Sciences et Techniques (Paris)*, *26*(1–3), 47–55.

Semichenko G.D. (1997) *Zol–gel protsess v keramicheskoy tekhnologii* [*Zo-gel process in ceramic technology*]. Kharkov: Khark. politekhn.

Sevim, İ. (2017). Intercalation of vermiculite in presence of surfactants. *Applied Clay Science*, *146*, 7–13. doi:10.1016/j.clay.2017.05.030

Shalygina, T. A., Rudenko, M. S., Nemtsev, I. V., Parfenov, V. A., Voronina, S. I., & Simonov, I. D. (2021). Influence of the Filler Particles' Surface Morphology on the Polyurethane Matrix's Structure Formation in the Composite. *Polymers*, *13*(22), 3864. doi:10.3390/polym13223864 PMID:34833164

Shambayati, S., Schreiber, S. L., Blake, J. F., Wierschke, S. G., & Jorgensen, W. L. (1990). Structure and basicity of silyl ethers: A crystallographic and ab initio inquiry into the nature of silicon-oxygen interactions. *Journal of the American Chemical Society*, *1990*(112), 697–703. doi:10.1021/ja00158a031

Shameem, M. M., Sasikanth, S. M., Annamalai, R., & Raman, R. G. (2021). A brief review on polymer nanocomposites and its applications. *Materials Today: Proceedings*, *45*(Part 2), 2536–2539. doi:10.1016/j.matpr.2020.11.254

Shin, S.-J., Lee, J.-K., Ha, H.-Y., Hong, S.-A., Chun, H.-S., & Oh, I.-H. (2002). Effect of the catalytic in preparation method on the performance of polymer electrolyte membrane fuel cells. *Journal of Power Sources, 106*(1–2), 146–152. doi:10.1016/S0378-7753(01)01045-X

Shu, G., Zhao, J., Zheng, X., Xu, M., Liu, Q., & Zeng, M. (2019). Modification of Montmorillonite with Polyethylene Oxide and Its Use as Support for Pd0 Nanoparticle Catalysts. *MDPI Polymers, 11*(5), 755. doi:10.3390/polym11050755 PMID:31035647

Siemens, H. A. G. (1902) Pat. 128253 Germany C., I. 1902. 448.

Sinha Ray, S., Yamada, K., Okamoto, M., Ogami, A., & Ueda, K. (2003). New polylactide/layered silicate nanocomposites. High performance biodegradable materials. *Chemistry of Materials, 15*(7), 1456–1465. doi:10.1021/cm020953r

Smirnov V.M. (2003) *Sovremennyye vozmozhnosti upravleniya nanostrukturirovaniyem tverdykh veshchestv i materialov Khimiya tverdogo tela [Modern possibilities for controlling solid state chemistry].* Khimicheskiye problemy sozdaniya novykh materialov. [Chemical problems of creating new materials], 57–87. Sankt–Peterburgskiy Gosudarstvennyy universitet.

Smith, G. N., Brok, E., Schmiele, M., Mortensen, K., Bouwman, W. G., Duif, C. P., Hassenkam, T., Alm, M., Thomsen, P., & Arleth, L. (2021). The microscopic distribution of hydrophilic polymers in interpenetrating polymer networks (IPNs) of medical grade silicone. *Polymer, 224*, 123671. doi:10.1016/j.polymer.2021.123671

Soheil, S., Abdolahifar, H., & Goudarzian, N. (2015). Preparation and Modification of Polystyrene/Clay Nanocomposite Using Ethylene Octene Copolymer and SBS Block Copolymer. Conference: Conference of Polymer Processing Society At: Graz, Austria

Song, C., & Villemure, G. (2001). Electrode modification with spin-coated films of mesoporous molecular sieve silicas. *Microporous and Mesoporous Materials, 44*, 679–682. doi:10.1016/S1387-1811(01)00249-9

Srebnik, S., & Lev, O. (2003). Theoretical investigation of imprinted crosslinked silicates. *Journal of Sol-Gel Science and Technology, 26*(1–3), 107–113. doi:10.1023/A:1020701721529

Strawhecker, K. E., & Manias, E. (2000). Structure and properties of poly(vinylalcohol)/Na+–montmorillonite nanocomposites. *Chemistry of Materials, 12*(10), 2943–2949. doi:10.1021/cm000506g

Sukhyy, K. M., Privalko, E. G., Privalko, V. P., & Burmistr, M. V. (2010). Thermoelasticity and Stress Relaxation Behavior of Synthetic Rubber/ Organoclay Nanocomposites. John Wiley & Sons.

Sukhyy, K. M., Tomilo, V. I., Burmistr, O. M., Burmistr, M. V. (2007). Modelyuvannya vlastyvostej nanokompozytiv na osnovi poliamidu–6 i organomodyfikavanykh sharuvatykh sylikativ [Modeling of the properties of nanocomposites based on polyamide-sh and organomodified layered silicates.]. *Voprosy khimii i khimichnoi tekhnologii [Question of chemistry and chemical technology]. 6,* 76–95.

Sukhyy, K. M., Belyanovskaya, E. A., Kozlov, YaN., Kolomiyets, E. V., & Sukhyy, M. P. (2014). Structure and adsorption properties of the composites 'silica gel–sodium sulphate', obtained by sol–gel method. *Applied Thermal Engineering*, *64*(1-2), 408–412. doi:10.1016/j.applthermaleng.2013.12.013

Sukhyy, K. M., Belyanovskaya, E. A., Sukhyy, M. P., & Tomilo, V. I. (2018). Modification technology of montmorillonite by polyionenes. *Journal of Chemistry and Technologies.*, *26*(1), 1–8. doi:10.15421/081801

Sukhyy, K. M., Belyanovskaya, E. A., Sukhyy, M. P., Tomilo, V. I., & Lytovchenko, R. D. (2018). Obtaining technology for nanocomposites based on polyamide-6 and organomodified montmorillonite. *Journal of Chemistry and Technologies*, *26*(1), 9–19. doi:10.15421/081802

Sukhyy, K. M., Gomza, Yu. P., Belyanovskaya, E. A., Klepko, V. V., Shilova, O. A., & Sukhyy, M. P. (2015). Resistive humidity sensors based on proton-conducting organic-inorganic silicophosphates doped by polyionenes. *Journal of Sol-Gel Science and Technology*, *74*(2), 472–481. doi:10.100710971-015-3622-7

Tadanaga K, Katata N, Minami T (1997) Super-Water-Repellent Al2O3 Coating Films with High Transparency J Am Ceram Soc 80, 1040-1042.

Tam, D. K. Y., Ruan, S., Gao, P., & Yu, T. (2012). 10 - High-performance ballistic protection using polymer nanocomposites. In *Woodhead Publishing Series in Textiles, Advances in Military Textiles and Personal Equipment* (pp. 213–237). Woodhead Publishing. doi:10.1533/9780857095572.2.213

Tandon, G. P., & Weng, G. J. (1984). Preparation, mechanical and rheological properties of a thermoplastic polyolefin (TPO)/organoclay nanocomposite with reference to the effect of maleic anhydride modified polypropylene as a compatibilizer. *Polymer Composites*, *4*, 327–333. doi:10.1002/pc.750050413

The Science and Technology of Rubber Fourth Edition. (2013), Editor(s): James E. Mark, Burak Erman, C. Michael Roland, Academic Press, 333

Theng, B. K. G. (2012). *Formation and Properties of Clay-Polymer Complexes* (2nd ed.). Elsevier.

Tretiakov, Yu. D. (2003). Protsessy samoorganizatsii v khimii materialov [Self organization processes in materials chemistry]. [in Russian]. *Uspekhi Khimii*, *72*(8), 731–762.

Truc, T., Thuy, T., Oanh, V., Hang, T., Nguyen, A., & … . (2019). 8-hydroxyquinoline-modified clay incorporated in an epoxy coating for the corrosion protection of carbon steel. *Tribology - Materials, Surfaces & Interfaces. Maney Publishing.*, *14*, 26–33.

Tseng, C., Wu, J., Lee, H., & Chang, F. (2001). Preparation and crystallization behavior of syndiotactic polystyrene-clay nanocomposites. *Polymer*, *42*(25), 10063–10070. doi:10.1016/S0032-3861(01)00568-7

Unnikrishnan, L., Mohanty, S., Nayak, S., & Singh, N. (2012). Synthesis and characterization of polysulfone/clay nanocomposite membranes for fuel cell application. *Journal of Applied Polymer Science*, *124*(S1), E309–E318. Advance online publication. doi:10.1002/app.34355

Upadhyay, A., & Singh, R. (2014). Use of Artificial Neural Network and Theoretical Modeling to Predict the Effective Elastic Modulus of Composites with Ellipsoidal Inclusions. *Open Access Library Journal, 1*(07), 1–14. doi:10.4236/oalib.1100903

Urban, P. M., Funke, A., Muller, J. T., Himmen, M., & Docter, A. (2001). Catalytic processes in solid polymer electrolyte fuel cell systems. *Appl. Catalysis. General, 221*(1–2), 459–470. doi:10.1016/S0926-860X(01)00819-5

Vaia, R. A., Sauer, B. B., Tse, O. K., & Giannelis, E. P. (1997). Relaxations of confined chains in polymer nanocomposites: Glass transition properties of poly(ethylene oxide) intercalated in montmorillonite. *Journal of Polymer Science. Part B, Polymer Physics, 35*(1), 59–67. doi:10.1002/(SICI)1099-0488(19970115)35:1<59::AID-POLB4>3.0.CO;2-Q

Valášek, P., Müller, M., & Ružbarský, J. (2014). Using Recycled Rubber Particles as Filler of Polymers. *Applied Mechanics and Materials, 616*, 260–267. doi:10.4028/www.scientific.net/AMM.616.260

Valverde, G., Macedo, J. G., & Cruz, D. (2003). Photoconductivity in Mesostructured Thin Films *J Sol-Gel. Sciences et Techniques (Paris), 26*, 605–608.

van Es, M., Xiqiao, F., van Turnhout, J., & van der Giessen, E. (2001). Comparing polymer–clay nanocomposites with conventional composites using composite modeling. Specialty polymer additives: principles and applications, 245 –260.

Vanderzande, C. (1998). *Lattice Models of Polymers (Cambridge Lecture Notes in Physics)*. Cambridge University Press. doi:10.1017/CBO9780511563935

Verbeek, J. (2003). The Influence of Interfacial Adhesion, Particle Size and Size Distribution on the Predicted Mechanical Properties of Particulate Thermoplastic Composites. *Materials Letters, 57*(13-14), 1919–1924. doi:10.1016/S0167-577X(02)01105-9

Verwey, E., & Overbeek, T. (1948). *Theory of the Stability of Lyophobic Colloids*. Elsevier.

Vollet, D. R., Donatti, D. A., & Ibanez Ruiz, A. (2002). A SAXS study of the nanostructural characteristics os TEOS–derived sonogels upon heat treatment up to 1100 °C. *Journal of Non-Crystalline Solids, 306*(1), 11–16. doi:10.1016/S0022-3093(02)01058-X

Vora, R. H., & Lau, K. S. Y. (2022) Subchapter 10.4 - Research and development of high-performance and thermomechanically stable Ceramer (polymers–ceramic composites materials) and 'CeramImide' nanocomposites material formulations based on fluoro-polyimides for the advanced industrial applications. In H. Dodiuk (ed), Plastics Design Library, Handbook of Thermoset Plastics (Fourth Edition), 517-585. William Andrew Publishing.

Wang, M.-J. (1998). Effect of Polymer-Filler and Filler-Filler Interactions on Dynamic Properties of Filled Vulcanizates. *Rubber Chemistry and Technology, 71*(3), 520–589. doi:10.5254/1.3538492

Wang, Z. M., Hong, H., & Chung, T. C. (2005). Synthesis of Maleic Anhydride Grafted Polypropylene with High Molecular Weight Using Borane/O_2 Radical Initiator and Commercial PP Polymers. *Macromolecules, 38*(22), 8966–8970. doi:10.1021/ma0516182

Ward, I. M., & Sweeney, J. (2004). *An Introduction to the Mechanical Properties of Solid Polymers* (2nd ed.). Wiley.

Wilde, G. (2021). *Nanostructured Materials*. Elsevier.

Wu, Z., Lee, K., Lin, Y., Lan, S., & Huang, L. (2003). Effects of surface-active substances on acid–base indicator reactivity in SiO_2 gels. *Journal of Non-Crystalline Solids*, *320*(1-3), 168–170. doi:10.1016/S0022-3093(03)00025-5

Xu, M., & An, B. (2022). Plastic deformation of polymer matrix composites reinforced with beaded fibers. *International Journal of Solids and Structures*, *256*, 111985. doi:10.1016/j.ijsolstr.2022.111985

Yang, P., Lü, M., Song, C., Gu, F., Liu, S., Xu, D., Yuan, Y., & Cheng, X. (2002). Luminescence of CaS and MnS nanocrystallites co-activated sol–gel derived silica xerogel. *Journal of Non-Crystalline Solids*, *311*(1), 99–103. doi:10.1016/S0022-3093(02)01600-9

Yang, Y., Nogami, M., Huang, Z., & Li, Y. (2013). Engineering Nanostructures of Inorganic Materials for Optical and Chemical Applications. *Journal of Nanomaterials*, *2013*, 157869. doi:10.1155/2013/157869

Yano, K., Usuki, A., Okada, A., Kurauchi, T., & Kamigaito, O. (1993). Synthesis and properties of polyimide–clay hybrid. *J. Polym. Sci. Part A.*, *31*(10), 2493–2498. doi:10.1002/pola.1993.080311009

Yeatman E. (2017). *Sol-gel fabrication for optical communication components: prospects and progress*. Imperial. doi:. doi:10.1117/12.279834

Yebra-Rodriguez, A., Alvarez-Lloret, P., Cardell, C. A., & Rodriguez-Navarro, B. (2011). Influence of processing conditions on the optical and crystallographic properties of injection molded polyamide-6 and polyamide-6/montmorillonite nanocomposites. *Applied Clay Science*, *51*(4), 414–418. doi:10.1016/j.clay.2010.12.031

Yilmazer, U., & Farris, R. J. (1983). Mechanical behavior and dilatation of particulate-filled thermosets in the rubbery state. *Journal of Applied Polymer Science*, *28*(11), 3369–3386. doi:10.1002/app.1983.070281106

Yu, Z., Zhu, Z., & Wang, S.-Q. (2005). Synthesis and rheological properties of polystyrene/layered silicate nanocomposite. *Polymer*, *104*, 785–796.

Zanurin, A., Johari, N. A., Alias, J., Mas Ayu, H., Redzuan, N., & Izman, S. (2022). Research progress of sol-gel ceramic coating: A review. *Materials Today: Proceedings*, *48*(Part 6), 1849–1854. doi:10.1016/j.matpr.2021.09.203

Zhang C., Cui Y., Lin S. & Guo J. (2022). Preparation and applications of hydrophilic quaternary ammonium salt type polymeric antistatic agents. *e-Polymers, 22*(1), 370-378.

Zhang, G., Wu, T., Lin, W., Tan, Y., Chen, R., Huang, Z., Yin, X., & Qu, J. (2017). Preparation of polymer/clay nanocomposites via melt intercalation under continuous elongation flow. *Composites Science and Technology*, *145*, 157–164. doi:10.1016/j.compscitech.2017.04.005

Zhong, Y., Zhu, Z., & Wang, S. (2005). Synthesis and rheological properties of polystyrene/layered silicate nanocomposite. *Polymer*, *104*(9), 785–796. doi:10.1016/j.polymer.2005.02.014

Zhu, Y., Iroh, J., Rajagopolan, R., Aykanat, A., & Vaia, R. (2022). Optimizing the Synthesis and Thermal Properties of Conducting Polymer–Montmorillonite Clay Nanocomposites. *Energies*, *15*(4), 1291. doi:10.3390/en15041291

Zulðqar, S., & Sarwar, M. I. (2009). Synthesis and Characterization of Aromatic–Aliphatic Polyamide Nanocomposite Films Incorporating a Thermally Stable Organoclay. *Nanoscale Research Letters*, *4*(5), 391–399. doi:10.100711671-009-9258-1 PMID:20596518

Zulfiqar, S., & Sarwar, M. I. (2008). Inclusion of aramid chains into the layered silicates through solution intercalation route. *Journal of Inclusion Phenomena and Macrocyclic Chemistry*, *62*(3-4), 353–361. doi:10.100710847-008-9478-5

About the Authors

Kostyantyn M. Sukhyy was born on 1976. Gained a PhD in 2002 and SciD in 2014 from Ukrainian State University of Chemical Engineering. Acquired the academic titles of associate professor in 2004 and professor in 2015. Has been the Rector of Ukrainian State University of Chemical Engineering since 2019. Scientific interests: development of methods for the synthesis of new polymer quaternary ammonium salts, polymer nanocomposites and polymeric electrolytes in solid state.

Elena A. Belyanovskaya was born on 1972. Gained a PhD in 2003 from Ukrainian State University of Chemical Engineering. Acquired the academic title of associate professor in 2015. Has been the associate professor of the Department of Power Engineering of Ukrainian State University of Chemical Engineering since 2014. Scientific interests: processes and devices for adsorptive heat energy storage and conversion, utilization of low-grade energy resources.

Index

K

M

N

O

P

Q

R

S

T

U

V

W

X

www.ingramcontent.com/pod-product-compliance
Lightning Source LLC
LaVergne TN
LVHW061543290926
849162LV00004B/890

* 9 7 8 1 7 9 9 8 9 7 2 8 6 *